W9-DIF-643

NEW LESSONS IN ARC WELDING

Published as an educational service to the welding industry

This book may be ordered from any dealer or representative of The Lincoln Electric Company, or through any recognized book dealer in the world or direct from:

THE LINCOLN ELECTRIC COMPANY
CLEVELAND 17, OHIO

LINCOLN ELECTRIC CO. OF CANADA, LTD.
Leaside, Toronto 17, Ontario, Canada

LA SOUDURE ELECTRIQUE LINCOLN
Grand Quevilly, (SM), France

LINCOLN ELECTRIC CO., (AUSTRALIA) PTY., LTD.
P.O. Box 22, Padstow, New South Wales

EXPORT REPRESENTATIVES
The Armco International Corporation,
Middletown, Ohio, U.S.A.

Copyright, 1957
THE LINCOLN ELECTRIC COMPANY

FIRST EDITION, (Lessons in Arc Welding) AUGUST, 1940
Reprinted, November, 1940
Reprinted, February, 1941
SECOND EDITION, MAY, 1941
Reprinted, December, 1941
Reprinted, March, 1942
Reprinted, June, 1942
Reprinted, November, 1942
Reprinted, April, 1943
Reprinted, July, 1945
Reprinted, March, 1946
Reprinted, November 15, 1946
Reprinted, January, 1947
THIRD EDITION, JUNE, 1947
Reprinted, June, 1948
Reprinted, April, 1949
Reprinted, February, 1950
NEW LESSONS IN ARC WELDING, MARCH, 1951
Reprinted, May, 1953
Reprinted, December, 1955
Reprinted, November, 1956
SECOND EDITION, AUGUST, 1957
Reprinted, April, 1958
Reprinted, August, 1959
Reprinted, August, 1960
Reprinted, August, 1961
Reprinted. August, 1962
Reprinted, April, 1963
Reprinted, December, 1963
Reprinted, August, 1964
Reprinted, January, 1965
Reprinted, July, 1965
Reprinted, December, 1965
Reprinted, September, 1966
Reprinted, March, 1967
Reprinted, November, 1967
Reprinted, July, 1968
Reprinted, January, 1969
Reprinted, March, 1970
Reprinted, June, 1971
Reprinted, April, 1972
Reprinted, September, 1972
Reprinted, October, 1973
Reprinted, June, 1974
Reprinted, February, 1975
Reprinted, December, 1975
Reprinted, April, 1976
Reprinted, October, 1977
Reprinted, June, 1978
Reprinted, December 1978
Reprinted, August 1979

Printed in U.S.A.

LB-10
Library of Congress
Card Catalog No. A58-2951

INTRODUCTION

This edition of NEW LESSONS IN ARC WELDING is the 4th revision of a text resulting from an enlargement of the lessons used in the Lincoln Arc Welding School. This school has been in operation since 1917. The methods and material presented here are not the ideas of any one person. They are the result of information gathered from the staff of the Lincoln Arc Welding School, weldors in the field, welding teachers, field engineers, and the comprehensive engineering and research facilities of The Lincoln Electric Co.

Arc welding has progressed tremendously in the past 10 years. New electrodes, machines and processes have been developed. This book presents concisely the latest technical information on the practice of arc welding.

The text is divided into three areas. Lessons 1.1 to 1.47 contain the basic information necessary to become an all-around weldor. Lessons to learn the skills are arranged according to their probable need and degree of difficulty. Some ability to perform each job should be learned before attempting the next. Information lessons are presented as they are needed to adequately understand the process, equipment or materials encountered in the welding jobs.

Lessons 2.1 to 6.4 contain information and operating procedures on specific machines and electrodes manufactured by The Lincoln Electric Company. These are representative of machines and materials encountered by the weldor in his work the world over. The name or trade mark of The Lincoln Electric Company is your guarantee of the highest quality in machine, accessory and electrode design, manufacturing and service.

The Welding Application Supplement in Section VII contains information and procedure data for minimizing welding costs, supplementary information to help weldors better understand the welding process and its use.

Further and more detailed technical information on the field of arc welding and its application may be obtained from the latest edition of the "Procedure Handbook of Arc Welding Design and Practice" or any of the specialized books as described in Part 7.6 of Section VII.

NEW LESSONS IN ARC WELDING will be useful to anyone engaged in the welding industry.

The student or beginning weldor will find step-by-step procedures for performing basic welding operations, as well as information on welding equipment, supplies and processes. The lessons may be used as instruction sheets to supplement teacher demonstrations and classwork, or as self-instruction units.

The experienced weldor may refresh his memory on technical phases of the welding field, pipe and alloy welding, or those seldom-encountered jobs and repairs.

The welding foreman or job shop operator will find that new electrodes and machines have been developed that will do special jobs as well as run-of-the-mill jobs better, faster and at reduced cost.

The designer or draftsman will find information on welding symbols, types of joints, welding design and electrode specifications to aid in planning and designing jobs using welded fabrication.

The Lincoln Electric Company is interested in the weldor and in helping him increase his knowledge and skill toward maximum job proficiency. Correspondence is invited with weldors or persons concerned with welding in regard to problems they may have.

Acknowledgement

The Lincoln Electric Company is grateful to William Sellon, Bemidji State College, Bemidji, Minnesota many of whose ideas on format and instruction objectives, are included in this new edition and whose writing has done much to improve the lesson section.

The Lincoln Electric Co.
Cleveland 17, Ohio
July, 1957

TABLE OF CONTENTS

SECTION I—ARC WELDING THEORY AND TECHNIQUES

SECTION II—ARC WELDING MACHINES

SECTION III—MILD STEEL ELECTRODES

SECTION IV—STAINLESS STEEL ELECTRODES

SECTION V—HARDSURFACING ELECTRODES

SECTION VI—CAST IRON, NON-FERROUS, AND WASH-COATED MILD STEEL ELECTRODE

SECTION VII—SUPPLEMENTARY DATA

SECTION 1

ARC WELDING THEORY AND TECHNIQUES

LESSON 1.1

Object:
To study the arc welding process.

General Information

This lesson provides a general explanation of the arc welding process, so that future lessons, each covering only one small part of the process, may be understood in relation to each other. This lesson should be understood clearly and be referred to frequently.

What Is Arc Welding?

Arc welding is a method of joining two pieces of metal into one solid piece. To do this, the heat of an electric arc is concentrated on the edges of two pieces of metal to be joined. The metal melts and, while these edges are still molten, additional melted metal is added. This molten mass cools and solidifies into one solid piece.

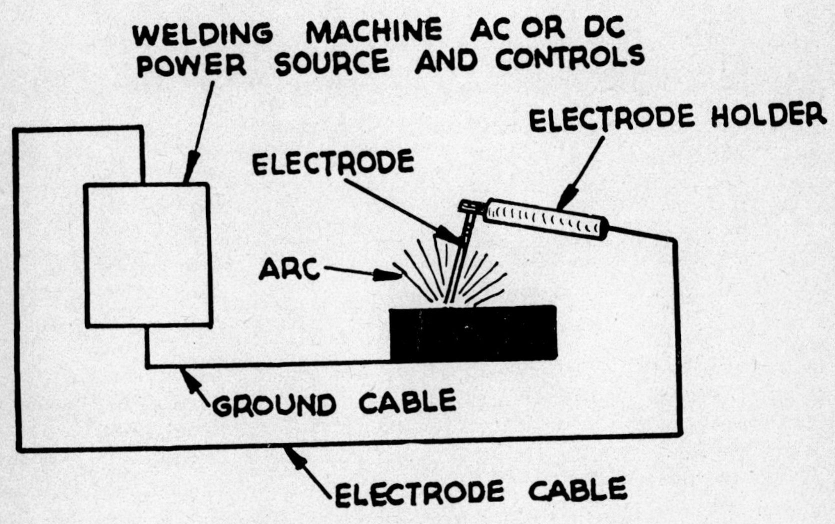

Fig. 1. The Welding Circuit.

The electric arc is made between the tip end of a small metal wire, the electrode, which is clamped in a holder and held in the hand. A gap is made in the welding circuit (see Figure 1) by holding the tip of the electrode 1/16″-1/8″ away from the work or base metal being welded. The electric current jumps this gap and makes an arc which is held and moved along the joint to be welded, melting the metal as it is moved.

Arc welding is a manual skill requiring a steady hand, good general physical conditions and good eyesight. The operator controls the welding arc and, therefore, the quality of the weld made.

What Happens in the Arc?

Figure 2 illustrates the action which takes place in the electric arc. It closely resembles what is actually seen during welding.

The "arc stream" is seen in the middle of the picture. This is the electric arc created by the electric current flowing through air between the end of the electrode and the work. The temperature of this arc is about 6000°F. which is more than enough to melt metal. The arc is very bright, as well as hot, and cannot be looked at with the naked eye without risking painful, though temporary, injury.

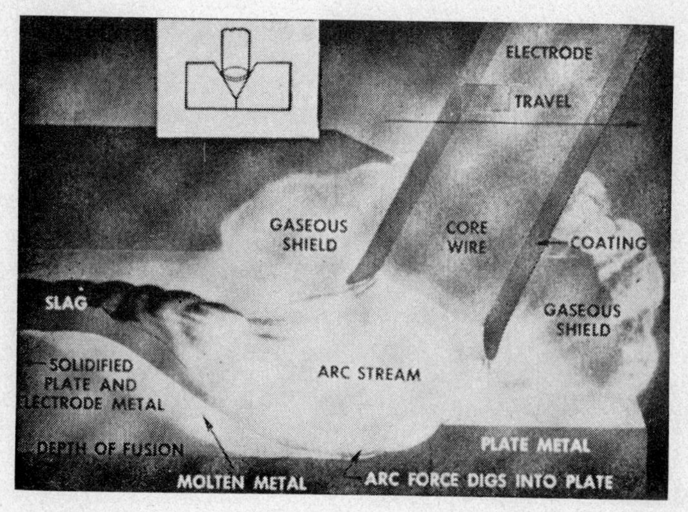

Fig. 2. The Welding Arc.

The arc melts the plate, or "base", metal and actually digs into it, much as the water through a nozzle on a garden hose digs into the earth. The molten metal forms a molten pool or "crater" and tends to flow away from the arc. As it moves away from the arc, it cools and solidifies. A slag forms on top of the weld to protect it during cooling.

The function of the electrode is much more than simply to carry current to the arc. The electrode is composed of a core of metal wire around which has been extruded and baked a chemical coating. The core wire melts in the arc and tiny droplets of molten metal shoot across the arc into the molten pool. The electrode provides additional "filler" metal for the joint to fill the groove or gap between the two pieces of the base metal. The coating also melts or burns in the arc. It has several functions. It makes the arc steadier, increases the arc force, provides a shield of smoke-like gas around the arc to keep oxygen and nitrogen in the air away from the molten metal, and provides a flux for the molten pool, which picks up impurities and forms the protective slag. The principal differences between various types of electrodes are in their coatings. By varying the coating, it is possible to greatly alter the operating characteristics of electrodes.

The Arc Welding Circuit

The operator's knowledge of arc welding must go beyond the arc itself. He must know how to control the arc and this requires a knowledge of the welding circuit and the equipment which provides the electric current used in the arc. Figure 1 is a diagram of the welding circuit. The circuit begins where the electrode cable is attached to the welding machine and ends where the ground cable is attached to the welding machine. Current flows through the electrode cable to the electrode holder, through the holder to the electrode and across the arc. From the work side of the arc, the current flows through base metal to the ground cable and back to the welding machine. The circuit must be complete for the current to flow, which means that it is impossible to weld if the cables are not connected to the machine or to either the electrode or work.

The several types of welding machines include motor-generators, engine-driven generators, transformers, rectifiers and combination transformer and rectifiers. Each type has its place and purpose, as will be shown in later lessons, but the basic function of each is the same—providing a source of controlled electric power for welding. This controlled electric power has the characteristic of high amperage at low voltage. The high amperage is required to provide sufficient heat at the arc. The voltage must be low enough to be safe for handling and yet high enough to maintain the arc. The welder (machine) permits the weldor (operator) to control the amount of current he uses. This, in turn, controls the amount of heat at the arc. Some welders also permit the operator to select either a forceful or soft arc and to control its characteristics to suit the job.

The Weldor's Job

A good weldor does more than simply hold the arc. He must, first of all, be able to select the correct size and type of electrode for each job. He must know which machine to use for each job and be able to set the current and voltage controls properly. He must be able to manipulate the electrode and arc so as to make a satisfactory weld under varying conditions. In addition, the weldor must have a knowledge of joint preparation, positioning the work, distortion, and many other factors which enter into the final result of a good weld. He must be a mechanic and a craftsman. Nearly anyone can "stick two pieces of metal together", but becoming a *good* weldor requires study, training and practice.

The Possibilities of Welding

Though not a new process, the arc welding industry is still rapidly expanding. There is room in the industry for every qualified weldor who desires to do his part in promoting the industry.

The expansion of the welding industry results from the cost reductions it makes possible in all types of metalworking plants. In most applications, arc welding is less expensive than riveting or bolting. Welding, when applied to proper designs, can also produce substantial savings over cast machinery. The possibilities of expansion in the welding industry are immense, but to recognize these possibilities to the fullest requires the best efforts of everyone in the industry. You can add to the expansion and help to secure your own future by consistently doing the best possible welding in the least possible time.

LESSON 1.2

Object:
To study safe and efficient work habits in the welding shop.

General Information

Before using the arc welder you should know certain precautions and information about operating and handling your equipment, accessories and tools. To do the job easily, properly and safely, there are some simple rules to observe. A safe worker is one who knows his equipment and materials and proceeds with correct work habits. A careless workman endangers not only himself, but those working around him.

Understand your welder. That means more than knowing where the "ON" and "OFF" switch is. Know the type of machine you are using and how it produces the welding current. Welders require very little maintenance to keep them in top working condition. Follow the operators manual and do not neglect that periodic check. Keep welders dry at all times, as moisture damages windings. Turn your welder on when you are ready to practice or start a job, and turn it off when you are finished.

Keep welding accessories in good condition. Cables should be of adequate size for the amperage they are to carry, have no insulation breaks, and be fastened securely to the welder, ground clamp and electrode holder. Loose connections cut down on welder efficiency and produce an unpredictable welding arc. The ground should fasten securely to the work or table at a point free from spatter, dirt or grease. The electrode holder jaws should be filed or brushed periodically to make good contact with the electrode. Keep cables out of the way, where they are not stepped on, tripped over, run over by vehicles or exposed unnecessarily to grease and oil.

Wear good protective equipment. Use a headshield or a handshield held close to the face with number 10 or 11 filter glass to protect the eyes and face from the heat and bright rays of the arc and spatter during welding. Sun glasses or gas welding goggles are not adequate protection. Caution the persons around you before striking an arc. Any light leaks in the shield should be covered. If the headshield does not have a "flip-up" filter glass, safety goggles should be worn to protect the eyes while chipping.

If a direct "flash" is encountered from the arc a rather painful, but not permanent, burning of the eyes results. This condition may be relieved or lessened by an application of cool boric acid or a few drops of 5% argyrol every 5 hours. Aspirin will relieve pain and headache and permit rest, helpful in promoting recovery.

Wear chrome tanned leather gloves and apron, and adequate work clothing to protect the skin against heat, spatter and "sunburn" of the arc rays. Ankle height shoes, turned down cuffs and closed pockets help ward off sparks. Flammable materials, such as "kitchen" matches and cellophane wrappers should be removed from pockets. A cap and leather sleeves and shoulders are necessary for all-position welding.

Remove flammable materials from the welding area before starting to weld. Sparks will travel for some distance, especially at high amperages. Welding small or movable items is best done on a metal table in a fireproof booth. A portable fire extinguisher should be accessible nearby.

Zinc (galvanizing), tin, brass, and lead fumes are especially toxic to breathe and must be ventilated out or blown away during welding. Mild steel flux fumes are not toxic, although like any smoke they may become uncomfortable if not properly ventilated.

Keep your tools away from the welding area, if they are not being used. Weld spatter will mark and roughen finished tool and machine surfaces.

Do not weld empty containers unless you know what they have contained. Tanks or containers which have held flammable material should be thoroughly cleaned and tested. If you intend to work on this type of welding it is well to obtain the booklet "Recommended Procedure to be Followed in Preparing for Welding or Cutting Certain Types of Containers Which Have Held Combustibles" from the American Welding Society, 33 West 39th St., New York, N. Y.

The floor or ground should be dry in the welding area. In case of water or moisture, welding should be done on a platform or rubber mat. Avoid welding with wet gloves.

Test pieces for heat before picking them up by holding an open palm over them. Hot metal should be handled with pliers or tongs, not welding gloves. Leather gloves will become stiff and uncomfortable if overheated.

Treat cuts or burns promptly to avoid infection by keeping a first aid cabinet in the shop. When in doubt about an injury, or in case of serious injury, consult a physician.

Plan your work. This is especially important to the something-new-every-day type of weldor in the job shop and on maintenance work. You can save yourself loss of time, materials and temper, if some simple rules are followed *before* starting to work.

1. Make or have a dimensioned drawing if the job is new construction.
2. Plan a step-by-step procedure to follow when doing the job. There may be more than one way of doing a job, but be sure yours is a workable way. Beware of the "shortcut" you have not tried before. You can "paint yourself into a corner" with an electrode holder as well as a paint brush.
3. Have all necessary materials on hand. The inability to secure certain parts or materials may necessitate a change in design or plans, best accomplished before starting work.
4. Be certain you can perform each part of the job. If a new electrode or type of weld is necessary, practice its use on scrap material.
5. Measure twice and cut once.

Review Questions

1. Why is moisture or water dangerous to both the welder and the operator?
2. What is caused by loose connections?
3. What shade of filter glass is worn for welding?
4. Must a filter glass be worn when observing as well as when welding?
5. When must safety goggles be worn?
6. What painful effects may the arc have on the weldor?
7. Which welding fumes are toxic?
8. Why is it necessary to know what an empty container held before welding?
9. How is a piece of metal checked to find out whether or not it is hot?

LESSON 1.3

Object:
To study the arc welding machines and accessories

General Information

The success of welding as a metal-joining process rests on the fact that a good weld is as strong or stronger than the plate in which it is made. This success has been established through the years by the gradual development of welding machines, accessories and electrodes that satisfy the complex requirements of the arc process.

Arc welding requires a continuous supply of electric current, sufficient in amount (amperes) and of proper voltage to maintain an arc. This current may be either alternating (AC) or direct (DC), but it must be provided through a source which can be controlled.

Several different types of welding machines are available for producing satisfactory welding current. Alternating current is produced in special welding transformers (Fig. 1). Direct current is produced in either electric motor-generator sets (Fig. 2), rectifier sets, or engine-driven generator sets. Combination welders, producing both AC and DC are basically transformer-rectifier sets.

Fig. 1. AC Transformer Welder

Fig. 2. DC Motor Generator Welder

Welding machines of all types are rated according to their current output. They range from 100 ampere machines, used for welding at home on the house power circuit, to 1200 amperes or more for use with automatic equipment. This rating is set by conscientious manufacturers in accordance with standards established by the National Electrical Manufacturers Association. These standards are established on a conservative basis, requiring a rating well below the maximum overload capacity of the machine so that it will provide safe operation efficiently over a long period of time. Ratings are given with a percentage duty cycle. The duty cycle of a welder is the percentage of a ten minute period that a welder can operate at a given output current setting. If a welder is rated 300 amperes at a 60% duty cycle, it means that the machine can be operated safely at 300 amperes welding current for 6 out of every 10 minutes. If this duty cycle is reduced in actual operation, the maximum permissible current is increased. At 35% duty cycle, a 300 ampere machine could be operated at 375 amperes.

Transformer welders are available for operation on single phase power lines. They transform high voltage–low amperage input current to a low voltage—high amperage welding current. Current controls must be provided on the transformer to permit its use for welding. There are two basically different types of current control systems used on transformer welders. The first type of current control is a continuous, crank operated control. The other is a tap type control in which the electrode leads are plugged into different jacks, or taps, to obtain different settings. Both types of control accurately regulate the welding current. With either, the operator can select just the welding heat he wants.

Rectifier sets are basically 3 phase or single phase transformers to which have been added selenium or other rectifiers to change the output current from alternating to direct current. These machines have the basic control and output characteristics that are inherent in transformers.

Direct current generators consist of an armature rotating in an electrical field. Current is generated in the armature and is taken off for use through a commutator. The armature is rotated either by an electric motor or an internal combustion engine. When generators are engine operated they are independent of electric power and may be operated in the field where power is not available.

Combination welders, producing both AC and DC current, are the most versatile of all types of welders (Fig. 3). They are basically a single phase transformer and a rectifier from which, by turning a switch, either alternating

Fig. 3. AC & DC Welder

or direct welding current is available. With DC, polarity can be switched to either electrode positive or electrode negative. On both AC and DC, the open circuit voltage can be set independently of the current to vary the arc characteristics. These machines can be adapted to individual job requirements, combining large AC capacity with a smaller DC capacity, or any other combination that is required.

Several factors must be evaluated when selecting a welding power source. The size or rated output of a machine required for a given job depends on the thickness of the metal to be welded and the amount of welding to be done. There is no need to buy more capacity than will be required by the job. Be sure, however, to check the duty cycle. Machines with a low duty cycle should be used only for maintenance or intermittent welding. 60% duty cycle is normal for industrial welders. Continued operation of a machine beyond its rated capacity will shorten its service life.

In selecting the welder, another consideration is the power source available. Fortunately, welders are made to be used with all types of power, but certain types are available for only given conditions. Motor-generator sets are generally available for only 3 phase AC power, but can be ordered to different cycles and voltages. They are also available for DC power. AC machines are generally available for only single phase AC power in various cycles, with or without power factor correction in the machine. Fortunately, in most manufacturing situations, the source of power does not present a limiting factor on the selection of a welder. The decision can be made on the basis of which is the most efficient and economical machine for a given job.

However, where service is through a 3 KVA transformer on residential or rural lines an industrial type AC welder cannot be used. It will be necessary to have a limited-input type transformer welder. This is designed for a limited power input, so that no more than a specified maximum amount of input current (37.5 amps) is drawn during operation.

Also, consider what type machine will make the job easiest to do and enable better welding to be done at lower costs. There is one best way to do every welding job.

Input power to the welder should be planned and installed only by qualified electricians. Lines should be of the proper size and properly fused, according to local requirements and correct standards.

Further information concerning specific types of welding machines is covered in Lessons 2.1 to 2.7.

Current carrying cables, cable lugs, electrode holder, ground clamp, weld cleaning devices, and protective equipment are essential for each welding ma-

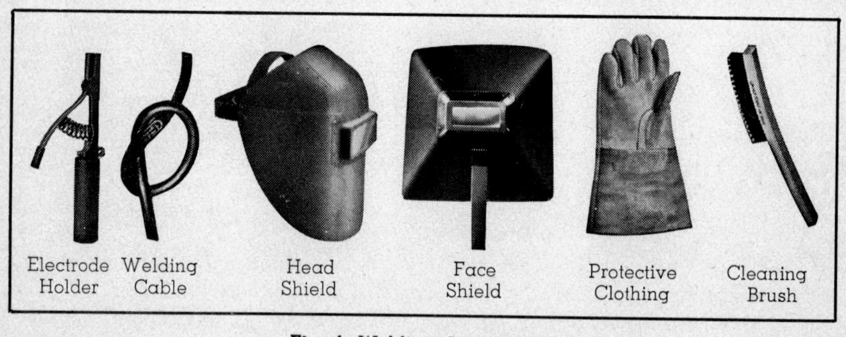

Fig. 4. Welding Accessories

chine and operator. These are called accessories (Fig. 4). For maximum safety and efficiency, good quality accessories must be used.

The welding current is conducted from the source of power to the electrode holder by an insulated copper or aluminum cable. An extra-flexible cable is used between the holder and the welding machine. This cable is designed expressly for welding and derives its flexibility from the thousands of very fine, almost hair-like wires enclosed in a durable paper wrapping which allows the conductor to slip readily within its insulation when the cable is bent. The high-grade insulation also contributes to flexibility. Wear resistance is provided this cable by an extra tough, braided cotton reinforcing and by the special composition of the covering which provides a smooth finish, highly resistant to abrasion. For grounding the welding circuit, a somewhat less flexible, but equally wear-resistant, cable is used.

The size (thickness) of the cables used in welding varies, depending upon the capacity of the machine and the length of cable required. Cable size is selected carefully because of its bearing on weld production and efficiency.

The cable is connected to the welder by means of a copper terminal lug. This lug is soldered or clamped to the end of the cable and fits on the welder terminal posts.

The electrode holder grips the electrode during the welding process. This holder should be reasonably light, well insulated, and sturdy enough to withstand the wear of continual handling. The holder should be the correct size for maximum machine output. A spring grip holder for quick insertion or release of the electrode is best.

A ground clamp fastened to the work or the table on which the work is mounted completes the welding circuit. A spring-pressure ground clamp is the quickest and easiest method of grounding work. Magnetic grounds are available, and may be necessary for a secure ground on broad surfaces where it is difficult to use ground clamps.

A shield for the face and eyes is necessary for protection from arc rays and heat, and the spatter of molten metal. This shield may have a head band for wearing on the head or a handle attached for holding in the hand. The head shield is most commonly used for welding, as both hands are usually needed for the welding process. The shield should have a number 10 or 11 filter glass for general purpose welding. The more expensive filter glass should be protected by a clear cover-glass which may be replaced when covered with arc spatter.

Gloves and apron should be of chrome tanned leather to protect the hands, body and clothes from heat and spatter. If welding is done in positions other than flat, leather shoulders and sleeves must be worn.

Safety goggles should be worn under the shield to protect the weldor's eyes when chipping hot slag, grinding metal for joint preparation and cleaning metal with the wire brush.

A chipping hammer and wire scratch brush should be used to thoroughly clean the beads. A grinder with a wire wheel is handy for cleaning welds and removing rust and scale from the base metal.

A special welding table should be used in the shop for welding practice and for work on small jobs. A workbench soon becomes spattered and burned. This table should be steel for fire-proofness and ease of grounding work. It may be easily made in the shop.

Further protection for other persons working in the shop may be provided around the welding area by a booth or shield of fire-proof canvas, sheet metal or transite. This might be a permanent or portable unit, depending upon the work being done.

Review Questions

1. What is the strength of a weld as compared to the base metal?
2. What type of current supply is required for welding?
3. What two types of welders produce DC welding current?
4. What type of welding machine produces AC welding current?
5. What organization sets standards for rating welders?
6. What is the duty cycle of a welder?
7. Can a welder be used above capacity for short periods?
8. What are the electrical characteristics of welding power, as contrasted with input power?
9. What is an "arc booster"?
10. What is the best all-around welder for doing light and heavy jobs in all positions?
11. What type of welder must be used when no electrical power is available?
12. Is it necessary to use a special cable to carry the welding current?

LESSON 1.4

Object:
To study the uses and characteristics of the metal electrode.

General Information

The electrode might be called the weldor's most important tool. It is through the electrode that the weldor is able to handle, concentrate, and vary the characteristics of the intense heat of the electric arc to make it usable for purposes of welding. The success of a weldor depends upon his ability to understand the action of the electrode, and to select and use the correct one for the job.

The electrode universally used for manual welding has a core of wire or rod covered with a baked chemical coating. The core provides filler metal for the joint being welded, and the chemical coating provides a shield for the weld as it is made. The process of "Shielded Arc Welding" gets its name from the action of this type of electrode. These electrodes are made with a variety of different core wires, coating and diameters.

A good weldor should be familiar with the common types of electrodes available. When selecting an electrode, it is necessary to use one which is made for welding with the type of current (AC or DC) being used, for the particular base metal being welded, and for the purpose to which the finished job is intended. Lessons 5.1–6.4 describe in detail the operation and application of all types of Lincoln electrodes.

The coating on electrodes has several functions and the chemicals are varied, depending upon the desired weld. Some of the common chemicals used in the coatings are cellulose (cotton or wood pulp) for a gas-shielding agent, titanium dioxide (rutile) as a slag former, ferro-manganese as a deoxidizing agent, asbestos to produce arc force and slag, and sodium silicate (water, glass) to bond the various chemicals and act as a fluxing agent. They are evenly applied to the core wire by the process of extrusion, similar to the way toothpaste is squeezed out of a tube. The coating largely determines the operating characteristics of the electrode.

The same core wire is used for most ferrous electrodes–low carbon steel. It is the addition of alloy in the coating which makes the difference in the physical properties of the deposited metal.

The chemical coating is called a flux, because it has a mixing and cleansing effect on the weld. This flux has three distinct purposes in metal arc welding.

Parts of the flux coating are melted and vaporized in the heat of the arc (Lesson 1.1, Figure 2). This forms a smoke or blanket of gas which protects both the droplets of metal projected across the arc and the pool of molten metal from the surrounding air. The air contains oxygen and nitrogen which combine readily with molten steel, and even more readily with the small droplets as they are projected across the arc gap, to form oxides and nitrides. If these are allowed to form, the finished weld is weakened by being porous and brittle. Its tensile strength and impact resistance are substantially reduced.

Other parts of the flux melt and mix with the weld metal, gather impurities and float them to the top of the molten pool in the form of slag. This slag protects the hot metal from the air, keeps it from solidifying too rapidly, so that gases may escape, and partly influences the shape of the bead. After the bead solidifies and cools down, the brittle slag may be chipped off leaving a clean, bright weld.

During the welding process, the flux coating projects over the end of the core wire and influences the action of the arc, similar to the way a shotgun barrel directs the pattern and distance of a load of shot. It stabilizes and directs the force of the arc and the droplets of molten filler metal. Hence, the coating makes the arc steadier and welding easier.

It is through the coating, also, that electrodes have been made for satisfactory operation with alternating welding current. An AC arc tries to go out 120 times a second on 60 cycle current. This makes arc stability a problem. The problem can be overcome by adding certain chemicals to the coating. These chemicals, when burned in the arc, produce special "ionized" gases which keep the arc going. It follows, then, that any AC electrode will work on DC, but that not all DC electrodes will work on AC.

Electrodes should be stored where it is dry. Excessive moisture can destroy the correct action of the coating. Storing open boxes of electrode in a closed cabinet heated by a heater coil or light bulb is good practice. Wet electrode may be dried out before using.

Welding has not always been as easy as you find it today using the shielded electrode. Before coated electrodes were produced commercially, welding was done with a bare or lightly coated wire. Bare electrodes are still used on some applications and you may run into them. They are used where maximum weld strength is not essential, when filling large grooves, filling cavities in castings, and where complete slag removal is difficult. Because they have no flux and cannot produce a shielded arc, the arc is less forceful and will short out easily as droplets of electrode bridge the gap. This makes them more difficult to use than a shielded electrode. The weldor must carry a consistently short arc. Without the shielding power of the flux, the resultant welds tend to be about one-third weaker in tensile strength and lower in impact resistance, due to porosity and impurities. They can be satisfactorily run only on a DC welder using electrode negative polarity.

Review Questions

1. Why is the electrode considered a tool?
2. Why is the name "Shielded Arc" used for a type of metal electrode?
3. Why is it necessary to have the correct electrode when making a weld?
4. Are the same chemicals used in all electrode coatings?
5. What are three purposes for putting flux on an electrode?
6. What elements in the air are harmful when mixed with the molten metal?
7. What important characteristic does an AC electrode flux have?
8. Why should the flux be left on the weld deposit during cooling?
9. What is the disadvantage of weld deposits made with bare electrodes?

LESSON 1.5

Object:

To strike and establish an arc.

Equipment:

Lincoln "Idealarc" or DC or AC welder and accessories.

Material:

Mild steel plate 1/8" or thicker; 5/32" "Fleetweld 5" (E6010) for DC or "Fleetweld 35" (E6011) for AC.
(See Lessons 3.1 and 3.2 for properties and application of these electrodes. If a limited input AC welder is used, "Fleetweld 180", Lesson 3.5, should be used.)

General Information

The basis of arc welding is the continuous electric arc. This arc is maintained when the welding current is forced across a gap between the electrode tip and the base metal. A weldor must be able to strike and establish the correct arc easily and quickly. There are two general methods of striking the arc, scratching and tapping.

The scratching method is easier for beginners and when using an AC machine. The electrode is moved across the plate inclined at an angle, as you would strike a match. As the electrode scratches the plate an arc is struck. When the arc has formed, withdraw the electrode momentarily to form an excessively long arc (Fig. 1), then return to normal arc length.

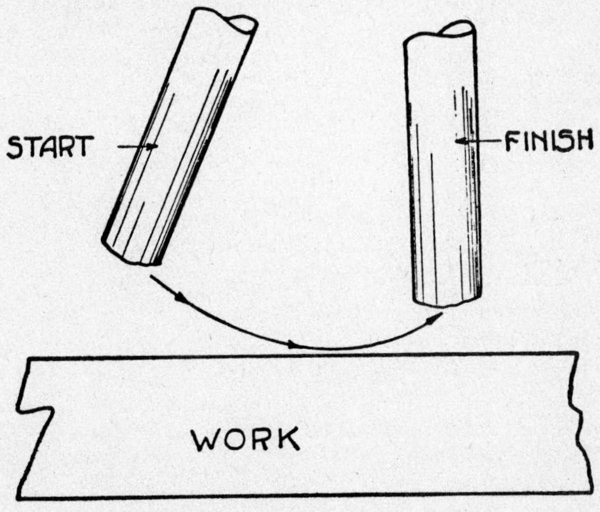

Fig. 1. "Scratch" Method of Arc Starting

In the tapping method the electrode is moved downward to the base metal in a vertical direction. As soon as it touches the metal it is withdrawn momentarily to form an excessively long arc (Fig. 2), then returned to normal arc length.

The principal difficulty encountered in striking the arc is "freezing", that is, the electrode sticks or fuses to the work. This is caused by the current melting the electrode tip and sticking it to the cold base metal before it is withdrawn from contact. The extra high current drawn by this "short circuit" will soon

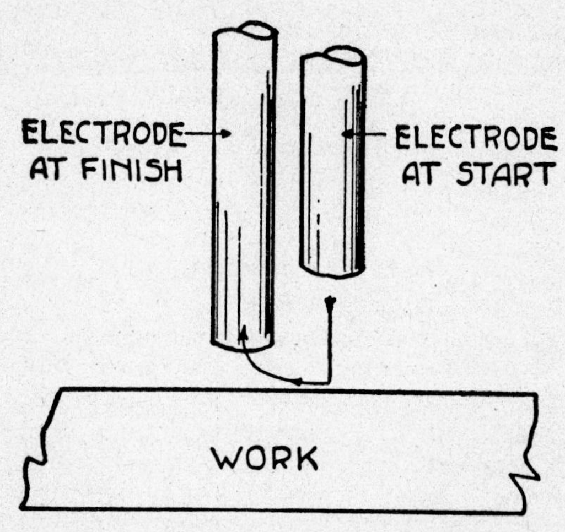

Fig. 2. "Tapping" Method of Arc Starting

overheat an electrode and melt it or the flux, unless the circuit is broken. Giving the electrode holder a quick snap backward from the direction of travel will generally free the electrode. If it does not, it will be necessary to open the circuit by releasing the electrode from the holder.

Caution: Never remove the shield from the face if the electrode is frozen. Free the electrode with the shield in front of the eyes, as it will "flash" when it comes loose.

Job Instructions

Job A: Strike the arc by the scratching method

1. Brush the base metal free of dirt and scale.
2. Position metal flat on metal table top or plate.
3. Attach the ground securely to work or table.
4. Set the amperage at 130 to 145 for 5/32" electrode, DC electrode positive.
5. Place the bare end of the electrode in the holder so that it is gripped securely at a 90 degree angle to the jaws.
6. Turn the welder "ON".
7. Assume a natural position and grasp the holder firmly but comfortably by using either one or both hands. Using both hands helps to steady the electrode and reduce fatigue. To use both hands rest the left elbow on the work table and with the left hand steady the right hand by holding the right wrist.
8. Hold the electrode above the plate and move it down until it is about an

inch above the plate. Hold it perpendicular to the plate, inclined at an angle of 20 to 25 degrees in the direction of travel (Fig. 3, Position 1).

9. Place the shield in front of your eyes.

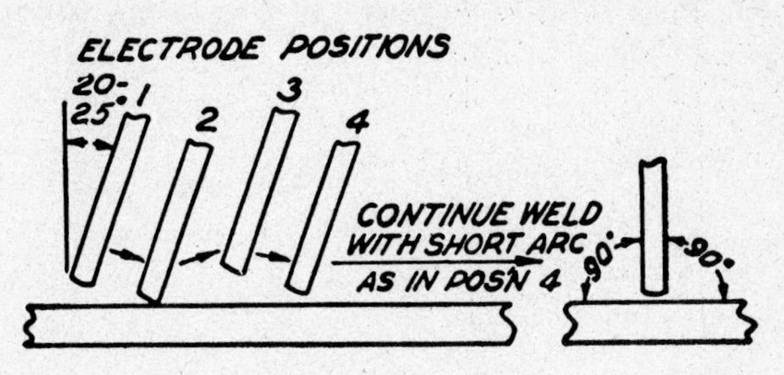

Fig. 3. Striking Sequence.

10. Strike the arc like a large match by gently and quickly scratching the electrode on the metal with a wrist motion as shown in Position 2. A sudden burst of light will be produced on contact with the plate.
11. Withdraw the electrode to form an excessively long arc, about 3⁄16", as in Position 3. This long arc is held only a second or two after which a normal arc length of 1⁄16" to 1⁄8", shown in Position 4, is assumed. This long arc prevents the large drops of metal which pass across the arc at this time from shorting out the arc, and causing it to "freeze". It also established the crater, eliminates excessive build-up of filler metal and helps to tie in more smoothly with the previously deposited bead.
12. Practice starting the arc, holding it, and breaking it, until you are able to easily strike the arc on first try. From the standpoint of time, the complete motion represented by Positions 1 to 4 takes place in two or three seconds.

Job B: Strike the arc by the tapping method.

1. Follow steps 1 to 7 in Job A, using the same amperage setting.
2. Hold the electrode above the plate in a vertical position, and lower it until it is about an inch above the point you wish to strike the arc (Fig. 2).
3. Place the shield in front of your eyes.
4. Touch the electrode gently to the plate by a downward motion of the wrist. With the first burst of light, quickly withdraw it to form a long arc, about 3⁄16". Hold the long arc for a second or two, then assume a normal arc length, 1⁄16" to 1⁄8".
5. Incline the electrode 20 to 25 degrees in the direction of travel as in Job A.
6. Practice striking the arc, holding it and breaking it.

Review Questions

1. What happens when electric current is forced across an open gap?
2. What two methods are used to strike an arc?
3. Which method is easier for a beginner?
4. Which method works better on an AC welder?

5. When an electrode "freezes" what should be done?
6. Should you remove your shield to see better when freeing a "frozen" electrode?
7. Should base metal be cleaned before welding?
8. Is it correct to use either one or both hands on an electrode holder when welding?
9. What should be done as soon as the arc appears at the tip of the electrode?
10. What is normal arc length?
11. What electrode angle is used?

LESSON

1.6

Object:
To run a straight bead in the downhand position.

Equipment:
Lincoln "Idealarc" or DC or AC welder and accessories.

Material:
Mild steel plate 3⁄16" or thicker; 5⁄32" "Fleetweld 5" (E6010) for DC or "Fleetweld 35" (E6011) for AC.

General Information

The bead is a continuous deposit of weld metal formed by the metallic arc on the surface of the base metal. It is this bead or series of beads, composed of a fused mixture of base metal and filler metal, that forms the weld.

Spend sufficient time on these jobs to become proficient in holding the proper arc length and electrode angle. Move the electrode along the plate at the correct speed, so as to secure smooth, uniform beads with adequate penetration as indicated in Figs. 1 and 6A. During welding observe the appearance of the bead and the characteristics of the arc. See how the arc digs into the metal for penetration, how it fills the crater and builds up the bead. Learn to recognize a good bead while you are making it. Keep your eye on the back of the crater as the arc force deposits and builds up the bead, so that you can quickly vary the arc length, electrode angle or speed of travel to correct a poor bead.

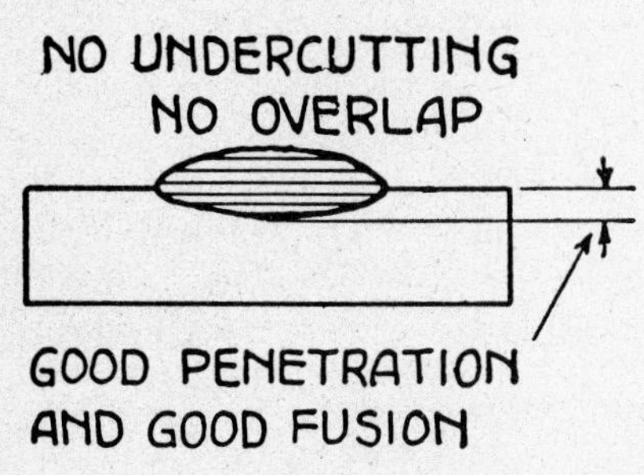

Fig. 1.

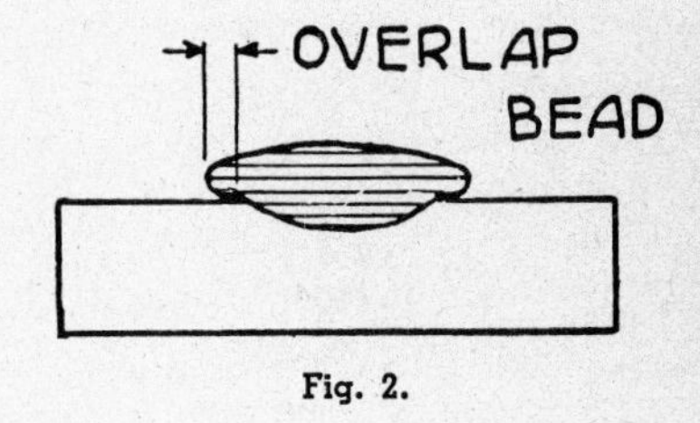

Fig. 2.

Normal arc length should be slightly less than the diameter of the electrode, and is usually considered to be 1⁄16" to 1⁄8". Judging arc length by fractions of an inch is difficult. Correct arc length will be developed by proper judgment of the weld deposit. With too long an arc there will be a noticeable increase of spatter; penetration will be poor; overlap will be noticeable; sound of the arc will be more of a hiss than a crackle; the metal will melt off the electrode in large wobbly drops and the slag will be difficult to remove from the completed bead.

When the rate of travel is too fast, the bead will be thin and stringy with poor penetration. If the rate of travel is too slow, weld metal will pile up and roll over with excessive overlap (Fig 2).

Correct amperage setting for any given electrode is important to secure proper shape of bead, proper penetration and a minimum of spatter. When amperage is set too high the bead will be flat with excessive spatter and some porosity, and the electrode becomes overheated. If amperage setting is too low difficulty is experienced in striking the arc and maintaining correct arc length. The weld metal piles up with excessive overlap and poor penetration.

Job Instructions

Job A: Run short beads.

1. Clean base metal and position flat on the table.
2. Check ground connection to table or work.
3. Set amperage at 130 to 145 for 5⁄32″ electrode, DC, electrode positive.
4. Hold the electrode perpendicular to the base metal inclined at a 20 to 25 degree angle in the direction of travel.
5. Strike and establish the arc as in Lesson 1.5.

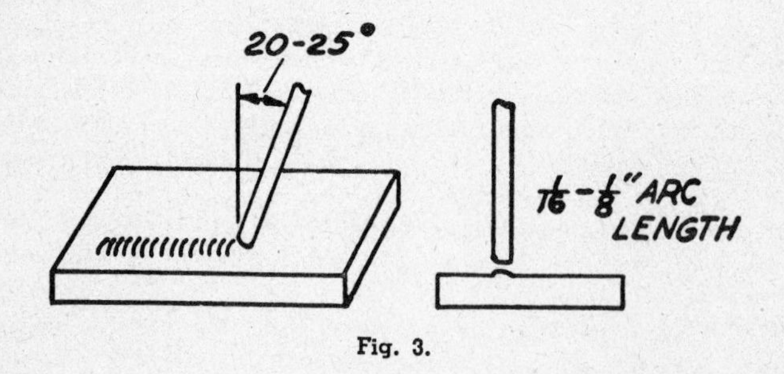

Fig. 3.

6. Maintain a normal arc length, 1⁄16″ to 1⁄8″ (Fig. 3), and move the electrode across the plate at a uniform rate. A right-handed weldor normally works from left to right.
7. Observe the back of the molten puddle, or crater, as the arc builds up the bead. Allow the arc force to penetrate the base metal and deposit filler metal. Correct speed will be indicated by the proper shape and size of the bead (Fig. 6).
8. Make beads one to two inches long and extinguish the arc by withdrawing the electrode.
9. Restrike the arc and run another bead.
10. Move over the plate increasing the length of the beads until you are able to stop and start as desired. Practice until you can make uniform beads 3 or 4 inches long.
11. Clean the slag off each bead by chipping with the chipping hammer and brush clean with the wire brush. Slag is removed more easily if the weld is allowed to cool a short time. Always chip slag away from you.
12. Examine the bead for shape, penetration and uniformity. Compare with Fig. 6.

Job B: Run long beads with correct rate of deposition.

1. Follow steps 1 to 7 in Job A.

2. Run parallel beads about 12" in length (Fig. 4).
3. Run beads toward you, away from you, and from the right and the left.
4. Chip off the slag and inspect the bead for shape, penetration and uniformity. There should be no overlapping or undercutting along the edge of the deposit.
5. Check the length of the weld with the length of the electrode used. Correct speed will produce approximately 1 inch of bead for each inch of electrode consumed (for non-iron powder electrodes). An electrode should be used so that about 1½" remains. Longer stubs cause excessive waste, and shorter stubs may damage the electrode holder.

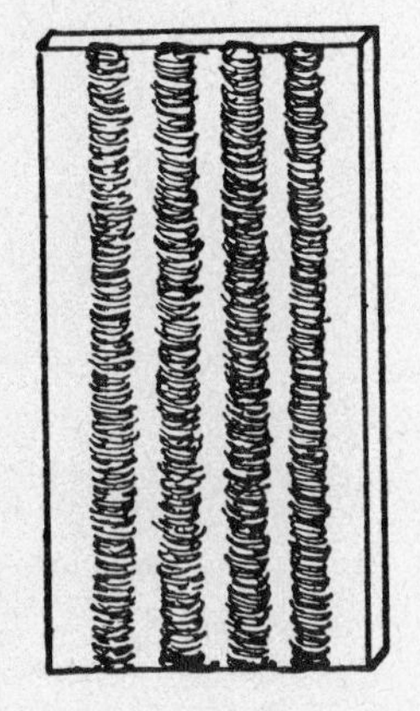

Fig. 4.

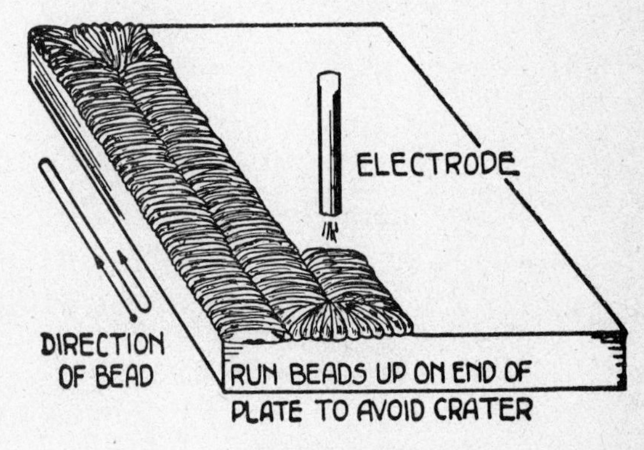

Fig. 5.

6. After you are able to run straight beads, weld in one direction to the end of the plate, move slowly to one side and reverse the direction of travel (Fig. 5). Keep the correct electrode angle when changing direction of travel. Try the same job welding toward and away from you.
7. Study the plate shown in Fig. 6 and the accompanying table. Compare with your beads.

Review Questions

1. Why is it important to be able to run a sound bead?
2. What electrode angle is used when running a bead?
3. Will moving the electrode too fast cause poor penetration?
4. Can a good weldor strike beads in any position and run in any direction?
5. How is a weld cleaned?
6. A 14" electrode should produce approximately what length of bead?
7. How long should the stub of an electrode be when it is thrown away?
8. What is penetration?
9. What are four important points to check when running a bead?

STUDY THIS PICTURE AND TABLE

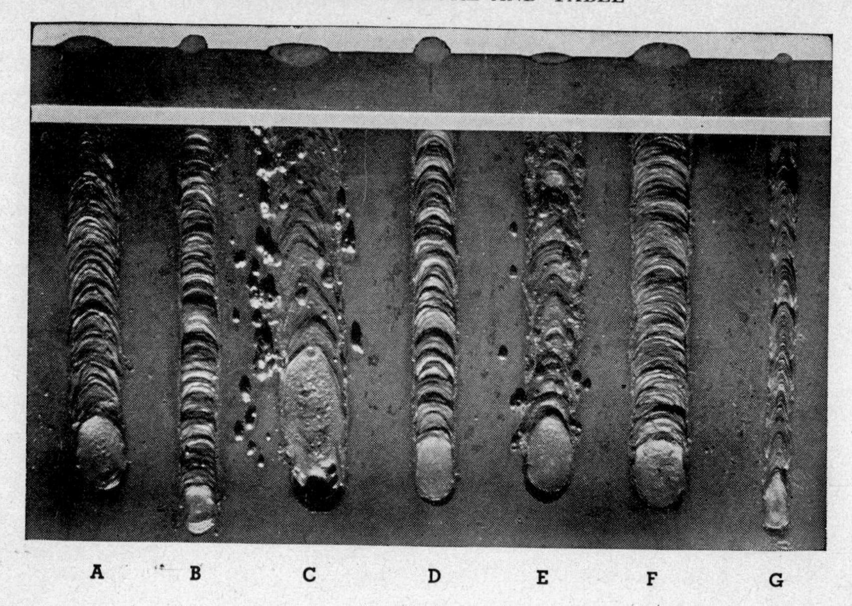

Fig. 6. Plan and elevation views of welds made under various conditions. (A) Current, voltage and speed normal. (B) Current too low. (C) Current too high. (D) Voltage too low. (E) Voltage too high. (F) Speed too low. (G) Speed too high.

Operating Variables	RESULTING WELD CHARACTERISTICS			
	Arc Sound	Penetration—Fusion	Burn Off of Electrode	Appearance of Bead
A. Normal Amps., Normal Volts, Normal Speed	Sputtering hiss plus irregular energetic crackling sounds	Fairly deep and well defined	Normal appearance	Excellent fusion—no overlap
B. Low Amps., Normal Volts, Normal Speed	Very irregular Sputtering Few crackling sounds	Not very deep nor defined	Not greatly different from above	On top of plate. Note there is not the overlap such as is on bare rod
C. High Amps., Normal Volts, Normal Speed	Rather regular explosive sounds	Deep-long crater	Shielded arc coating is consumed at irregular high rate	Broad rather thin bead—Good fusion
D. Low Volts, Normal Speed, Normal Amps.	Hiss plus steady sputter	Small	Coating too close to crater. Touches molten metal and results in porosity. Rod freezes.	Sits upon plate but not so pronounced as for low amps. Somewhat broader
E. High Volts, Normal Speed, Normal Amps.	Very soft sound plus hiss and few crackles	Wide and rather deep	Note drops at end of electrode. Flutter and then drop into crater	Wide—Splattered
F. Low Speed, Normal Amps., Normal Volts	Normal	Crater normal	Normal	Wide bead—overlap large. Base metal and bead heated to considerable area.
G. High Speed, Normal Amps., Normal Volts	Normal	Small, rather well-defined crater	Normal	Small bead—undercut. The reduction in bead size and amount of undercutting depends on ratio of high speed and amps.

LESSON 1.7

Object:
Determining the amount of current required for the job.

General Information

The actual procedure for setting a particular welding machine to the desired current is described for each machine in Section II. However, before you can set the machine, you must first know how much welding current you need.

First, select an electrode. A general rule of electrode selection is that the electrode diameter should not exceed the thickness of the base metal. This rule is only the starting point. Here are some other considerations:

1. Welding speed is the largest factor in welding costs. Use the largest electrode and greatest amperage possible.
2. Position of welding—3/16" electrode is maximum size for good puddle control on vertical, horizontal and overhead welding.
3. Joint preparation—reaching the bottom of a narrow vee may prevent the use of larger electrodes on first pass.
4. Fit-up—burn-through may occur with large electrodes on first pass. Back-up strips permit larger electrodes to be used.
5. Machine capacity—amperage rating on machine may not be sufficient to use largest electrode possible for a given job.

For any given size of electrode, there is an amperage range specified by the manufacturer. This range takes into account the use of this electrode under all types of welding conditions. Somewhere within that range there is an optimum setting for each job with its combination of conditions. For example: a vee joint may require less amperage than a square butt joint; a thin or small article will require less amperage than a thick or heavy article; welding vertically up will require lower amperage for puddle control than vertically down; and better weldors can use higher amperages than the beginner.

During welding practice, or when encountering a new job, it is advisable to set the amperage about the middle of the range recommended for a given electrode. Run a few beads, then observe the arc and examine the completed bead (Lesson 1.6, Figure 6). You will be able to see whether or not the amperage setting is correct, too high, or too low. At first, selecting the correct amperage for a new job will be quite slow and may involve several adjustments. Experience will develop the judgment to select the correct amperage quickly for each job and will also develop the ability to work successfully with maximum amperage settings for each type of electrode, thus obtaining faster welding speeds and reducing welding costs.

Review Questions

1. What factor is most important in the cost of welding?
2. What are the disadvantages of using small electrodes?
3. What general rule governs electrode selection for downhand welding?
4. What is maximum size electrode for overhead welding?
5. Why can't the manufacturer designate a specific amperage for each electrode?
6. What amperage should be selected when starting on a new job?
7. What two factors give maximum welding speed?

LESSON 1.8

Object
To restart a continuous bead and fill the crater.

Equipment:
Lincoln "Idealarc" or DC or AC welder and accessories.

Material:
Mild steel plate 3/16″ or thicker; 5/32″ "Fleetweld 5" (E6010) for DC or "Fleetweld 35" (E6011) for AC.

General Information

During welding it is often necessary to interrupt a continuous bead, as when finishing one electrode and starting another. Restarting a bead must be done without a depression or a lump to spoil the bead uniformity. After a little practice you will soon know how long to hold and direct the arc into the crater before continuing on with the bead. Holding the arc too long, or directing it too far back on the bead will cause a lump. Not holding it long enough or not directing it back far enough into the crater will cause a depression in the bead.

When craters occur at the end of the joint there are no further beads to cover them. Unfilled craters cause stress points and are the weakest spot in the completed weld. They must be filled by one of two methods to bring them up to the height of the weld bead.

Job Instructions

Job A: Restart a continuous bead.

1. Clean the base metal and position flat.
2. Check ground connection.
3. Set amperage at 130 to 145 for 5/32″ electrode, DC, electrode positive.
4. Hold electrode perpendicular, inclined 20 to 25 degrees in the direction of travel.
5. Strike an arc and run a straight bead for 2 to 3 inches. Extinguish the arc by drawing the electrode tip away. This leaves a bead which tapers off into a crater.
6. Chip the slag out of the crater and for at least 1/2″ back on the bead.
7. Restrike the arc about 1/4″ ahead of the crater, move the arc back into the

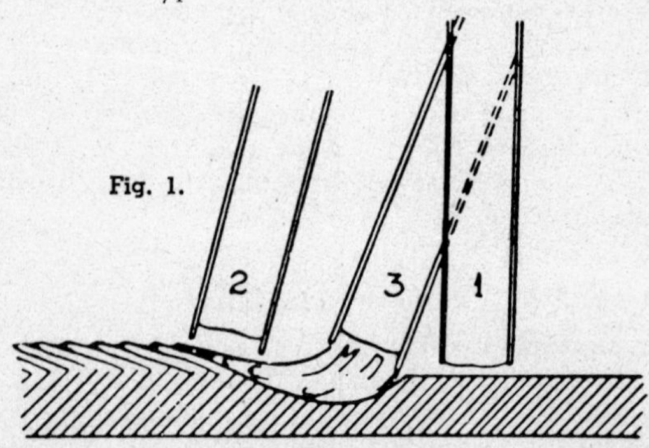

Fig. 1.

crater and continue ahead with the bead (Fig. 1). By striking the arc ahead of the crater any marks of the arc are covered by the bead.

8. Repeat, welding in the other directions, until a uniform bead is obtained.

Fig. 2.

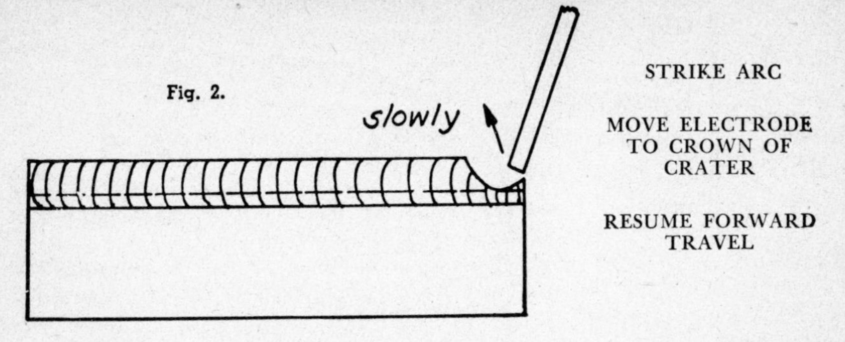

Job B: Fill the crater at the end of a weld.

1. To fill the crater as a bead is brought to the end of the plate, draw the electrode up slowly and backward over the completed weld. Slowly drawing up the electrode permits the crater to be filled with metal. Moving back over the weld will minimize the crater by placing it on top of the bead ¼″ to ½″ from the end (Fig. 2).
2. If the crater starts to burn through or spill over the end of the plate, extinguish the arc, chip out the crater, and restrike the arc. Remove the crater by building up in this manner.
3. The other method to overcome the crater depression is to break the arc about 1″ to 2″ from the end of the plate, jump the electrode tip to the end, restrike the arc and weld back toward the finished bead. Remember to incline the electrode in the direction of travel when reversing the direction. Continue welding over the crater and extinguish the arc as the top of the two beads run together. This also puts the crater on the top of the bead (Fig. 3).

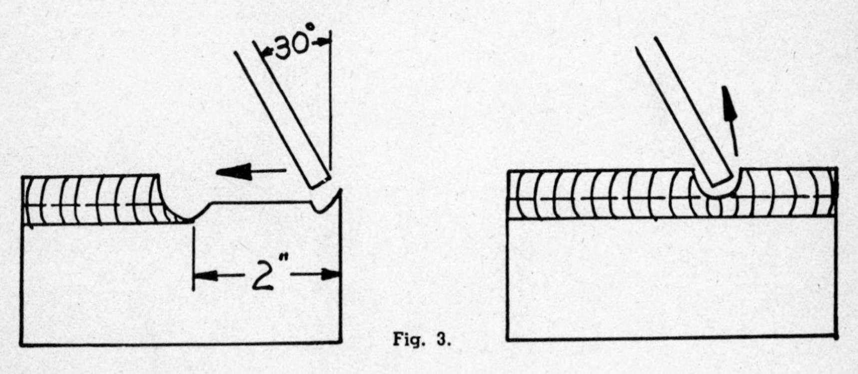

Fig. 3.

Review Questions

1. What causes a crater?
2. What are two defects in the bead caused by incorrectly restarting a bead?
3. Why is the arc restruck ahead of the crater?
4. Why defects may be caused by unfilled craters?
5. What should be done if the crater starts to burn through the end of the base metal?
6. What two methods may be used to fill a crater?

LESSON 1.9

Object:
To run a bead with a weaving motion.

Equipment:
Lincoln "Idealarc" or DC or AC welder and accessories.

Material:
Mild steel plates 3/16" or thicker; 5/32" "Fleetweld 5" (E6010) for DC or "Fleetweld 35" (E6011) for AC.

General Information

Weaving is an oscillating motion, back and forth, crosswise to the direction of travel. These motions are used to float out slag, deposit a wider bead, secure good penetration at the edges of the weld, or allow gas to escape and avoid porosity. The weave pass will be used later, on welding and hardsurfacing in all positions.

To make a wide bead, it is necessary to move the electrode from side to side, at the same time moving forward to advance the bead; such a weave is shown in A, Fig. 1.

Some may find it easier to weave in a crescent motion, as shown in B. The purposes accomplished by both these motions are substantially the same and their use is largely a matter of preference. Weave B is probably the most popular. The "figure 8" motion shown in C or the circular motion shown in D is preferred by some weldors.

Weave E makes a hesitation at each side of the weave to allow a slight build-up or working of the metal into the edges of the joint.

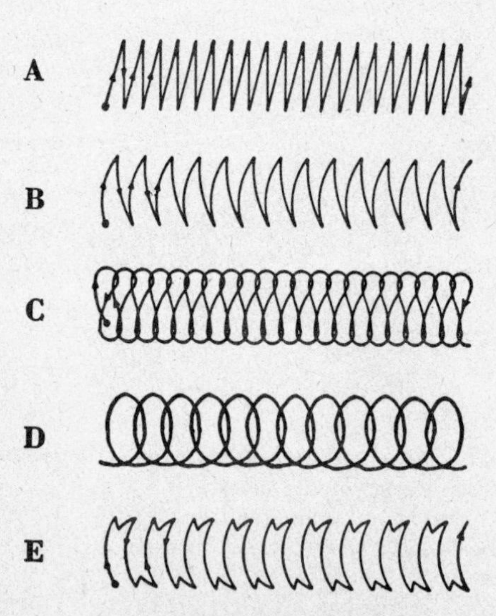

Fig. 1. Weaving motions.

Job Instructions

1. Clean the base metal and position flat.
2. Set amperage at 130 to 145 for 5/32" electrode.
3. Hold electrode perpendicular, inclined 20 to 25 degrees in the direction of travel.
4. Strike the arc and carry the bead using the same arc length as when running straight beads. Practice the motions shown in A and B in the form of a bead. Make beads 1/2" to 3/4" wide. Bead width should not exceed 6 times the diameter of the electrode.
5. Run weave passes in all four directions.
6. Movements C, D and E may be practiced if desired until you can obtain the same smooth type of bead deposited by A and B.
7. You may find that you will develop a weave slightly different from those shown which works easily for you. Practice these weave motions so they will become free and natural and can be done automatically when required.

Review Questions

1. What is a weave pass?
2. Why are weave passes used?
3. Is there more than one type of correct weaving motion?
4. What is the maximum width of a weave pass?
5. Can a weave pass be used only in the downhand position?

LESSON 1.10

Object:
To run a bead with a whipping motion.

Equipment:
Lincoln "Idealarc" or DC or AC welder and accessories.

Material:
Mild steel plate 1/4" or thicker; 5/32" "Fleetweld 5" (E6010) for DC or "Fleetweld 35" (E6011) for AC.

General Information

Whipping is an oscillating motion lengthwise with the direction of the bead (Fig. 1). A similar motion may be used to obtain two opposite results, keeping the puddle "hot" or keeping it "cool".

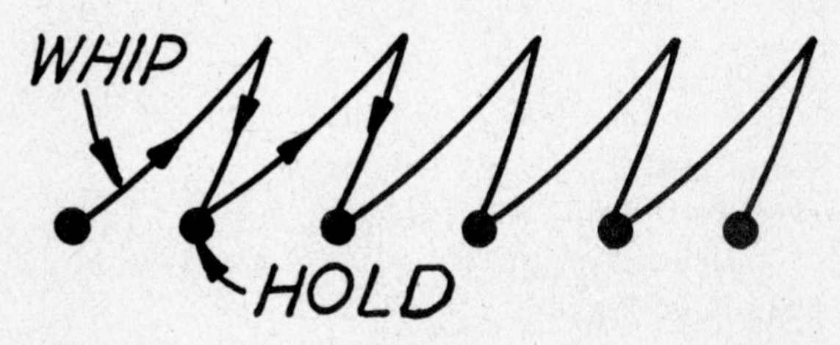

Fig. 1.

Whipping may be used on downhand work to keep the puddle "hot" or in a fluid state to obtain good penetration with even ripples and uniform build-up.

For vertical and overhead work or in joints where burn-through is a problem, the whipping motion is used to keep the puddle "cool" and prevent it from sagging or running down. In the case of thin metal, it keeps the puddle from penetrating too deep.

Job Instructions

Job A: Run a bead with a whipping motion to keep a puddle "hot".

1. Clean the base metal and position flat.
2. Set amperage at 130 to 145 for 5/32" electrode.
3. Hold electrode perpendicular, inclined 20 to 25 degrees in the direction of travel.
4. Assume a position that permits you to see behind and ahead of the arc so that when a faulty action occurs it can be corrected immediately.
5. Strike the arc and carry the bead with a normal arc length during the entire whipping motion.
6. The whipping motion to keep the puddle "hot" or fluid should be about 5/16" forward and 1/8" to 1/4" back toward the crater, depending upon the size of the bead desired. When the backward motion is completed, hesitate in the electrode motion. Penetration is obtained on the forward motion and

build-up of the bead is obtained on the "hesitation" of the backward motion. The longer the hesitation, the larger the weld deposit. The length of the backward motion controls the ripple appearance of the weld. Except for the hesitation, the motion is rapid.

7. Practice the motion using different lengths of return strokes, until you can build up a heavy or light bead.

Job B: Run a bead with the whipping motion used to keep a puddle "cool".

1. Follow steps 1 to 4 in Job A.
2. Strike the arc and carry the bead with a normal arc length on the backward stroke, and a long arc, 3/8" to 1/2" at the hesitation point of the forward stroke.
3. Move the arc ahead with a forward stroke of approximately 3/8", hesitate, holding the long arc, then move backward approximately 1/4" assuming a normal arc length (Fig. 2). The long arc on the forward stroke reduces penetration as well as the amount of metal deposited, and allows the weldor to see how the puddle is solidifying. Shortening the arc on the backward stroke allows a normal deposition of metal. A longer hesitation will allow the crater to solidify more. Returning the arc to the partially solidified crater will not cause slag to be trapped in the weld, because it has not had time to completely solidify.

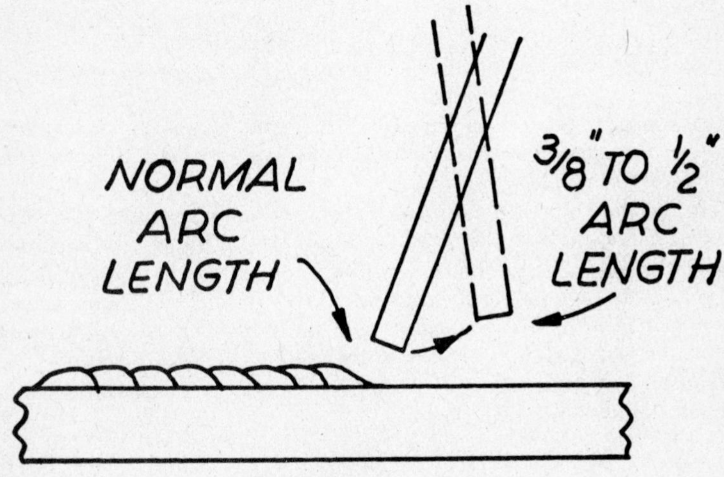

Fig. 2.

Review Questions

1. What is the difference between whipping and weaving?
2. What are the two reasons for using a whipping motion?
3. When keeping the metal fluid, what motion provides penetration?
4. Why is a long arc used on the forward stroke to keep the puddle "cool"?
5. Is the hesitation movement in the same place on each type of whipping motion?

LESSON 1.11

Object:
To build a pad.

Equipment:
Lincoln "Idealarc" or DC or AC welder and accessories.

Material:
Mild steel plate 4x2½x¼ and bar stock 1½ to 3 inches in diameter; 5/32" "Fleetweld 5" (E6010) for DC or "Fleetweld 35" (E6011) for AC.

General Information

Padding is a common welding application. It is often necessary to build up metal surfaces with one or more layers of weld deposit. Rebuilding a worn surface or repairing a machining error are two such applications. This work may be done on either flat or curved surfaces by depositing overlapping straight beads or weave passes.

Job Instructions

Job A: Padding on a flat surface.

1. Clean the base metal and position flat.
2. Set amperage at 130 to 145 for 5/32" electrode.
3. Hold electrode perpendicular, inclined 20 to 25 degrees in the direction of travel.
4. Run a straight bead along the edge of the plate. A weave pass may be used, but on the edge more heat control is necessary to keep from burning off the corner of the plate.
5. Chip the bead free of slag before running succeeding passes. This must be done for each pass, so that excess slag will not be trapped in the deposit.
6. Run a second bead parallel to the first, and overlapping it about one third. Use the weaving motion (Lesson 1.9). Make certain that complete fusion is obtained with the plate as well as with the previous bead. Beads should be the same height with no depression or "valley" between the two.
7. As succeeding beads are run, a comparatively smooth surface of weld metal should be obtained (Fig. 1).

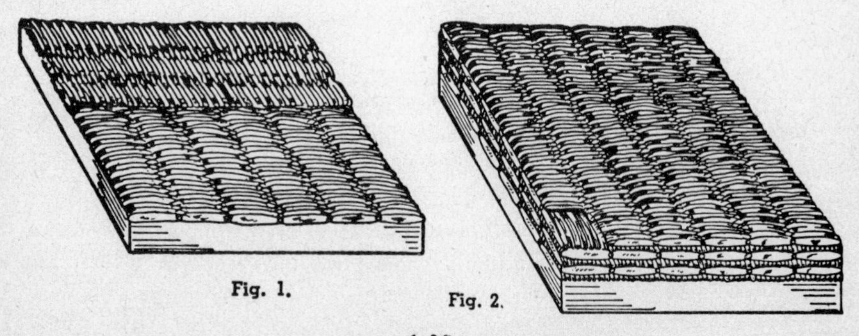

Fig. 1.

Fig. 2.

8. After the first layer has been deposited the oxide and scale should be completely removed from the surface by using a chipping hammer and brush. Inspect completed layer for smoothness and penetration.
9. Run the second layer of passes at right angles to the first layer. This is called lacing (Fig. 1).
10. Straight overlapping beads without weaving may be laid when padding. Sound passes by either method produce a sound pad. If the pad is being made for practice, alternate layers of weaving and straight passes might be used.
11. Build up the pad to a height of 3/4" to 1" (Fig. 2). As the plate is built up, larger sized electrode may be used. Control the electrode deposits so that the pad is not only dense, but the edges are built up square and straight. The last layer of passes should be the same length and width as the first layer.
12. To check a pad for dense build-up, it may be sawed through. Check visually for pin-holes, pores and slag inclusions. Do not quench in water, but allow it to cool normally, so that it will saw easier. If a hacksaw is not available, cut through with the arc (Lesson 1.40) and grind the surface.
13. To provide a more thorough method of visual inspection, etching may be used. Grind off the cut surface and etch with a dilute solution of nitric acid. You can then get a comparison of the weld deposit with the base metal. Observe the lines of fusion between the beads, layers, and plate. Check for porosity. *Caution:* Nitric acid causes stains and severe burns. Extreme care must be taken when storing and using acid. Use one part acid to three parts water. Always pour the acid into the water when diluting. Apply the solution with a glass stirring rod. After a few seconds wash off in warm running water.

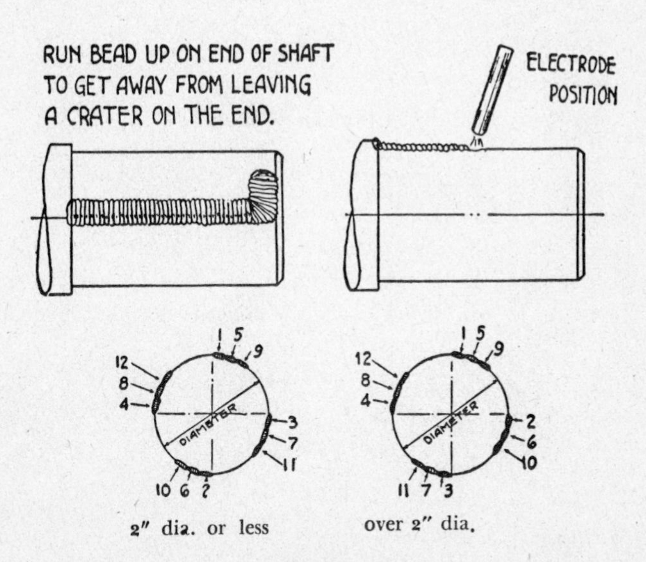

Fig. 3.

Job B: Build up a shaft by padding.

1. Turn the shaft diameter 1/16" to 1/8" under the finished size.
2. Place the shaft in a flat position on vee blocks or a roll positioner, so that it will be steady and may be turned when desired. When doing an actual padding job on a shaft the type of metal should be known, so that the correct electrode and procedure may be used.
3. Set amperage at 135 to 165 for 5/32" electrode (base metal is thick).
4. Weld parallel to the axis of the shaft if the weld area is close to the end. Place one bead and turn the shaft either 90 or 180 degrees (depending upon the diameter) to place the next bead. The two types of sequence shown in Fig. 3 may be used. This is done to minimize warping the shaft.
5. If the place to be welded is some distance from the end, it is advisable to weld around the shaft (Fig. 4). This will minimize distortion. Turn the

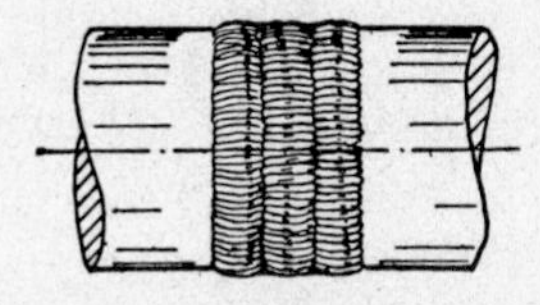

Fig. 4.

shaft while welding so that the weld metal is always on a slight upward incline for best penetration.
6. Before starting a second pass using either sequence and method of welding, remove all slag and oxide from previous pass. Be certain beads are properly fused together, as well as to the shaft.

Review Questions

1. What is padding?
2. What are some typical applications of padding?
3. Is it necessary to clean each pass before running the next?
4. What is lacing?
5. Is it necessary to use only a weave pass when padding?
6. How may a practice pad be tested?
7. How should nitric acid be mixed with water?
8. Is there a definite pass sequence to be followed when building up a shaft?
9. Why is this sequence necessary?
10. Should the pads be quenched in water after welding?

LESSON 1.12

Object:
To show the types and positions of welded joints.

General Information

There are numerous types of welded joints, and various positions in which they are welded. Figure 1 shows a variety of these joints as they may appear on welding jobs.

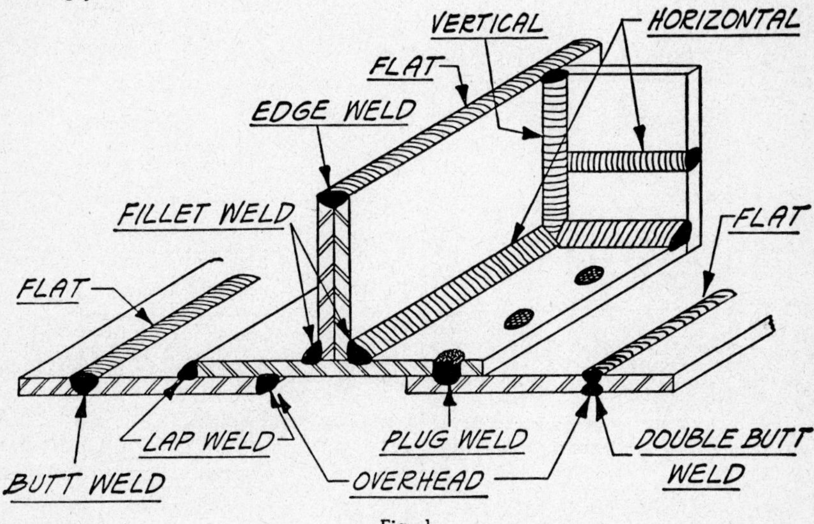

Fig. 1.

Figure 2 is a diagram showing the four basic welding positions. It is possible to weld any type of joint in any of the four positions, but whenever it is possible joints are placed in the flat position. Welding in the flat position is much faster and easier than any of the other positions.

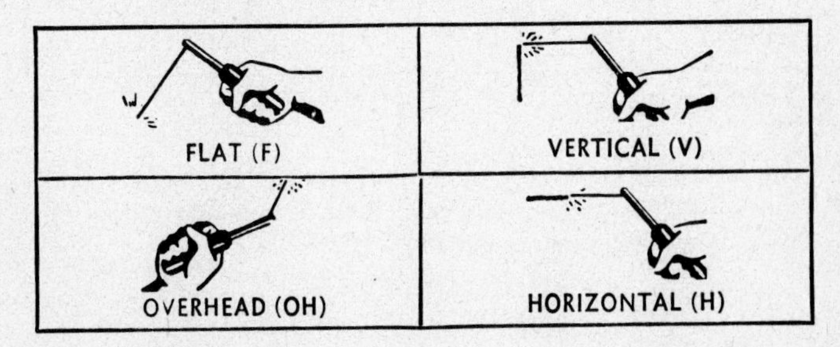

Fig. 2.

A diagramed summary of basic types of joints and the position in which they are commonly welded is shown in figure 3. The welding position (F, V, H, or OH) is shown in the box in the upper right corner of each joint diagram.

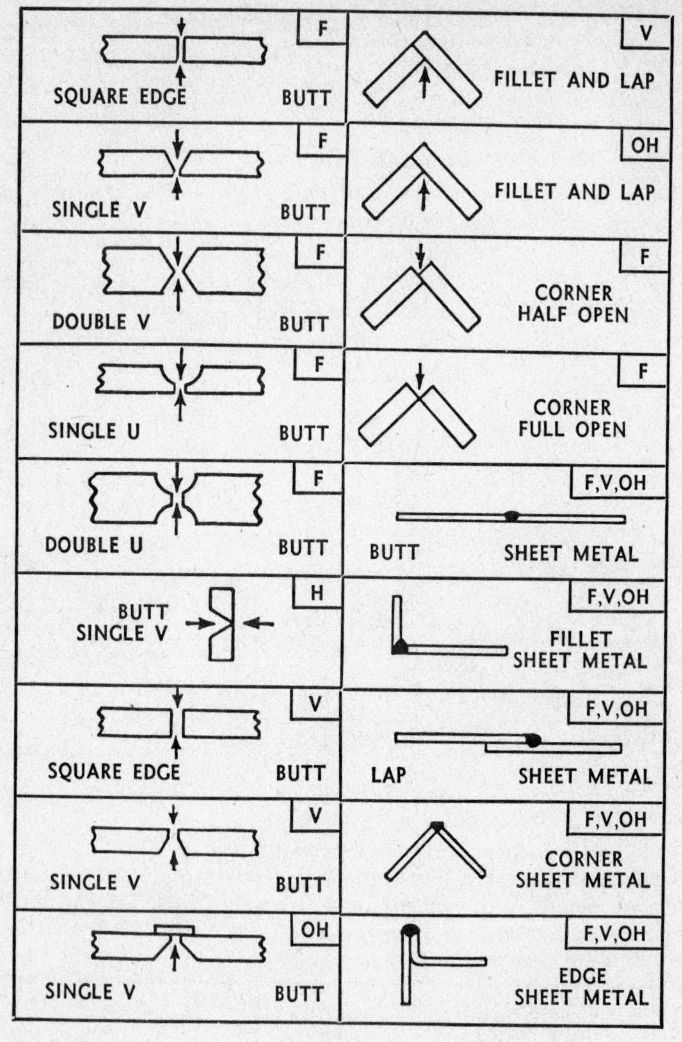

Fig. 3.

Additional information on the edge preparation and set-up procedure for most of these joints will be given in succeeding lessons and detailed procedures for welding them are included in Section VII.

LESSON
1.13

Object:
To study and demonstrate the effects of correct and incorrect polarity.

Equipment:
Lincoln "Idealarc" or DC welder and accessories.

Material:
Scrap plate 3/16" or thicker; 5/32" "Fleetweld 5" (E6010) and two 1/4" carbon electrodes.

General Information

The terms "straight" and "reversed" polarity are used around the shop. They may also be expressed as "electrode negative" and "electrode positive" polarity. The latter terms are more descriptive and will be used throughout the book.

Polarity results from the fact that an electrical circuit has a negative and a positive pole. Direct current (DC) flows in one direction, resulting in a constant polarity. Alternating current (AC) flows half the time in one direction and half the time in the other, changing its polarity 120 times per second with 60 cycle current.

A weldor should know the meaning of polarity, and recognize what effect it has on the welding process. It is the ability to adjust polarity that lends DC welding its versatility. With few exceptions, electrode positive (reversed polarity), results in deeper penetration. Electrode negative (straight polarity), results in faster melt-off of the electrode and, therefore, faster deposition rate. The effect of different chemicals in the coating may change this condition. The high cellulose coated mild steel rod, such as "Fleetweld 5" is recommended for use on positive polarity for general welding. Some types of shielded electrodes function on either polarity, though some operate on only one polarity.

Positive polarity also has the ability to break up an oxide film that coats certain non-ferrous metals. For this reason it is used when welding copper alloys, aluminum and magnesium.

The use of the AC transformer type welder necessitated the development of an electrode which would work on either polarity, due to the constant changing of polarity in the AC circuit. Though AC itself has no polarity, when AC electrodes are used on DC they usually operate best on one specific polarity. The coating on the electrode designates which polarity is best and all manufacturers specify on the electrode container what polarity is recommended.

For proper penetration, uniform bead appearance and good welding results, the correct polarity must be used when welding with any given metallic electrode. Incorrect polarity will cause poor penetration, irregular bead shape, excessive spatter, difficulty in controlling the arc, and overheating and rapid burning of the electrode.

Most machines are clearly marked as to what the terminals are, or how they can be set for either polarity. Some machines have a switch to change polarity, whereas on others it is necessary to change the cable terminals. If there is any question as to whether or not the correct polarity is being used, or what polarity is set on a DC machine, there are two easily performed experiments that will tell you. The first is to use a carbon electrode which will work correctly only on negative polarity. The second is to use "Fleetweld 5" electrode, which works outstandingly better on positive polarity than on negative polarity.

Job Instructions

Job A: Determine polarity by using the carbon electrode.

1. Clean the base metal and position flat.
2. Shape the points of the two carbon electrodes on a grinding wheel, so they are identical with a gradual taper running back 2 or 3 inches from the arc tip.
3. Grip one electrode in the electrode holder close to the taper.
4. Set amperage at 135 to 150.
5. Adjust to either polarity.
6. Strike an arc (use shield) and hold for a short time. Change arc length from short to long, affording an observation of the arc action.
7. Observe the arc action. If the polarity is negative (straight) the arc will be stable, easy to maintain, uniform and conical in shape. If the polarity is positive (reverse), the arc will be difficult to maintain and will leave a black carbon deposit on the surface of the base metal.
8. Change the polarity. Strike an arc with the other electrode and hold for a similar length of time. Observe the arc action as before.
9. Examine the ends of the two electrodes and compare. The one used on negative polarity will burn off evenly keeping its shape. The electrode used on positive polarity will quickly burn off blunt.

Job B: Determine polarity by the metallic electrode (E6010).

1. Clean base metal and position flat.
2. Set amperage at 130 to 145 for 5/32" electrode.
3. Adjust to either polarity.
4. Strike an arc. Hold normal arc length and standard electrode angle and run a bead.
5. Listen to the sound of the arc. Correct polarity, with normal arc length and amperage, will produce a regular "crackling" sound. Incorrect polarity, with normal arc length and amperage setting will produce irregular "crackling" and "popping" with an unstable arc.
6. See above for characteristics of arc and bead when using metallic electrode on correct and incorrect polarity.
7. Adjust to the other polarity and run another bead.
8. Clean beads and examine. With the wrong polarity, the electrode negative, you will get many of the bad bead characteristics shown in Lesson 1.6.
9. Repeat several times, until you can quickly recognize correct polarity.

Review Questions

1. What is the polarity of the electrode on "straight" polarity?
2. What is the polarity of the work on "reversed" polarity?
3. Can you control polarity on AC current?
4. What polarity must be used with a bare electrode?
5. At which polarity setting is the highest rate of electrode melt-off exhibited?
6. What type of coating is on "Fleetweld 5" electrode?
7. What effects does positive polarity have on certain non-ferrous metals?

LESSON 1.14

Object:
To study arc blow and welding with AC and DC current.

General Information

Weldors who have both AC and DC welding current available must learn which current to use on each job. Also, the question of which type is best frequently comes up during consideration of buying a new welding machine, and it is to the weldor's credit when he can assist in the decision.

As will be shown later, much of the decision rests on the presence or absence of "arc blow". So let's understand arc blow before considering the AC-DC question further.

What Is Arc Blow?

Arc blow occurs when the arc refuses to go where it's supposed to, blows wildly forward or back, and spatters badly. (See Figure 1.) Those weldors who have already encountered arc blow need no description, for trying to weld with severe arc blow is a difficult task which is long remembered. In less severe cases, arc blow makes it difficult to control the molten pool and slag.

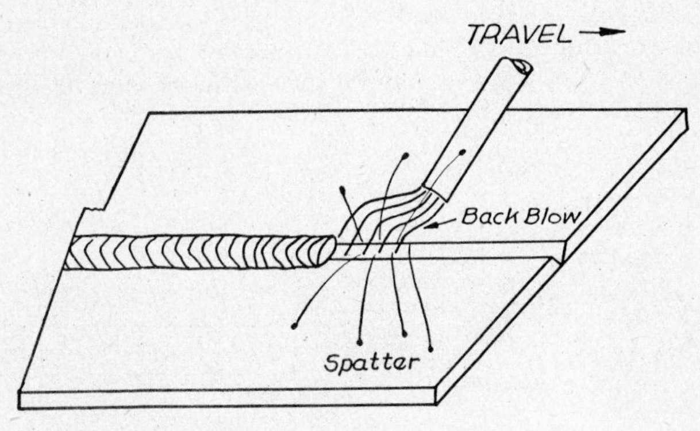

Fig. 1.

Arc blow is most frequently encountered at the start and finish of joints, and in corners and deep grooves, particularly when high amperages are being used in welding thick plates. It makes welding very difficult, reduces speed, and lowers weld quality. When the arc blows opposite to the direction of travel, as in Figure 1, it is called "back blow"; when it blows with the direction of travel, it is called "forward blow".

What Causes Arc Blow?

Arc blow is caused by a magnetic force acting on the arc, making the arc "blow" from its normal path. Every wire or conductor which carries current is surrounded by lines of magnetic "flux", or "force", as seen in Figure 2. These lines prefer to travel through steel than air, never touch each other, and exert a force when bunched together. This force is proportional to the amount of current in the conductor. Their normal pattern is a series of concentric rings

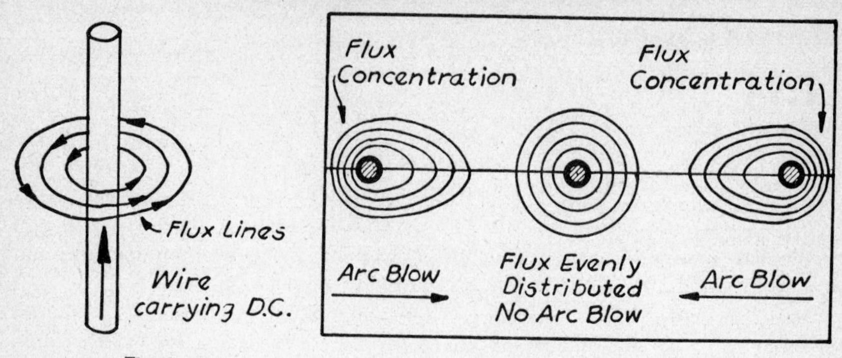

Fig. 2.

Fig. 3.

around the conductor as seen in the center portion of Figure 3.

The normal pattern of the flux is disturbed on the ends of the joint where the flux bunches up in the steel, instead of going into the air beyond the end of the joint. A new pattern, such as that seen on either side of Figure 3, results in concentrations on the ends of the joint. In these areas of concentration, the lines are bunched together and exert a force on the arc which pushes, or blows, it.

A similar situation occurs when the conductor is bent as in Figure 4. Again the lines are forced into a bunch and push the arc. This is called "ground effect" because the direction of this blow can be changed by moving the ground.

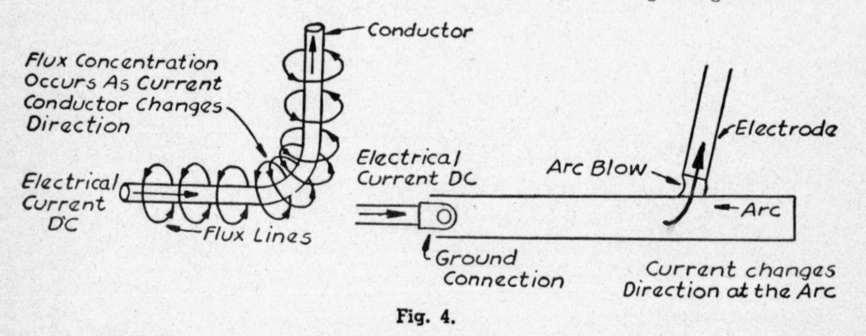

Fig. 4.

In welding, both of these situations occur simultaneously, though it should be noted that "ground effect" is less forceful than end concentrations and becomes even less noticeable as the size of the base metal increases. Figure 5 illustrates the effect on arc blow of various combinations of the two types of arc blow.

Similarly, though somewhat more complicated, arc blow is encountered in corners and deep vee joints. In each of these cases, the lines of flux bunch together and exert a force on the arc, causing it to blow and be erratic.

How to Reduce Arc Blow

To reduce arc blow, one must reduce the causes of arc blow; that is, reduce or counteract the strength of the force, or minimize flux concentrations. Here are several corrective steps which may help:

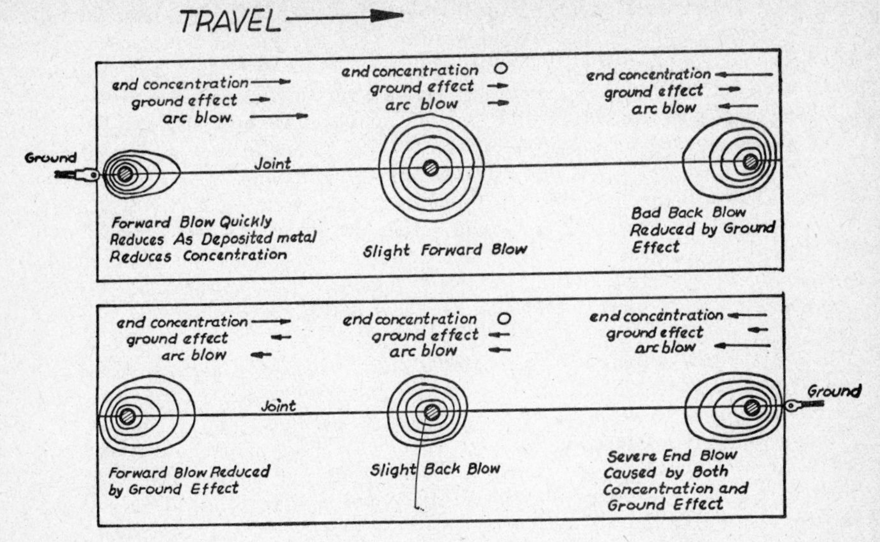

Fig. 5.

1. Reduce current.
2. Weld toward a heavy tack or toward a weld already made.
3. Use back-stepping on long welds.
4. Place ground connection as far from joint to be welded as is possible.
5. If back blow is the problem, place ground connection at start of weld and weld towards a heavy tack.
6. If forward blow causes trouble, place ground connection at end of weld.
7. Wrap ground cable around the work piece and pass ground current through it in such a direction that a magnetic field will be set up to neutralize the magnetic field causing the blow.
8. Hold as short an arc as possible to help the arc force counteract the arc blow.
9. If the machine being used is of the type producing both AC and DC, switch to alternating current.

The last step requires some explanation. AC markedly reduces the effect of arc blow by practically eliminating the strength of the flux. With AC, the current goes through zero 120 times a second. This means that the flux lines build-up and collapse 120 times a second. The result of this action is that arc blow is greatly reduced—to the point of elimination on many jobs.

AC or DC?

Now that we have an understanding of arc blow and realize that AC greatly reduces arc blow, let's continue the discussion of AC and DC.

The decision to use either AC or DC is based on which one will permit the operator to produce the best welds in the least time. Generally speaking, AC is best when using iron powder electrodes or where arc blow is a problem; DC is best for low current or on any application where arc blow is not likely a problem.

The electrode used may determine the type of current required. Jetweld iron powder type electrodes operate best on AC, even in the smaller sizes. With AC, they produce a smooth, evenly-shaped bead, while, with DC, they tend to produce a narrower and stringier bead. Most other electrodes operate better on DC. Some electrodes, such as most stainless, non-ferrous, low hydrogen in small sizes, and hardsurfacing electrodes, operate very poorly or not at all on AC.

When electrodes are used which operate on both AC and DC, arc blow is usually the deciding factor. Even when arc blow is not severe, it may be bad enough to reduce welding speeds substantially. Weldors tend to select a welding current that will produce the best weld and make welding easy. Invariably, when arc blow, even slight, is present, operators select a lower current with DC than with AC. Lower current means slower welding. Currents in the neighborhood of 250-300 amperes are sufficient to cause arc blow on most jobs. Therefore, AC should be used on most applications requiring currents this large or larger.

Below 250-300 amperes with electrodes other than iron powder, DC will usually be preferable. The electrodes operate better and make welding easier on critical jobs, such as sheet metal, vertical, overhead, etc.

In summary, when deciding between AC and DC, one must consider the job, the position of welding, the electrode to be used, and the currents involved. The chart below is a summary of preferred types of current for all sizes of the various steel electrodes. If, as is frequently the case, both types of welding are involved and only one machine is available, it is advantageous to have a combination AC and DC welder, which will provide both alternating and direct currents.

PREFERRED TYPE OF CURRENT FOR VARIOUS ELECTRODES

Electrode Type	Electrode Size					
	1/8	5/32	3/16	7/32	1/4	5/16
E-6010	DC+ ———	———	———	———	———	——→
E-6011	DC+ (AC) ———	———	——→	AC (DC+) ———	———	——→
E-6012	DC− (AC) ———	———	——→	AC (DC−) ———	———	——→
E-6014 E-6027	AC (DC−) ———	———	———	———	———	——→
E-6018	DC+ ———	———	——→	AC (DC+) ———	———	——→
E-6024	AC (DC+) ———	———	———	———	———	——→

Type of current in parenthesis may also be used.

Review Questions

1. Where is arc blow most frequently encountered?
2. Is arc blow dependent on the amount of current?
3. Is "ground effect" stronger than end concentrations of magnetic flux?
4. On a production job, what is the least desirable method of reducing arc blow? The most desirable?
5. Does AC reduce arc blow? How?
6. What are some electrodes which generally do not operate well on AC?
7. Do Jetweld iron powder type electrodes operate best on AC, or DC?
8. How much current is required to cause arc blow?
9. How does arc blow affect welding speeds?
10. Will moving the ground have any effect on arc blow?

LESSON 1.15

Object:
To study the effects of welding heat on metals.

General Information

Metals become larger when heated, and become smaller upon cooling. During the arc welding process, the arc heats the metal being welded, causing it to become larger, or expand. As the heat is removed, the surrounding metal and air cause a cooling effect upon the heated area, which results in the metal becoming smaller, or contracting. When this expansion and contraction is not controlled excessive distortion (warping) is likely to result. On the other extreme, if expansion and contraction is restrained or controlled too rigidly, severe stress and strain may result and impair the weld.

For every degree of temperature rise or fall there is a corresponding change in the size of the metal. Nothing we can do will change the laws of expansion and contraction. We can, however, recognize that changes will take place, figure out how it will effect the work upon which we are welding, and prepare for them.

The three following rules can be followed to aid materially in the prevention and control of distortion.

1. REDUCE THE FORCES WHICH CAUSE SHRINKAGE.
2. MAKE SHRINKAGE FORCES WORK TO REDUCE DISTORTION.
3. BALANCE SHRINKAGE FORCES WITH OTHER FORCES.

The following discussion gives examples of these rules. In many cases, the application of a single rule will be sufficient. Sometimes a combination of rules may be required.

1. REDUCE THE FORCES WHICH CAUSE SHRINKAGE

(a) Avoid overwelding. The addition of excess weld metal not needed to meet the service requirements of a joint is known as "overwelding".

Overwelding causes distortion (Fig. 1) and contributes nothing to the strength and performance of the joint. It is a waste of time and money. In certain cases it may even weaken the joint.

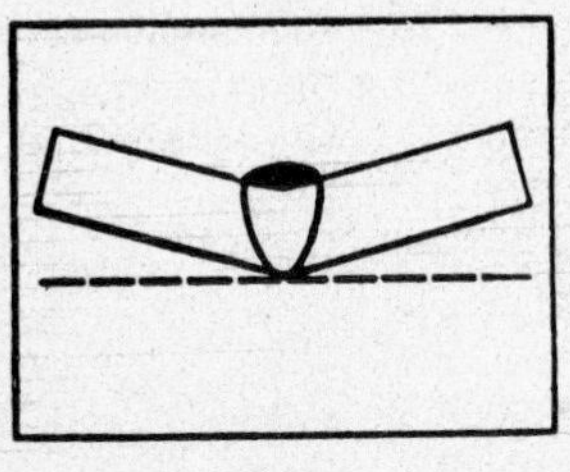

Fig. 1.

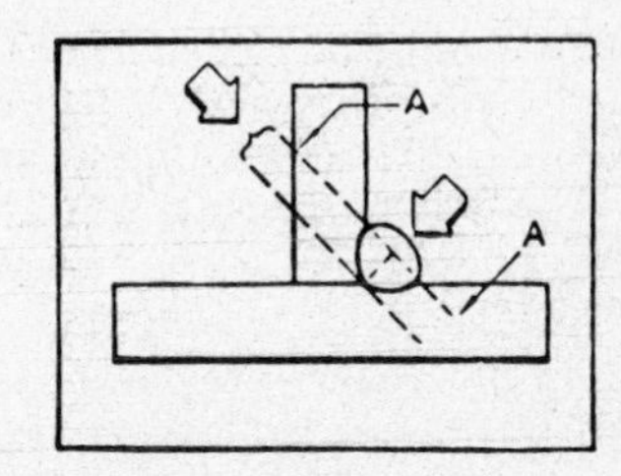

Fig. 2.

Deposit as little weld metal as possible and make intelligent use of the weld metal that is needed. The strength of a fillet weld for a T joint is determined by the throat size (Fig. 2). In this illustration there is an excess of weld metal above line A-A, which does not increase the strength, but obviously increases the shrinkage force, due to the contraction of the large molten puddle.

(b) Use proper edge preparation and fit-up. It is also possible to reduce shrinkage forces through proper edge preparation. The bevel should not exceed 90 degrees to obtain proper fusion at the root of a weld with a minimum of deposit. Proper fit-up is important. Plates should be spaced 1/32" to 1/16" apart (root spacing) to minimize the amount of deposit necessary for proper strength.

(c) Use few passes. Distortion in the lateral direction (across the weld) is a major problem. The use of one or two passes with large electrodes reduces distortion in this direction (Fig. 3). As a general rule, lateral distortion is approximately 1 degree per pass.

In some cases, however, distortion is in the longitudinal direction (lengthwise). Due to the greater ability of a small bead to stretch longitudinally, compared to a large bead, the number of passes should be increased rather than decreased.

There is inherent rigidity against the longitudinal bending of a plate, providing the plate is thick enough. Light gauge sheets have little rigidity in this direction and will buckle easily. Lateral distortion is more common because unless the two plates are restrained, they will move as shown in figure 1.

Fig. 3.

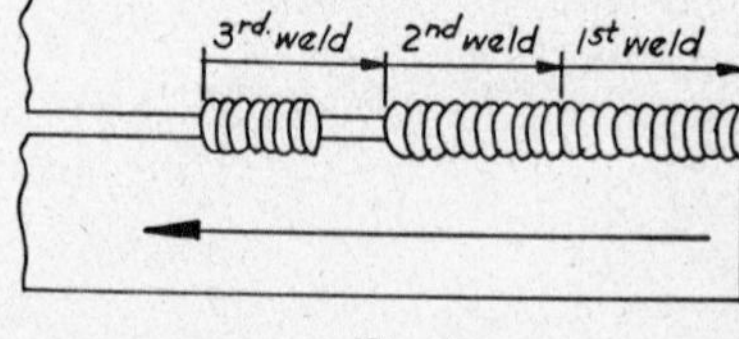

Fig. 4.

(d) Use intermittent welds. To further reduce shrinkage force by minimizing the amount of weld metal, intermittent welds may sometimes be used instead of continuous welds. It is often possible to use up to two-thirds less weld metal and still obtain the strength required. An example of this is when welding stiffeners for bulkheads and plates of all kinds. The use of intermittent welds also distributes the heat more widely throughout the structure.

(e) Use "back-step" welding method. If the job requires a continuous weld it is possible to reduce shrinkage forces by the "back-step" technique. With this technique, the general direction of welding progression may be from left to right, but each bead is deposited from right to left (Fig. 4). Expansion will be less and less with each bead because of the locking effect of each weld. The tendency of plates to spread is locked in by each step.

Where a continuous bead is laid in one direction, in many cases there is a tendency for the plates to spread and become locked in the spread position as the welding progresses. Welding speed is the determining factor here. As a general rule, the greater the speed, the more the amount of spreading. In some cases a speed can be found at which the plates will not separate at all.

2. MAKE SHRINKAGE FORCES WORK TO MINIMIZE DISTORTION.

(a) Locate parts out of position. A simple way to use the shrinkage force of weld metal to advantage is to obtain proper location of parts before welding. Fig. 5 shows a T being made with the vertical plates out of alignment before the weld is deposited. When the weld deposit shrinks it will pull the vertical plate to its correct 90 degree position.

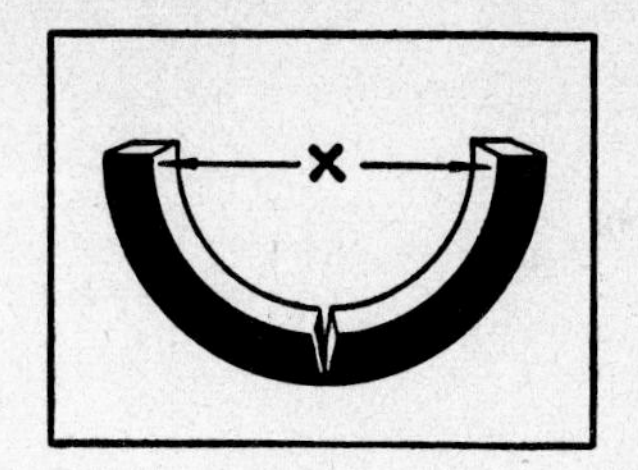

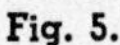
Fig. 5.

Fig. 6.

(b) Space parts to allow for shrinkage. Another method is to space parts before welding. Experience indicates how much space should be allowed for a given job, so the parts will be in correct alignment after welding. An example is the trunnion arms of a large searchlight as shown in Fig. 6. The distance between the two arms had to be accurately controlled. Correct spacing of the parts prior to welding allowed the arms to be pulled into correct position by the shrinkage forces.

(c) Prebend. Shrinkage force may be put to work by prebending or prespringing the parts to be welded. When the plates in Fig. 7 are sprung away from the weld side, the counter force exerted by the clamps overcomes most of the shrinkage tendency of the weld metal, causing it to yield. When the clamps are removed, there is still a slight tendency for the weld to contract. The contraction force pulls the plates into alignment.

3. BALANCE SHRINKAGE FORCES WITH OTHER FORCES.

Often the structural nature of parts to be welded is such as to provide sufficient rigid balancing forces to offset welding shrinkage forces. This is particularly true in heavy sections where there is inherent rigidity because of the arrangement of the parts. If these natural balancing forces are not present, it is necessary to balance the shrinkage forces to prevent distortion.

(a) Balance one shrinkage force with another. Proper welding sequence will place weld metal at different points about the structure. As one section of metal shrinks, it will counteract the shrinkage forces of previous welds. An example is the welding alternately on both sides of a butt weld (Fig. 8), or the pass sequence used when building up a shaft (Lesson 1.11).

Fig. 7.

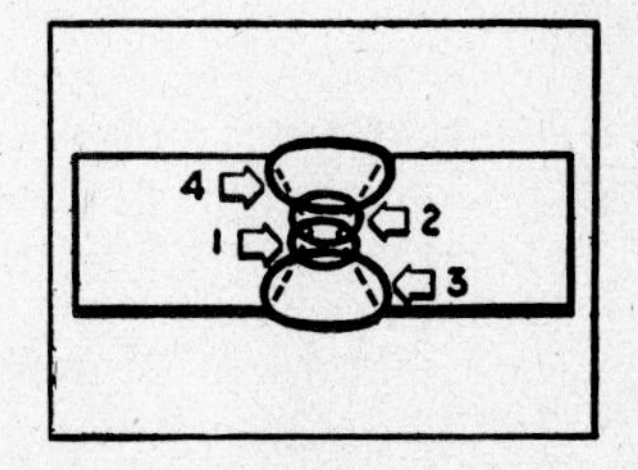

Fig. 8.

Another application of this principle is the staggering of intermittent welds applied in a sequence such as shown in Figs. 9a and 9b. Here the shrinkage force of weld No. 1 is balanced by that of weld No. 2; the shrinkage force of weld No. 2 is balanced by that of weld No. 3 and so on.

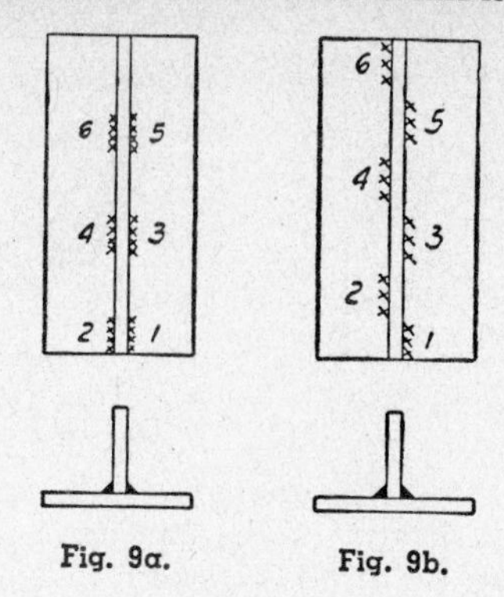

Fig. 9a. Fig. 9b.

(b) Peening. This is a mechanical working of the metal by means of hammer blows. Peening stretches the bead, counteracting its tendency to contract as it cools. Peening should be used with great care. Too much peening may damage the weld metal, work harden it excessively, or damage the base metal.

(c) Use jigs and fixtures. The most important method of avoiding distortion, and one in which Rule 3 is applied to the fullest extent, is the use of clamps, jigs, or fixtures to hold the work in a rigid position during welding. In this way, the shrinkage forces of the weld are balanced with sufficient counterforces to prevent distortion.

The balancing forces of the jig or fixture cause the weld metal itself to stretch, thus preventing the distortion.

For a more detailed discussion of this subject, see "Procedure Handbook of Arc Welding Design and Practice".

Review Questions

1. What happens when steel is heated?
2. What happens when steel is cooled?
3. Should shrinkage force be decreased or increased?
4. Does an excess of weld metal increase the shrinkage force?
5. What determines the strength of a fillet weld?
6. Is distortion less when less metal is deposited?
7. What bevel should be used to obtain proper fusion at the root of a weld with a minimum of weld deposit?
8. What is proper root spacing of plates for minimum use of weld metal?
9. Will use of larger electrodes and fewer passes reduce distortion in the lateral direction.
10. What can be done when longitudinal distortion is a problem?
11. Which is most common, lateral or longitudinal distortion?
12. Can work be spaced or located out of position and the shrinkage forces be made to pull them into position?
13. Can shrinkage forces be nullified by placing welds in such positions and sequence that they counteract each other?
14. What does peening do to the metal?
15. What is the most important method of avoiding distortion?

LESSON 1.16

Object:
To study electrode classification and identification.

General Information

It has been discussed previously (Lesson 1.4) that there are many different types of electrodes designed for specific jobs or categories of jobs. You may sometimes be confronted with the problem of identifying these different electrodes, or specifying and ordering electrode to be used for a job. It is necessary, then, to know what methods of electrode identification are available, how electrodes are classified, and for what specific purposes the electrode in the various classifications are used.

The easiest method of identifying a common electrode (aside from reading the manufacturer's specifications on the container, if they are in the correct container) is by the coating color and a color code (Fig. 1) which has been set up for the major electrode classifications.

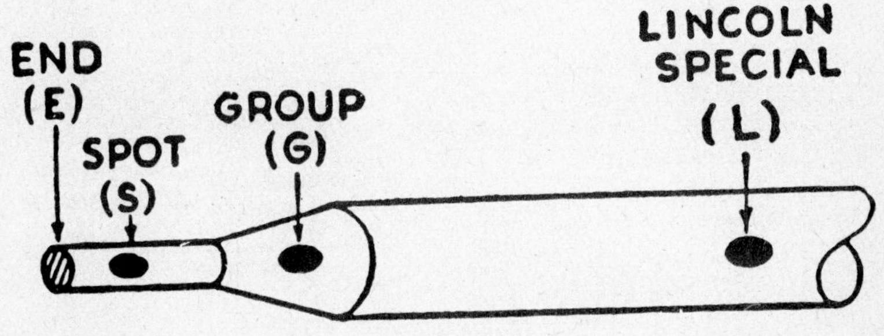

Fig. 1.

The code colors for electrode identification are established by the National Electrical Manufacturers Association (NEMA). The classifications are established by the American Welding Society (AWS). Classifications and code colors assist the user in proper selection and identification of electrodes. NEMA has also specified some color codes for electrodes having no AWS class.

Some manufacturers make several electrodes in the same class. Lincoln has added the Lincoln Special mark (see Fig. 1) to distinguish between these, such as "Fleetweld 7" and "Fleetweld 72". In addition, there are some electrodes which have no AWS class and no NEMA code. Many of these are also given special code colors by the manufacturer.

The AWS classification and marking of each electrode is the one under which that electrode is qualified. Although an electrode may meet the requirements of more than one class, it may only be listed under one classification. The "AWS classification numbering system" table shows what the classifications mean. When selecting an electrode, consider the following:

For Fast Freeze. Types E-6010 and E-6011 electrodes are designed primarily for fast freezing characteristics. This makes them excellent for all-position welding.

For Fast Follow. The electrodes used for high speed welding on sheet metal and light plate are particularly designed to have fast following characteristics. The E-6012 and E-6013 electrodes would usually fall into this category.

For Fast Fill. Those electrodes used for making heavy fillets and groove welds are designed to give fast filling deposits. Electrodes falling into this category are the iron powder electrodes, E-6014, E-6024 and E-6027.

For Hard-To-Weld Steels. The E-6016 and E-6018 electrodes are designed primarily for use on hard-to-weld steels. These electrodes are excellent for welding sulphur-bearing (free machining) steel, steels with high carbon content, and some medium alloy steels which are difficult to weld with other electrodes.

For more information on electrode selection see Section 7

A.W.S. CLASSIFICATION NUMBERING SYSTEM

Prefix "E" stands for electrode—designates arc welding.
Prefix "R" stands for rod—designates gas welding.
In mild and low alloy steel electrodes:

First two digits of four digit numbers and first three digits of five digit numbers indicate tensile strength:

E60xx..............................60,000 psi Tensile Strength
E70xx..............................70,000 psi Tensile Strength
E100xx..........................100,000 psi Tensile Strength, etc.

Next-to-last digit indicates positions:

Exx1x..............................Indicates all positions
Exx2x..............................Indicates H-Fillets or flat
Exx3x..............................Indicates flat position only

Last digit cannot be considered individually.
Last two digits together have significance as to polarity.

Exx10—D. C. Positive
Exx11—D. C. Positive or A. C.
Exx12—D. C. Negative or A. C.
Exx13—D. C. Negative or A. C.
Exx14—A. C. or D. C.
Exx15—D. C. Positive
Exx16—A. C. or D. C. Positive
Exx18—A. C. or D. C. Positive
Exx20—A. C. or D. C.
Exx24—A. C. or D. C. Either Polarity
Exx27—A. C. or D. C. Negative
Exx30—A. C. or D. C. Polarity

In stainless electrode such as E-347-15:

A. First three digits indicate type of stainless steel.
B. Last two digits indicate position and polarity.

Types of Coating:

E-6010 and E-6011 electrodes have a high organic (cellulose) type coating.
E-6012 and E-6013 electrodes have a high rutile (titania) type coating.
E-6016 and E-6018 electrodes have a low hydrogen (lime and sodium carbonate or lime-titania) type coating.
E-6020 and E-6030 electrodes have a high mineral (iron oxide and manganese oxide) type coating.
E-6014, E-6024 and E-6027 electrodes have a iron powder type coating.
In stainless electrodes:—E-xxx-15 types have lime coatings.
E-xxx-16 types have titania coatings.

Review Questions

1. Why is it necessary to know electrode identification and classification?
2. What is the easiest method of identifying an electrode?
3. Who established the electrode classifications?
4. Who established the color code identifying these classifications?
5. What are the four general categories into which all the classified electrodes will fall?
6. What electrodes should be used for overhead welding?
7. What electrode should be used on high carbon steel?
8. What do the first two numbers of the electrode classification stand for?

LESSON 1.17

Object:
To study the characteristics and uses of iron powder electrodes.

General Information

The latest development in the evolution of electrode coatings has been the addition of iron powder. When this is added in relatively large amounts, the speed of welding and the quality of the welds are increased. Extra metal in the form of iron powder is available for deposition, in addition to that of the core wire. These are sometimes called a "contact" rod, which is in part a misnomer, because not all iron powder electrodes are of the contact type.

The theory of iron powder electrodes is not new, and they have had limited production and use in Europe for nearly two decades. Problems of relatively high cost and slow speed have kept them from general use. The solving of these problems in America, about 1953, has marked an important milestone in the progress of arc welding.

The first American iron powder electrode on the market to prove the success of this type was "Jetweld" produced by the Lincoln Electric Co. "Jetweld" electrodes are used in the downhand position and for horizontal fillets and lap welds. They have brought new economies to welded fabrication through faster and easier welding.

More recently, iron powder electrodes suitable for all-position welding have become available. Research has proven that most electrode coatings lend themselves to the addition of iron powder. Undoubtedly the future will see even greater use of iron powder in electrode coatings.

Iron powder in electrode coatings almost automatically produces:

1. Higher rate of metal deposition.
2. Smooth, uniform weld appearance with well-shaped beads.
3. Very little spatter.
4. Welds that are self-cleaning or clean very easily.
5. Excellent arc stability and characteristics.

To the welding industry, probably the most important single advantage of the iron powder electrode is the high rate of metal deposition. Fig. 1 shows a comparison of the deposition rates between the iron powder-rutile electrode and one of the fastest E-6012 electrodes. This means that a weldor can lay more pounds of weld metal per day and produce considerably more weld footage,

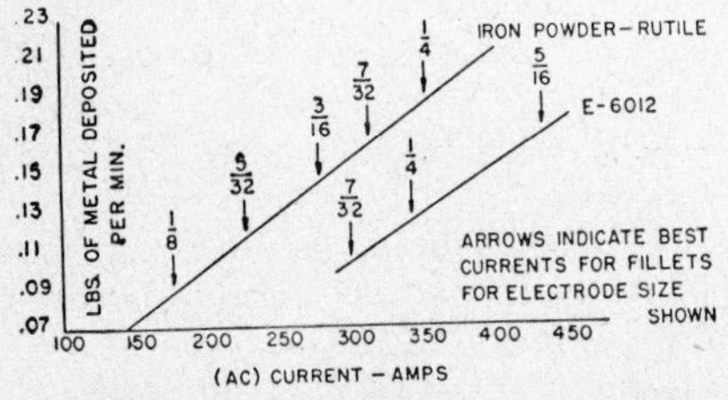

Fig. 1. Iron powder electrodes are capable of depositing weld metal approximately 50% faster than E-6012 electrodes.

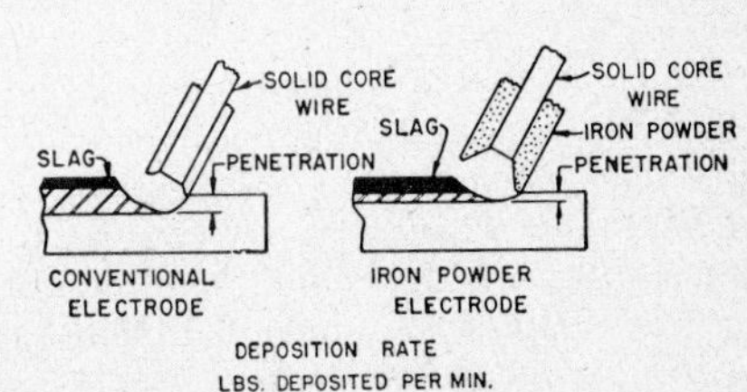

Fig. 2. The heavy coating of iron powder electrodes produces a crucible effect so that the electrical energy of the arc is more efficiently used in melting the core wire and coating.

even though smaller sizes of iron powder electrodes than E-6012 electrodes are used.

The higher rate of metal deposition, at a given current, which affords greater welding speed is the result of more efficient use of arc heat. A cross-section comparison of the electrodes as they are melted (Fig. 2) shows the deep crucible effect formed by the iron powder electrode due to its heavier coating. This crucible confines the heat of the arc, concentrates it, and utilizes it to melt the core wire and iron powder, instead of uselessly overheating the base metal and surrounding air. When depositing metal with the E-6012 electrode the deposited metal must come entirely from the core wire, while with iron powder electrodes, up to one-third of the metal comes from the coating. This metal, being in the powder form, melts with comparatively less arc heat than solid metal in the core wire. Because of this efficient use of arc heat, it is possible to lay down more metal with a 5/32" "Jetweld" electrode at 250 amperes than with a 7/32" E-6012 electrode at 300 amperes.

Iron powder electrodes, because of their heavier coating, have the ability to safely carry higher maximum amperages per size of electrode than E-6012 electrodes, as may be noted in Fig. 1. This means that more arc heat is available. Combining more heat with more efficient use of the heat, it is easy to understand how, size for size, iron powder electrodes will deposit 50% more metal per minute than E-6012 electrodes. Figure 3 shows why this is so important.

HORIZONTAL FILLET WELD

	PLATE SIZE	FILLET SIZE	ELECTRODE SIZE	CURRENT	ARC SPEED "/ MIN.
IRON POWDER E-6020	3/8"	1/4"	3/16"	270 AC	15.0
	3/8"	1/4"	1/4"	350 AC	12.0
IRON POWDER E-6012	3/8"	5/16"	1/4"	360 AC	14.0
	3/8"	5/16"	5/16"	400 AC	10.0

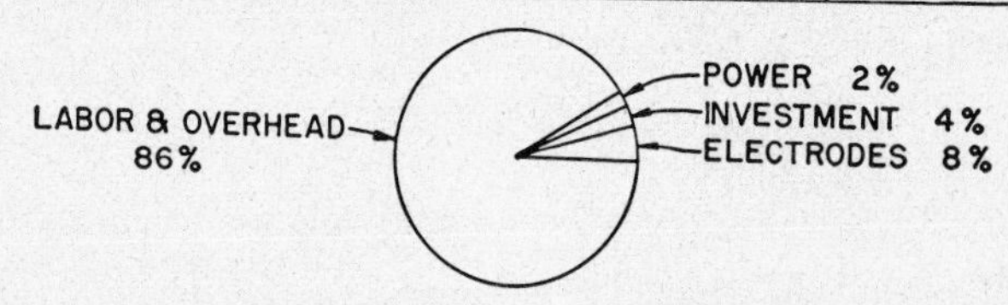

Fig. 3. A comparison between iron powder and E-6020 and E-6012 electrodes shows the higher speeds made possible through higher deposition rate. Large possible savings in labor and overhead cut into the major portion (86%) of welding cost.

Iron powder electrodes produce wide shallow beads as compared to the narrower, deeper penetrating beads of the E-6012 electrodes. This is an advantage, especially when welding thin sections or poor fit-up joints, though it may be a disadvantage when deep penetration is necessary.

Iron powder electrodes produce uniformly smooth beads, approaching the quality obtained by automatic welding. The most popular weld form in fabrication is the fillet weld used on T joints, inside corner joints and lap joints. On fillet welds the iron powder electrode produces a bead which is slightly concave to flat. This makes the most economical weld form with no metal wasted in the form of excessive convexity (Fig. 4), and the shrinkage forces are minimized (Lesson 1.15). There is no undercutting, spatter loss is nil at almost any con-

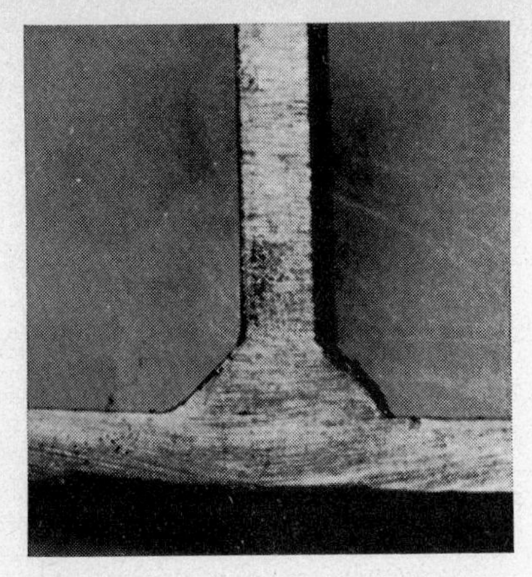

Fig. 4. The bead shape of a conventional fillet (right) compared to that of an iron powder electrode (left). With iron powder no metal is wasted in excessive convexity.

Fig. 5. High speeds are attained with a high standard of appearance. The top weld was made with a single iron powder electrode, and the bottom with a single E-6012 electrode. The iron powder electrode was 36% faster, with better appearance.

ceivable current, and the bead is smoother than by any other manual welding method (Fig. 5). Deep groove welds have good appearance and wash-in (Fig. 6). The weld bead is often self cleaning, and in any event, the slag may be easily removed. (Fig. 7).

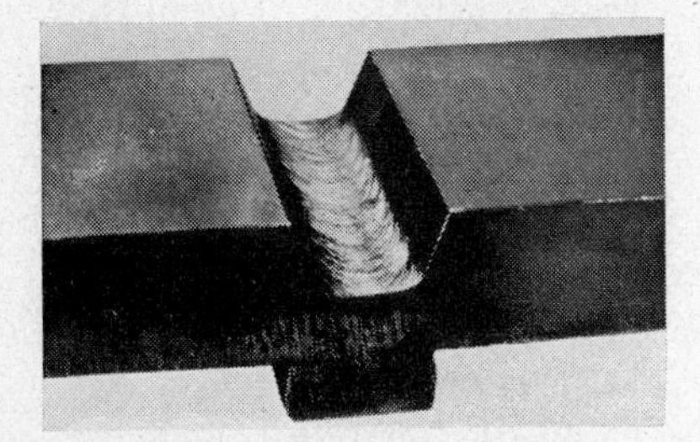

Fig. 6. Deep groove welds have excellent appearance and wash-in.

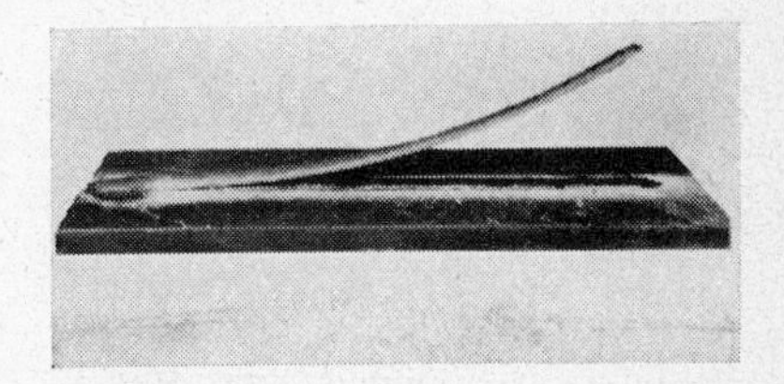

Fig. 7. Under favorable conditions slag is self-cleaning.

The nearly flat or concave beads produced by these electrodes are desirable when welding crack-sensitive steels. This results from: less admixture with the base metal, less total heat input due to increased welding speed and efficient use of current, and higher strength weld deposit. These factors affect shrinkage and the chemistry of the weld metal. In certain instances E-6012 and even some low-hydrogen electrodes (Lesson 1.38) have been successfully replaced on crack-sensitive applications by iron powder electrodes. The weld deposit exhibits excellent ductility. (Fig. 8).

Care must be used when selecting the amperage setting for iron powder electrodes. Excessive amperage settings may not be noticed due to the electrode's ease of operation, stable arc characteristics, lack of spatter and resistance of coating to heat deterioration. However, excessive current is to be avoided if weld quality and ease of cleaning are to be maintained. Fig. 9 shows a table of recommended amperage settings.

Both AC and DC operation may be used for iron powder electrodes, but AC is preferred for easiest operation. Also with AC higher welding currents may be used than with DC.

Fig. 8. Excellent ductility is demonstrated by the smooth iron powder weld compared to the conventional E-6012 weld which cracked when this plate was bent. Surface condition affects ductility.

RECOMMENDED CURRENTS				
ELECTRODE SIZE	CURRENT	POLARITY	FILLET SIZE	ARC SPEED "/ MIN.
5/32	225	AC	7/32"	15
5/32	200	DC +	7/32"	12.5
3/16	275	AC	1/4"	15
3/16	250	DC +	1/4"	12.5
1/4	350	AC	5/16"	14
1/4	300	DC +	5/16"	12

Fig. 9. Table of recommended currents.

The heavy coating of iron powder makes the electrode ideal for use with the contact or drag technique of welding. This generally produces the best and fastest results. Weldor fatigue is reduced by not having to hold the electrode a certain height above the work to obtain correct arc length, though some operators prefer to hold a short arc. Consult specific instructions for the application of any iron powder electrode used.

Review Questions

1. How did "iron powder" electrodes receive their name?
2. What was the first iron powder electrode on the American market?
3. What welding positions were recommended for the first iron powder electrodes?
4. Are there any all-position iron powder electrodes on the market?
5. Can iron powder be used in most types of electrode coating?
6. What are several advantages of iron powder electrodes?
7. What is the most important advantage to industry?
8. Can welding costs be cut with iron powder electrodes?
9. What welding amperages are used as compared to an E6012 electrode?
10. Does the iron powder in the coating unite with the weld deposit?
11. What type of penetration is obtained compared to the E6012 electrode?
12. What is the most-used form of weld in fabrication?
13. Can iron powder electrodes be used with certain hard-to-weld steels?
14. What type of welding current is preferred?
15. What type of electrode technique is recommended?

LESSON 1.18

Object:

To make a lap weld in the horizontal and downhand positions.

Equipment:

Lincoln "Idealarc" or AC or DC welder and accessories.

Material:

Mild steel plates, 1/4"; 5/32" "Jetweld 1" (E6024) and "Fleetweld 7" (E6012). (See Section III for properties and application of electrodes.)

General Information

The lap joint is welded with the bead made on the surface of one plate and the edge of the other. Fit work so there is no appreciable gap. Speed of welding, amperage and quality of weld vary directly with fit-up. "Fleetweld 7" is recommended for poor fit-up welds.

On practical applications most lap welds will be made in the horizontal position with both the base metal pieces flat. However, when the work can be tilted so that the joint is a trough for the molten pool of weld metal (see Fig. 2), much higher welding speeds can be obtained. Tilting the plates at least 10 degrees will usually be sufficient to speed up the welding appreciably.

The electrodes specified are the types recommended for lap welds to obtain maximum speed and minimum cost, although other types may be used. Consult Lessons listed above for further information on properties and application of these electrodes.

Job Instructions

Job A: Make a horizontal lap weld.

1. Clean base metal and position the two pieces of 1/4" plate on the table with a 2 inch overlap.
2. Set amperage at 130 to 145 for 5/32" "Fleetweld 7" or 215 to 235 for 5/32" "Jetweld 1".

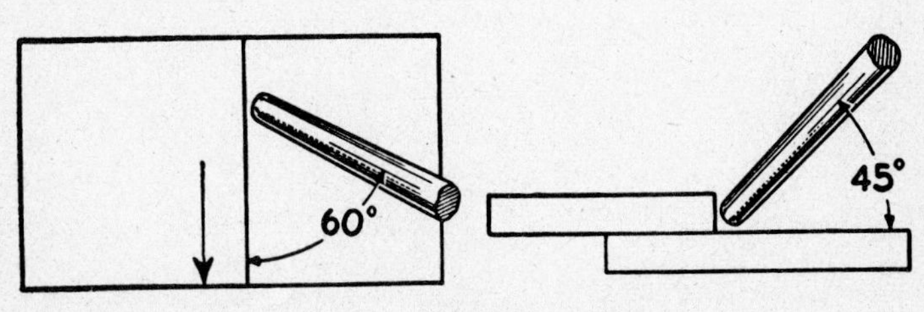

Fig. 1.

3. Hold electrode about 45 degrees from horizontal plate, inclined 25 to 30 degrees in the direction of travel. (Fig. 1)
4. Tack weld the joint at each end, so it will not move during welding.

5. Hold electrode lightly against both plates with the arc directed into the corner. Move ahead in a straight line.
6. Observe carefully the bead as it builds up. Change electrode angle or speed if bead sags or there is a tendency to undercut. Faster travel rate must be used with "Jetweld 1".
7. Clean the bead and examine it. It should be uniform and smooth without overlap or undercuts, penetrating evenly into each plate.
8. Break the plates apart if the weld is under 5 inches long, by placing one plate in a sturdy vise and hammering on the back of the other plate. The bead should have even penetration into each plate and completely into the corner.

Job B: Make a downhand lap weld.

Follow the same steps as Job A. Position the work as shown in Figure 2. Avoid undercutting the bottom plate.

If a welder of sufficient size is available, larger size electrode and higher amperages can be used, resulting in an increased welding speed. A slight weaving motion can be used to carry a larger bead on heavy plate.

Fig. 2.

Review Questions

1. What other weld is similar to the lap weld?
2. Why is it advisable to have a good fit-up?
3. What electrode should be used for poor fit-up joints?
4. What electrode movement is used when welding a horizontal lap weld?
5. What are the advantages of positioning a lap weld?
6. What electrode movement is used on a positioned lap weld?
7. What does the bead on a broken lap weld look like?

LESSON 1.19

Object:
To make a fillet weld in the horizontal or downhand position.

Equipment:
Lincoln "Idealarc" or DC or AC welder and accessories.

Material:
Mild steel plates 1/4" or thicker; 5/32" "Jetweld 1" (E6024) and "Fleetweld 7" (E6012).

General Information

Joining members or plates coming in at a 90 degree angle to each other with a fillet weld is the most commonly used joint in welded fabrication. The lap joint also uses the fillet type bead.

Most fillet welds are made in the horizontal position (Fig. 1). It is sometimes possible, however, to position the fillet downhand so the surface of the molten pool is horizontal. Higher amperages and larger electrodes may be used on the positioned welds, resulting in substantially higher welding speeds. Welding may be done easier on the positioned fillet without too much practice, because there is less tendency to undercut as is sometimes done to the vertical member of the horizontal fillet.

The arc is directed into the corner if the plates are of the same thickness. If the plates are of unequal thickness, the arc is directed more on the thicker plate, to heat both plates equally.

For maximum strength the leg of a fillet should be at least as thick as the base metal. On heavy plate the vertical member may be beveled for maximum penetration.

The electrodes specified are the types recommended for fillet welds to obtain maximum speed and minimum cost, although other types may be used.

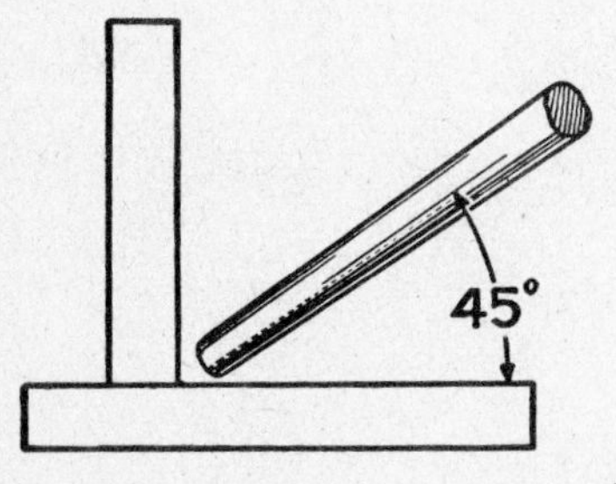

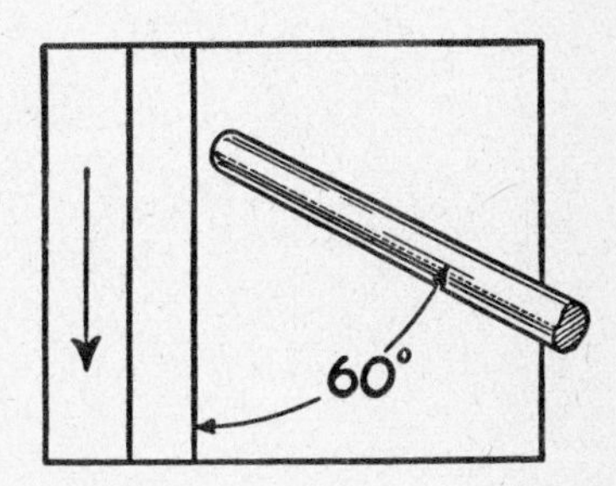

Fig. 1.

Job Instructions

Job A: Make a horizontal fillet weld.

1. Clean base metal and place plates as shown in Figure 1.
2. Set amperage at 130 to 145 for "Fleetweld 7" or 215 to 235 for "Jetweld 1".
3. Hold electrode angle as shown in Fig. 1.
4. Tack weld each end of the joint in position with a good fit-up.

5. Hold electrode lightly against both plates with the arc directed into the corner. Move the electrode ahead in a straight line.
6. Observe the bead carefully as it forms under the arc. There may be some tendency to undercut the vertical plate. Vary the electrode angle specified above slightly to get the correct bead shape. If travel speed and electrode angle are correct, the bead will not undercut. Travel faster with "Jetweld 1".
7. Clean the bead, and examine it for signs of overlap or undercut.
8. Break the weld to see if penetration is equal into both plates and complete into the corner.
9. Heavier horizontal fillet welds require more than one bead. Proper procedure is to use the bead sequence shown in Fig. 2.

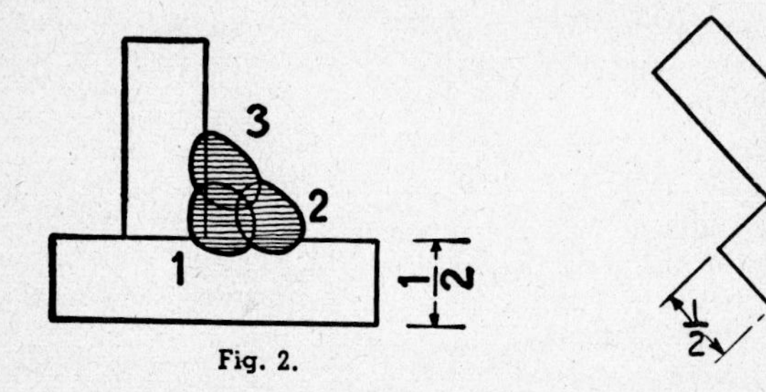

Fig. 2.

Fig. 3.

Job B: Make a downhand positioned fillet weld.

Use same steps as in Job A, except to position the weld (Fig. 3). Amperage may be set higher. If the top of the molten pool approaches a horizontal position, a slight weaving motion can be used to carry a larger bead. If subsequent passes are needed to bring the weld up to size, use the weaving technique with the bead sequence shown in Fig. 4. A larger electrode may be used with higher amperage settings for maximum speed.

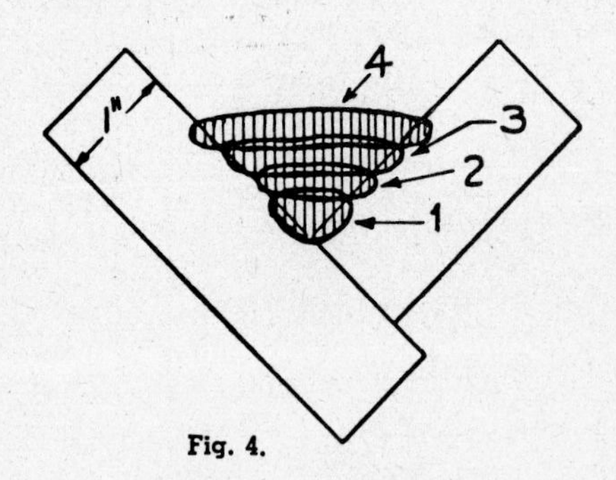

Fig. 4.

Review Questions

1. What other weld form is similar to a fillet weld?
2. What arc length is used on fillet welds?
3. How are plates of unequal thickness welded?
4. What are the advantages of positioning a fillet downhand?
5. Is it necessary to vary the specified electrode angle during welding?
6. What does a good fillet weld look like when broken?

LESSON

1.20

Object:

To make a butt weld in the downhand position.

Equipment:

Lincoln "Idealarc" or DC or AC welder and accessories.

Material:

Mild steel plates of various thickness; 5/32" and 3/16" "Fleetweld 5" (E6010) for DC, "Fleetweld 35" (E6011) for AC and "Jetweld 2" (E6027). (See Section III for properties and application of "Jetweld 2")

General Information

The joint made by placing the edges of two plates together and fusing them with the arc is called a butt weld.

Preparation for the butt weld depends upon the thickness of the metal, whether it will be welded from one or both sides, and the equipment available for preparing the edge.

A butt weld is most easily made by running beads on both sides of the joint, taking care that the penetration of the beads on each side meet for maximum strength (Fig. 2). Butt welds may also be run from one side only.

This is often the only way a weld can be made, as when doing pipe and tank welding. When welding from one side, be sure that the first pass achieves 100% penetration. Each bead must be cleaned well before subsequent passes are run. When welding from one side, better penetration is insured if metal 1/4" thick and over is beveled equally to form a 60 degree vee along the joint (Fig. 3). Make enough passes on either type of joint to bring the weld bead slightly above the surface of the base metal.

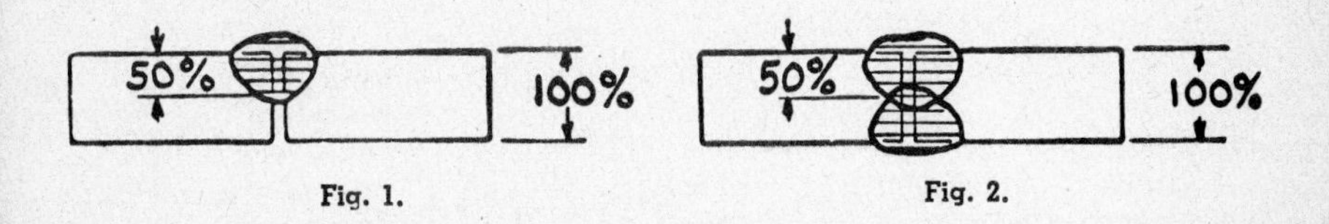

Fig. 1.　　　　Fig. 2.

Job Instructions

Job A: Make a square butt joint.

1. Clean base metal and place two 1/4" plates flat on the table.
2. Set amperage 130 to 145 for 5/32" "Fleetweld 5 or 35" electrode.
3. Hold electrode perpendicular to the plate, inclined 20 to 25 degrees in the direction of travel.
4. Tack weld the plates together with 1/32" to 1/16" root spacing (Fig. 1).
5. Run a straight bead to produce the weld in Fig. 1.
6. Break the weld apart; check for 50% minimum penetration.
7. For full penetration, turn the plate over and run a similar weld on the reverse side (Fig. 2).

Job B: Make a vee butt weld on heavier plate.

1. Prepare ⅜" plate with a 30 degree bevel (Fig. 3). If beveled plate is unavailable, use square edged plates placed at an angle (Fig. 4).
2. Clean base metal and position flat on the table.
3. Tack weld joint at each end with approximately ⅛" root spacing.

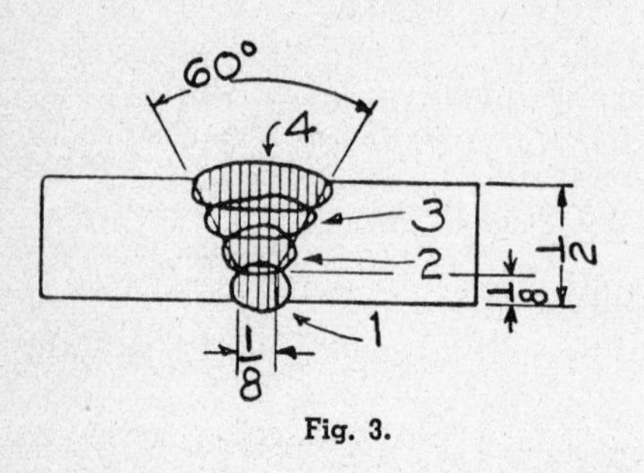

Fig. 3.

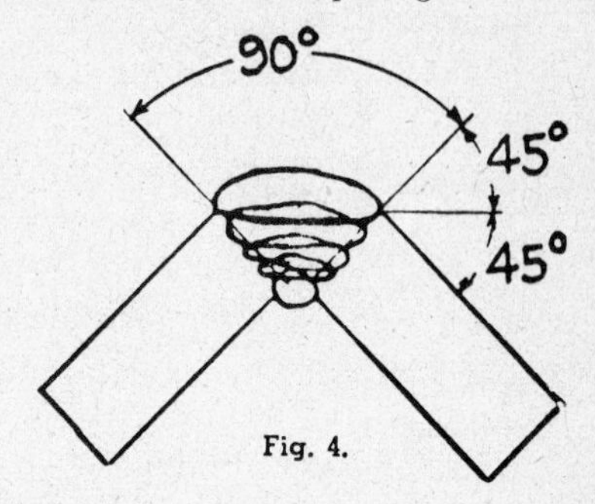

Fig. 4.

4. Lay root pass bead at the bottom of the vee with a 5/32" electrode.
5. Set amperage at 230 to 250 for "Jetweld 2" electrode.
6. Hold electrode perpendicular to the plate inclined 30 degrees in the direction of travel.
7. Chip off slag if it does not remove itself. Examine the weld for complete penetration and even fusion into each plate.
8. Run second, third and fourth beads, or enough to fill the vee with a slight reinforcement bead on the last pass. Larger electrode may be used with increased amperages for filler passes on heavier metal. Use a slight weaving motion for filler passes. Wash the deposit up on each plate, so that a slightly concave bead is formed. It is easier to chip, assures good penetration into each plate, and minimizes slag inclusions along the edge of the bead. When vee becomes wider than 3 times the electrode diameter, lay two beads, rather than weaving a wide one.
9. Practice making U-joint welds on ½" or heavier plate using procedure similar to that for vee joints. Use 3/16" or 7/32" "Jetweld 2" electrode.
10. A back-up strip may be used in making butt joints of the vee or U type. Larger electrodes and higher speeds can be used without danger of burn-through. Steel back-up strips become a part of the weld and can be removed only by machining. Copper strips ¼" thick or more may be used for back-up. They will not fuse with the weld and may be removed after welding. Root spacing when using "Jetweld 2" with a back-up strip should be the diameter of the electrode.

Review Questions

1. What are three edge preparations used for butt welds?
2. What is the easiest way of obtaining full penetration?
3. Why is full penetration important?
4. What precaution must be taken when making a weld from one side?
5. When welding from one side, how many passes are used?
6. Is it necessary to thoroughly clean each pass before running the next?
7. What are the advantages of using back-up strips?
8. What type of back-up strip may be used?

LESSON 1.21

Object:
To run a bead in the horizontal position.

Equipment:
Lincoln "Idealarc" or DC or AC welder and accessories.

Material:
Steel plates 3/16" or thicker; 5/32" "Fleetweld 5" (E6010) for DC or "Fleetweld 35" (E6011) for AC.

General Information

When welding out-of-position the electric arc counteracts gravity with the arc force created by the coating. The arc will deposit droplets of electrode metal into the crater in any position. However, gravity will influence the action of the molten metal once it is deposited in the crater. It is necessary, therefore, that the size of this molten pool be kept small and the force of the arc be used to help keep it in place. Electrode movement and arc length become very important. Electrodes should have "fast freeze" characteristics. The importance and effect of practice also become apparent when welding in these positions.

It is necessary at times to make horizontal welds, particularly in the field. These are welds on plate which is in the vertical position, but the joint runs parallel to the ground. An example of this is the girth seam in large storage tanks, and butt welds on vertical pipelines.

Job Instructions

1. Clean base metal and position vertically on the welding table.
2. Set amperage at 130 to 140 amps for 5/32" electrode.
3. Tack weld metal to a scrap plate (Fig. 1) or clamp it in position.

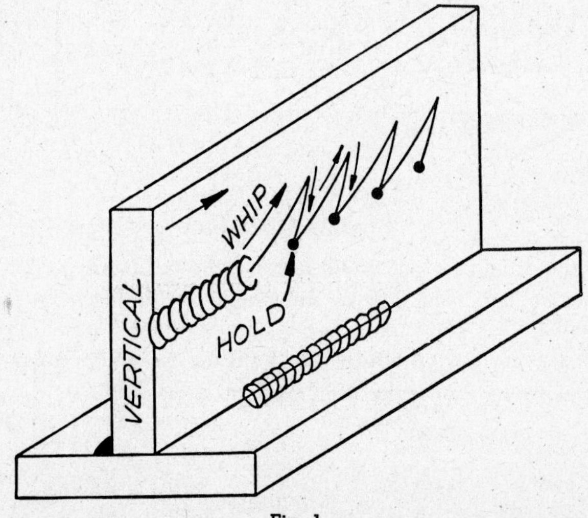

Fig. 1.

4. Hold the electrode angle about 5 degrees below perpendicular and inclined approximately 15 to 20 degrees in the direction of travel. (Fig. 1).
5. Strike an arc on the vertical plate and draw the bead along in a horizontal line, holding a short arc. Make the arc deposit molten metal on the vertical plate. While it is comparatively easy to maintain the arc, it is rather difficult

Fig. 2. Fig. 3.

to get a uniform, well-shaped bead. The molten metal has a tendency to sag or run down the plate, giving an appearance similar to Fig. 2. Reducing the current may assist in obtaining a bead of good shape.

6. If an irregular bead persists, a slight whipping motion such as shown in Fig. 1 may be used to assist in overcoming it, and obtaining the bead shown in Fig. 3. Shortening or "crowding the arc" at the top of the weave will be helpful in controlling the molten metal.
7. For wider beads, such as used on butt welds, a somewhat more extensive weaving motion is necessary, as shown in Fig. 4. The upward motion is rapid, and the metal is deposited on the downward motion.

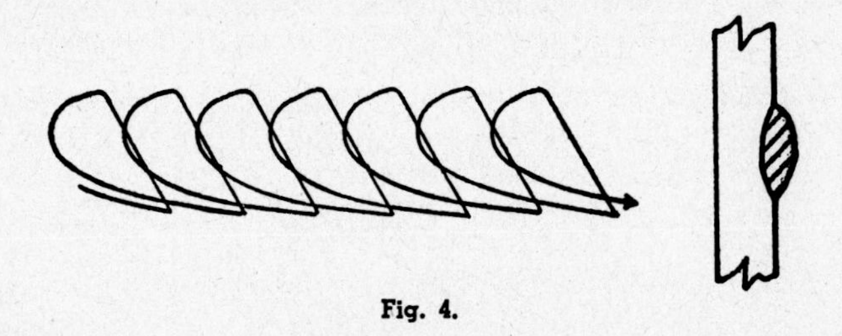

Fig. 4.

Review Questions

1. How are the droplets of electrode metal deposited in the crater?
2. What are two important factors influencing welds made in positions other than downhand?
3. What are two corrections which may be made for an irregular bead?
4. Can a weaving motion be used in horizontal welding?

LESSON 1.22

Object:
To make a butt weld in the horizontal position.

Equipment:
Lincoln "Idealarc" or DC or AC welder and accessories.

Material:
Mild steel plates ¼" or thicker; 5/32" and 3/16" "Fleetweld 5" (E6010) for DC or "Fleetweld 35" (E6011) for AC.

General Information

There are a number of horizontal joints which may be encountered in field work, such as butt, lap, fillet, edge and corner. The lap joint in Fig. 1 is similar to the horizontal lap weld (Lesson 1.18) and need not be discussed further. The lap joint shown in Fig. 2 is made like the overhead lap or fillet weld (Lesson 1.33). The technique used on the edge joint in the horizontal position is similar to running a horizontal bead, and welding a corner joint is similar to the horizontal vee butt weld. Since these types of joints are so similar to other types, the butt weld will be the only one discussed under the horizontal position.

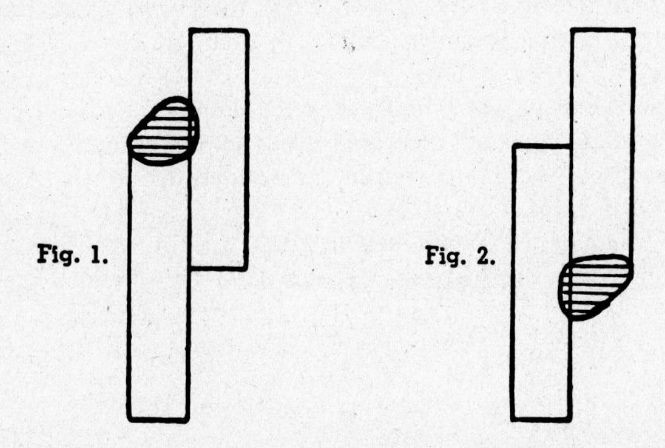
Fig. 1. Fig. 2.

The mastery of this weld is very important, because it is encountered frequently in pipe and tank welding.

Horizontal butt joints may be of three types, square, bevel one plate (Fig. 3), and bevel both plates (Fig. 4).

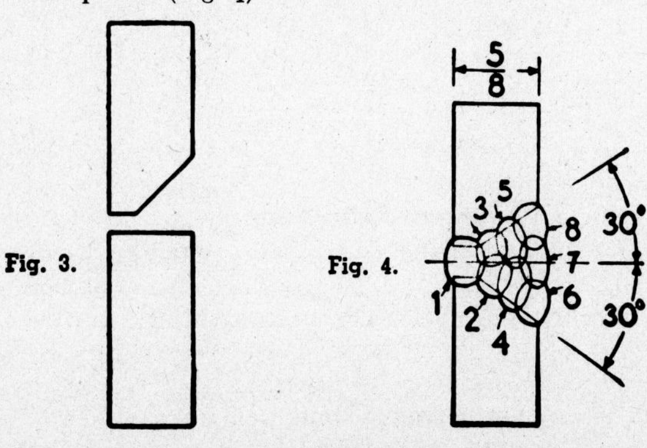

Fig. 3. Fig. 4.

Job Instructions

Job A: Make a square butt weld in horizontal position.

1. Tack weld two ¼" plates together in a square butt joint. Secure in a vertical position.
2. Set amperage at 130-145 for 5/32" electrode.
3. Hold the electrode perpendicular to the plate or 5 degrees under, inclined 5 to 10 degrees in the direction of travel.
4. Hold a short arc length. Run a bead on one side using a slight whipping motion. If spatter is excessive or puddle is difficult to control, reduce heat.
5. Examine the completed bead. Break apart for better inspection if desired. Bead should fuse equally into each plate slightly over half way through.
6. Repeat the job, welding from both sides for complete penetration.

Job B: Make a vee butt weld as shown in Figs. 3 or 4.

1. Set up as for Job A, using plate with 30 degree bevel.
2. Set amperage 130-145 for 5/32" electrode; 160-175 for 3/16" electrode.
3. Use 5/32" electrode for root pass, and 3/16" for filler and cover passes.
4. Hold the electrode angle as shown in Fig. 6, inclined 5 to 10 degrees in the direction of travel.
5. Make welds from both sides whenever possible. This allows a larger size electrode (3/16") to be used on the root passes to increase the welding speed. Fig. 7 indicates how ½" plate should be welded using 3/16" electrode and welding from both sides.
6. In horizontal welding, the beads are usually laid in straight lines without weaving (Fig. 5). For the last pass (cover pass), however, a weave may be made using the motion such as shown in Lesson 1.21, Fig. 4. Use a rapid upward motion, and deposit the metal on the downward movement.

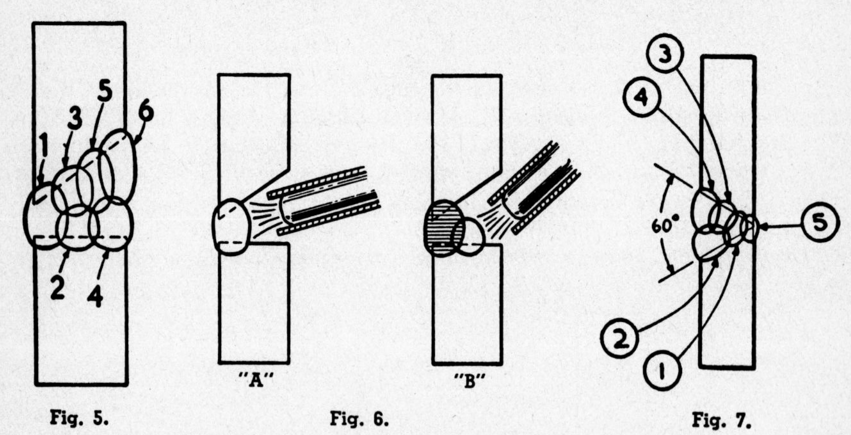

Fig. 5. Fig. 6. Fig. 7.

Review Questions

1. What other joint is welded similar to the horizontal butt weld?
2. What types of edge preparation may be used on horizontal butt welds?
3. How does the amperage setting for horizontal welding compare with down-hand technique?
4. What can be done to control excessive spatter?
5. What arc length is used on horizontal butt welds?

LESSON 1.23

Object:
To run a bead in the vertical position welding down.

Equipment:
Lincoln "Idealarc" or DC or AC welder and accessories.

Material:
Mild steel plate 3/16" or thinner; 5/32" "Fleetweld 5" (E6010) for DC or "Fleetweld 35" (E6011) for AC.

General Information

When welding in the field it is essential that a weldor be able to do vertical welding. There are two ways to make vertical welds; start at the top and weld down or start at the bottom and weld up. Vertical down welding is recommended on metals 3/16" or less in thickness. Down welding is usually found to be easier than vertical up welding. The cover pass on heavier metal is sometimes welded down to produce a smooth appearance. Vertical butt welds on horizontal transmission pipelines are generally welded down.

Not as much metal can be carried in a down pass. On heavy metals, therefore, it takes more passes to complete the joint. This can cause excessive distortion and is more time consuming.

Job Instructions

1. Secure the plate in a vertical position.
2. Set amperage at 130 to 145 for 5/32" electrode.
3. Hold the electrode pointing up at an angle of about 60 degrees with the plate (Fig. 1).

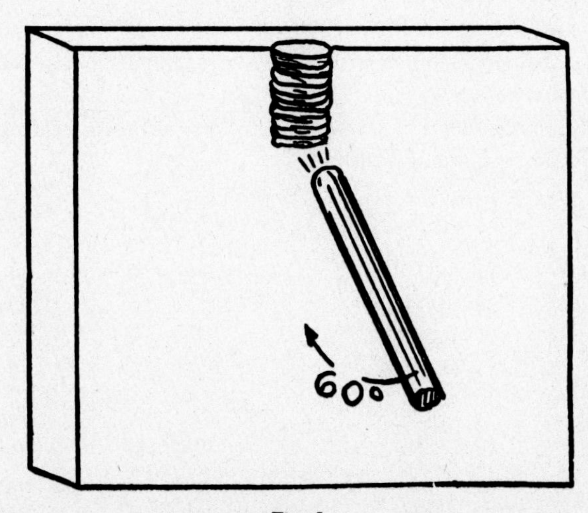

Fig. 1.

4. Strike the arc and hold a short, but visible arc. Draw the electrode down in a straight line. Move rapidly enough to keep the slag from running ahead of the molten pool. This will make a thin bead. If the slag runs ahead of the puddle, extinguish the arc, chip the slag from around the crater and ahead of it, and restart the bead.
5. Repeat the operation using a whipping motion as shown in Fig. 2A. This will give a narrow bead but one which is well-proportioned.
6. Use a wider weave, such as Fig. 2B and C. These weaves will be used later when making joints.

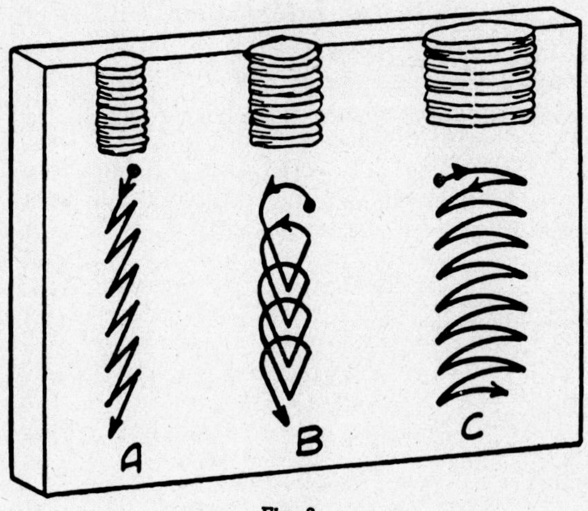

Fig. 2.

Review Questions

1. What two methods are used in vertical welding?
2. When is down welding used?
3. What is vertical up welding used for?
4. Which method is easier?
5. How fast should the electrode travel be on vertical down welding?
6. Can a weave be used welding down?

LESSON 1.24

Object:
To make a lap and fillet weld in the vertical position welding down.

Equipment:
Lincoln "Idealarc" or DC or AC welder and accessories.

Material:
Mild steel plates 3/16" or thinner; 5/32" "Fleetweld 5" (E6010) for DC or "Fleetweld 35" (E6011) for AC.

General Information

Lap and fillet welding in vertical positions are similar. The bead form is the same, as well as the arc length, electrode angle and electrode motion. For this reason they will be presented as a single lesson.

Job Instructions

Job A: Make a lap weld in the vertical position welding down.

1. Set amperage at 130 to 145 amps for 5/32" electrode.
2. Tack weld two plates together in a lap joint and secure in a vertical position.
3. Hold the electrode pointing upward 60 degrees from the vertical plate and directly into the corner, 45 degrees from the plate surface.

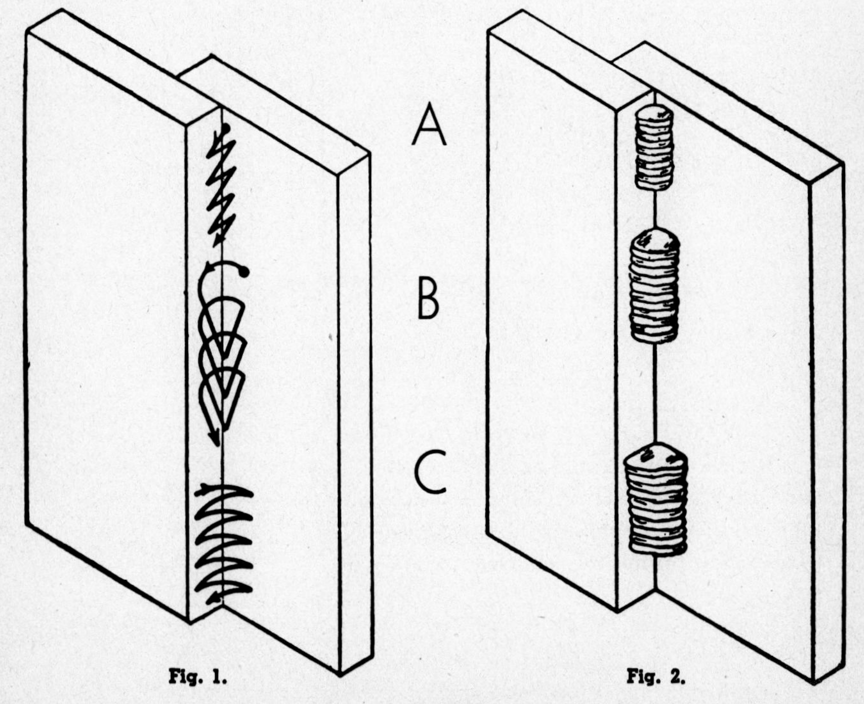

Fig. 1.

Fig. 2.

4. Strike the arc at the top and weld down keeping a short arc length. Use a straight bead or a whipping motion (Fig. 1A). Travel downward should be at such a rate that the slag does not run ahead of the crater.
5. Repeat, using weave B. This will result in a slightly heavier bead.
6. Repeat weave A or B, clean the bead well and lay a second bead over it, using weave C. Make certain you are getting penetration into the corner and evenly into each plate.

Job B: Make a fillet weld in the vertical position welding down.

1. Use amperage settings and electrode angle as in Job A.
2. Tack weld two plates in a tee joint, and secure in a vertical position.
3. Repeat the same passes as in Job A, making a fillet weld (Fig. 3). Practice particularly the heavy weave.

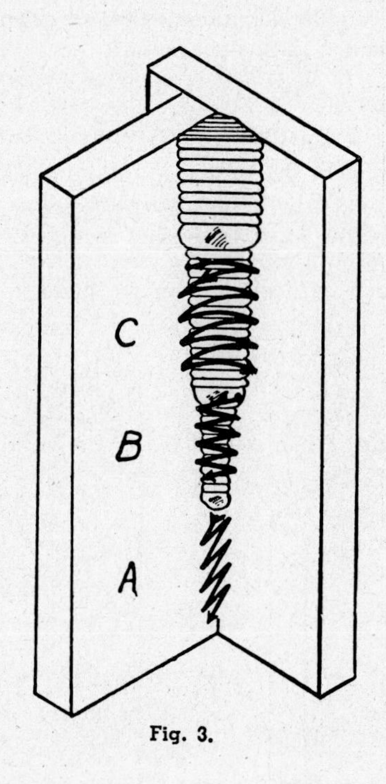

Fig. 3.

Review Questions

1. Is vertical welding encountered a lot in the field?
2. How is a heavier bead run in vertical down welding?
3. Why are the different weave patterns used?
4. How do you control the tendency to undercut?

LESSON 1.25

Object:
To make a butt weld in the vertical position welding down.

Equipment:
Lincoln "Idealarc" or DC or AC welder and accessories.

Material:
Mild steel plate 3/16" or thinner; 5/32" "Fleetweld 5" (E6010) for DC or "Fleetweld 35" (E6011) for AC.

General Information

The various edge preparations, spacing and procedure for welding butt joints is discussed in Lesson 1.20. As vertical down welding is generally used on metal up to 3/16" thick, only the square butt will be practiced here. Lesson 1.45 covers vee butt down welding of pipe.

In general, the same electrode angle, arc length and electrode motions are used in making vertical down butt welds as for fillet and lap welds.

Job Instructions

1. Set amperage at 130 to 145 for 5/32" electrode.
2. Tack weld two plates together with 1/16" root spacing (Fig. 1) and secure in a vertical position.

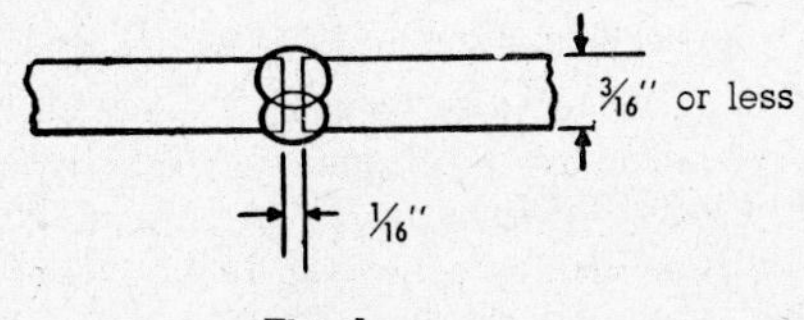

Fig. 1.

3. Hold electrode pointing directly into the joint, upward 60 degrees from the plate.
4. Strike the arc at the top and weld down, keeping a short arc length. Put in the first bead using the whipping motion (Lesson 1.24, Fig. 1A).
5. Clean the bead thoroughly and inspect it. Penetration should be adequate as on a downhand bead, and bead should be evenly fused into each plate.
6. Start at the top of the reverse side and put in the second bead, using the same electrode motion.

Review Questions

1. Why is the square butt joint used in vertical down welding?
2. Why is tack welding used on joints?
3. Why are the pieces spaced apart?
4. Why are larger electrode sizes not used on vertical down welding?

LESSON

1.26

Object:
To make a corner weld.

Equipment:
Lincoln "Idealarc" or DC or AC welder and accessories.

Material:
Steel plate 3/16" or 1/4"; 5/32" "Fleetweld 5" (E6010) for DC or "Fleetweld 35" (E6011) for AC.

General Information

Corner welds are used on a number of jobs, especially in making tanks and containers. They may be used on material of any thickness. No edge preparation is needed. The base metal should be tacked into position with a good fit-up. The square plate edges form a vee.

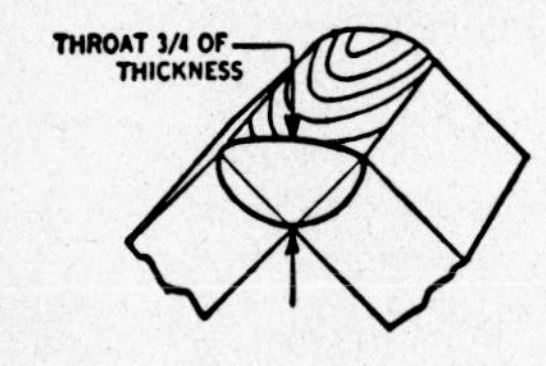

Fig. 1.

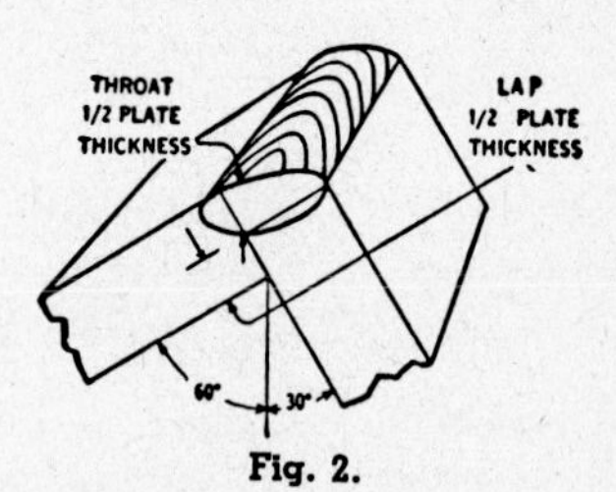

Fig. 2.

A full open joint (Fig. 1) may be used for maximum strength and the half open joint (Fig. 2) or closed joint may be used where less strength or only liquid tightness is needed. Corner welds on thin metal tend to warp or spread apart easier than butt welds and will need to be tack welded at closer intervals. Back-step welding will minimize distortion, as will the use of copper back-up strips placed in the corner. The procedure in making a corner weld is similar to a vee butt weld, except that less heat may be required. For maximum strength, a fillet weld is necessary on the inside.

Whenever possible these welds should be made in the downhand position, although horizontal and vertical positions may be used. When vertical welding is necessary, down welding is recommended.

Job Instructions

1. Set amperage at 125 to 140 for 5/32" electrode.
2. Tack weld base metal together. Root spacing of 1/16" (maximum) may be used for penetration if desired.
3. Hold electrode pointing directly into the root of the weld, inclined 10 to 15 degrees in the direction of travel.
4. Strike an arc and carry a short arc laying the bead into the vee with a slight weaving motion. Avoid undercutting along the top of the edges.
5. Clean the bead and examine the weld. Bead should be flush with the top of the vee with no undercutting or overlapping.

Review Questions

1. Where are corner welds commonly used?
2. In what position should corner welds be made?
3. Should vertical up welding be used?
4. What thickness of material is used for corner welds?
5. What are the three major types of corner joints?
6. What three precautions aid in minimizing distortion on corner welds?
7. Is root spacing used when setting up a corner weld?

LESSON

1.27

Object:

To make edge welds.

Equipment:

Lincoln "Idealarc" or DC or AC welder and accessories.

Material:

Sheet steel, 12 to 16 ga.; 3/32" "Fleetweld 7" (E6012) or "Fleetweld 35" (E6011).

General Information

Edge welds, sometimes called flange welds, are usually restricted to base metal 1/4" and under, and for joints not subjected to heavy loads. Around the small shop they will probably be used on metal 12 to 16 gauge, due to the difficulty of flanging heavier metal. The stiffening effect of the flanged edge minimizes distortion. Welding procedure is similar to running a bead on a flat plate. The only difficulty encountered will be keeping the bead evenly on the narrow edge. Both types of joints in Fig. 1 may also be welded with the arc torch or carbon arc by fusing the flanges together without additional filler metal.

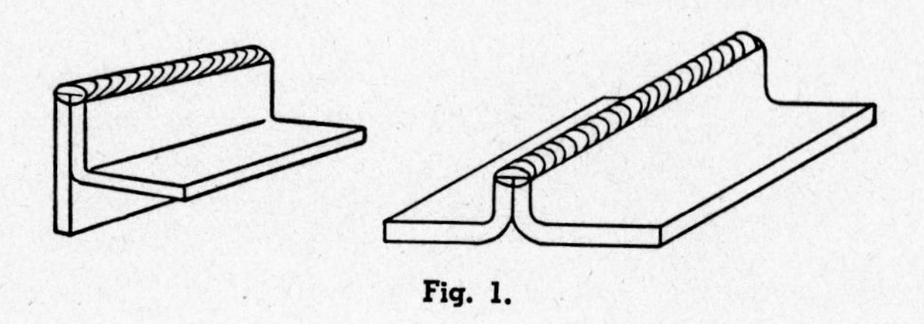

Fig. 1.

Whenever possible these welds should be made in the downhand position. When vertical welding is necessary, down welding should be used.

Job Instructions

1. Prepare flanges by turning over in a brake or with a hammer. For practice, narrow strips may be cut and put together with the edges up for welding to save the work of bending flanges.
2. Set amperage at 60 to 75 amps for 3/32" electrode.
3. Tack weld joint with no root spacing.
4. Hold electrode perpendicular to the joint and inclined 10 to 15 degrees in the direction of travel.
5. Strike the arc and move along rapidly holding a short arc.
6. Clean weld and examine bead. Weld should be fused into both pieces and be uniform in height without depressions burned down into the flange caused by irregular travel or excessive amperage.
7. Edge welds may be run on heavier metal, 1/8" to 3/16", using 1/8" electrode.

Review Questions

1. What size metal is usually used for edge welds?
2. What is another name for edge welds?
3. Are edge welds suitable for joints which receive heavy stresses?
4. What method of welding may be used besides the metallic arc?
5. What position is best for edge welds?
6. How should edge joints be welded vertically?

LESSON 1.28

Object:
To study the welding of thin gauge metals.

General Information

The use of thin gauge metal, generally termed sheet metal, around the shop is very common. Technically, sheet metal is any gauge less than 1/8" (approximately 11 ga. U.S.S.). Heavier material is called plate. Much of the work in the shop will be concerned with the thinner gauges of sheet steel, 14 gauge or less. Most sheet metal is mild steel with a coating of oxide (black iron) or zinc (galvanizing). Car and tractor bodies are usually mild steel with a primer and paint coating.

The general procedure for welding sheet metal will be the same as that discussed in the lessons for welding the standard joints. There are some problems, however, which are peculiar to sheet metal welding and must be taken into consideration before starting a job. Because of its thin gauge, the correct heat must be used when welding sheet metal. There is little room for error. Too little heat will not maintain the arc and too much will burn through the base metal. Distortion due to expansion and contraction is more of a problem in light metal than in heavier, more rigid metal. Lesson 1.15 should be studied before welding sheet metal.

Several precautions may be observed to make it easier to work with thin gauge metals. Joints should always have a good fit-up, and be carefully tack welded prior to welding. The lap weld is easily set up on sheet metal and provides extra thickness for burn-through protection. The flange type welds provide a more rigid joint to minimize warping (Lesson 1.27).

The metallic arc is recommended on metal as thin as 16 gauge, and may sometimes be used on 18 gauge. (Fig. 1) When using the metallic arc, a short arc should be held. There are some electrodes made for sheet metal welding. These have arc characteristics which produce a soft arc "spray" type of deposit. The E6013 and E6012 electrodes work well. The "deep penetration" type of electrodes should be avoided. Some welders are equipped with an attachment or adjustment to facilitate the welding of thin metal.

If gaps exist in joints they should be backed up with 1/4" copper strips to protect against burn-through. (Fig. 2) The weld metal will not stick to copper

ELECTRODE SIZES FOR SHEET STEEL

Size Gauge	Approximate Thickness in Fractions of an Inch	Approximate Thickness in Decimals of an Inch	Electrode Size	Amperage
11	1/8	.125	1/8"	90-100
12	7/64	.109375	1/8"	80-100
13	3/32	.09375	3/32"	45-65
14	5/64	.078125	3/32"	25-45
16	1/16	.0625	1/16"	20-30
18	3/64	.050	*Carbon arc	

*Carbon arc torch or single carbon is preferred for brazing metal of this thickness.

Fig. 1.

and its high conductivity assists in carrying away excess heat to reduce distortion. Steel back-up strips may be used, but they become a part of the joint and must be left on.

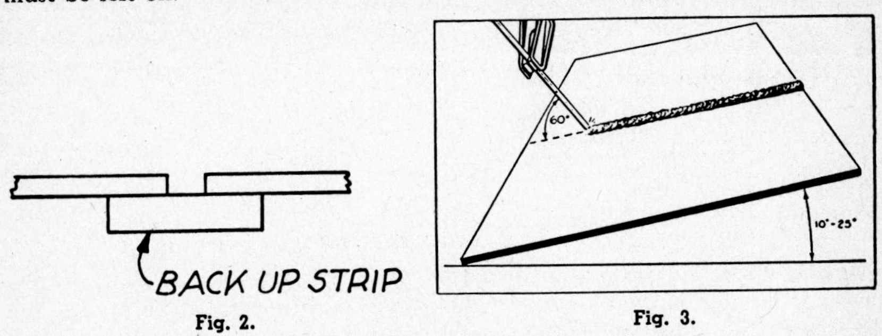

Fig. 2. **Fig. 3.**

Tipping the base metal 10 to 15 degrees (Fig. 3) and welding downward is advisable on light metals. Faster forward movement is possible.

The arc torch or carbon arc (Lesson 1.39) provides a method for fusing edge welds and brazing light metals. It is recommended for 18 gauge and thinner. The heat may be controlled more carefully than with the metallic arc. The base metal must be clean and free of oxide. Flux must be used when brazing to help clean the metal, retard oxidation during the process, and aid the bronze to flow and adhere to the base metal. The rod is fed into the flame area at an angle of 30 degrees from the joint. The arc flame is played on the joint and the rod, so that the base metal is heated to a dull red, the melting point of the rod. The metal from the rod should flow out as it contacts the base metal. The bronze will follow the heat of the arc along the base metal. If the bare rod and powdered flux is used, the flux is applied to the rod by heating and dipping it into the flux, or a little flux may be spread along the joint. Be careful not to burn the filler rod or overheat the base metal. A minimum of warp will result if the heat is held down and welding speed is rapid.

Galvanized metal must be welded or brazed outside, or in a well-ventilated area. The zinc fumes are toxic and should not be breathed into the lungs.

Review Questions

1. What thickness is sheet metal?
2. What base metal is commonly used in sheet metal?
3. What is galvanized iron?
4. What is black iron?
5. What are two major problems confronted by the sheet metal weldor?
6. What joint works well on sheet metal to provide extra thickness?
7. What is the thinnest gauge generally recommended for welding with the metallic arc?
8. What arc characteristics does a sheet metal electrode have?
9. What should be done with joints with poor fit-up?
10. What type of positioning works well when welding sheet metal?
11. Why is the use of the carbon arc recommended on very thin sheet metal?
12. Why is flux used when brazing?
13. What danger is there in welding galvanized metal?

LESSON 1.29

Object:
To run a bead in the vertical position welding up.

Equipment:
Lincoln "Idealarc" or DC or AC welder and accessories.

Material:
Mild steel plate 1/4" or thicker; 5/32" "Fleetweld 5" (E6010) for DC or "Fleetweld 35" (E6011) for AC.

General Information

Vertical up welding is generally used on plate 1/4" or thicker. Greater penetration is possible by vertical up welding, and more metal can be carried in each pass. This allows welds to be made with fewer passes, thereby reducing distortion and speeding up welding.

Amperage setting is less than for downhand welding and vertical down welding. The puddle must be kept small for easy control.

Some beginners find it easier to run beads on plate inclined 30 to 40 degrees and gradually increasing the angle until it is vertical.

Job Instructions

1. Set amperage 125 to 140 for 5/32" electrode.
2. Position plate inclined or vertical, and tack to a piece of scrap or clamp securely in place.
3. Hold electrode perpendicular with the plate pointing upward about 5 degrees (Fig. 1).

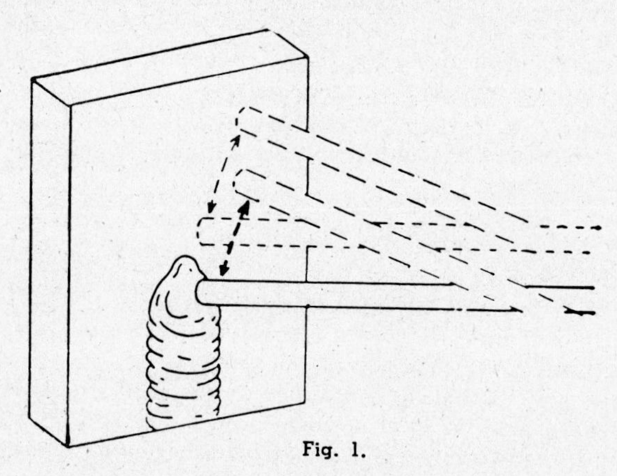

Fig. 1.

4. Strike the arc and establish a puddle holding a short arc.
5. Use the whipping motion to keep the puddle "cool", as practiced in Lesson 1.10. Move the electrode tip ahead of the puddle about 1/2" to 1" holding a long arc, hesitate, and return it to the puddle to deposit more metal with a short arc. Bead uniformity depends upon timing of the whipping motion. If the puddle is difficult to control or excessive spatter is obtained, it may

be necessary to reduce the amperage setting. Practice whipping motion until a uniform bead is obtained.

6. Run slightly wider beads using weave motions B, C and D (Fig. 2).

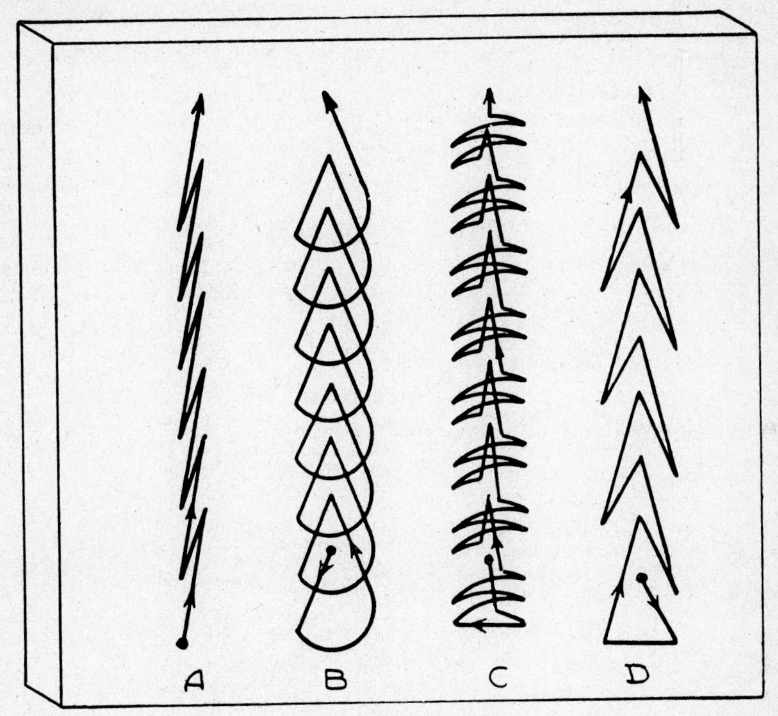

Fig. 2.

Review Questions

1. What size plate is generally welded vertically up?
2. Why is heavier plate not welded vertically down?
3. Which method of vertical welding can produce the heavier bead?
4. Why is the whipping motion used in vertical up welding?

LESSON

1.30

Object:

To make a lap and fillet weld in the vertical position welding up.

Equipment:

Lincoln "Idealarc" or DC or AC welder and accessories.

Material:

Mild steel plates 1/4" and 3/8"; 5/32" and 3/16" "Fleetweld 5" (E6010) for DC or "Fleetweld 35" (E6011) for AC.

General Information

The making of lap and fillet joints is similar, using the same weld form, electrode angle, and electrode motion. Maximum penetration on vertical joints in metal 1/4" or thicker is insured by welding up.

Job Instructions

Job A: Make a lap weld in the vertical position welding up.

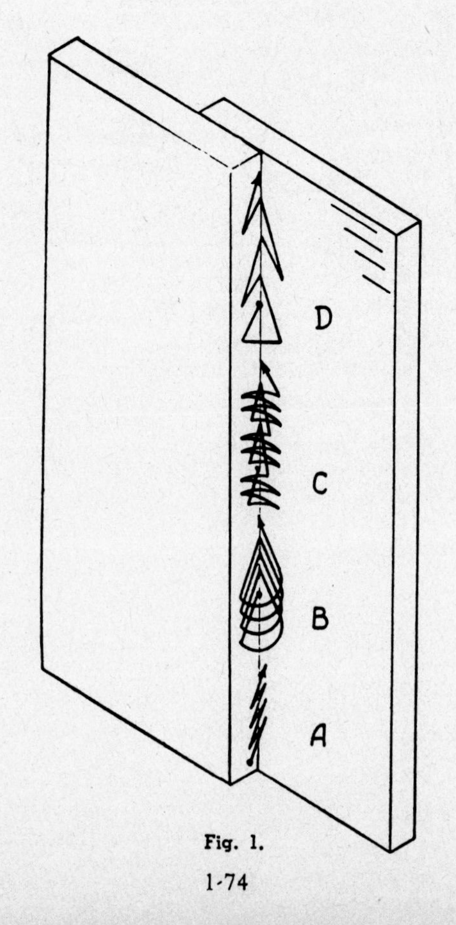

Fig. 1.

1. Set amperage 125 to 140 for 5/32" electrode.
2. Tack weld plates for a lap joint, and secure in a vertical position (Fig. 1).
3. Hold electrode pointing upward 5 degrees, directly into the corner.
4. Hold a short arc, and establish a puddle penetrating evenly into each plate.
5. Make first pass using the whipping technique shown in Fig. 1A. Whip the

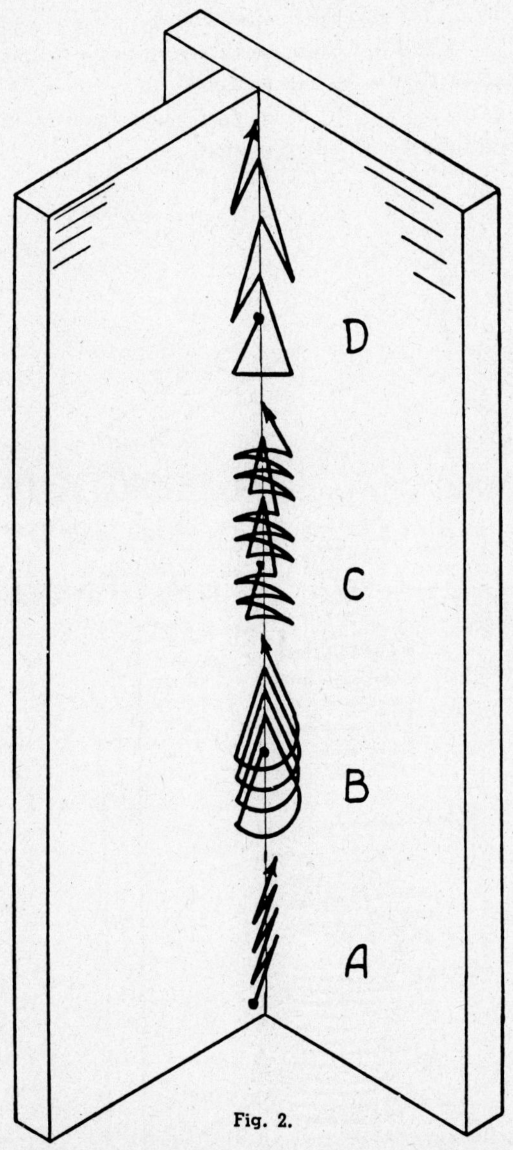

Fig. 2.

electrode tip upward from the crater about ½" to 1" holding a long arc, hesitate and return to the crater with a short arc to deposit more metal. Bead uniformity depends upon proper timing of whipping motions. If puddle is difficult to control, reduce amperage.

6. Practice weaves B, C and D to obtain a wider bead.

Job B: Make a fillet weld in the vertical position welding up.

1. Use same amperage setting, electrode angle, and electrode motion as for Job A.
2. Tack weld plates for a fillet weld and secure in a vertical position (Fig. 2).
3. Use whipping motion A to lay the first bead, cover using weaves B, C and D.

Job C: Repeat Jobs A and B, welding vertically up using 3/8" plate or thicker.

1. Set amperage 150 to 165 for 3/16" electrode.
2. Tack weld plates for lap or fillet weld; secure in vertical position.
3. Run the first bead using a whipping motion.

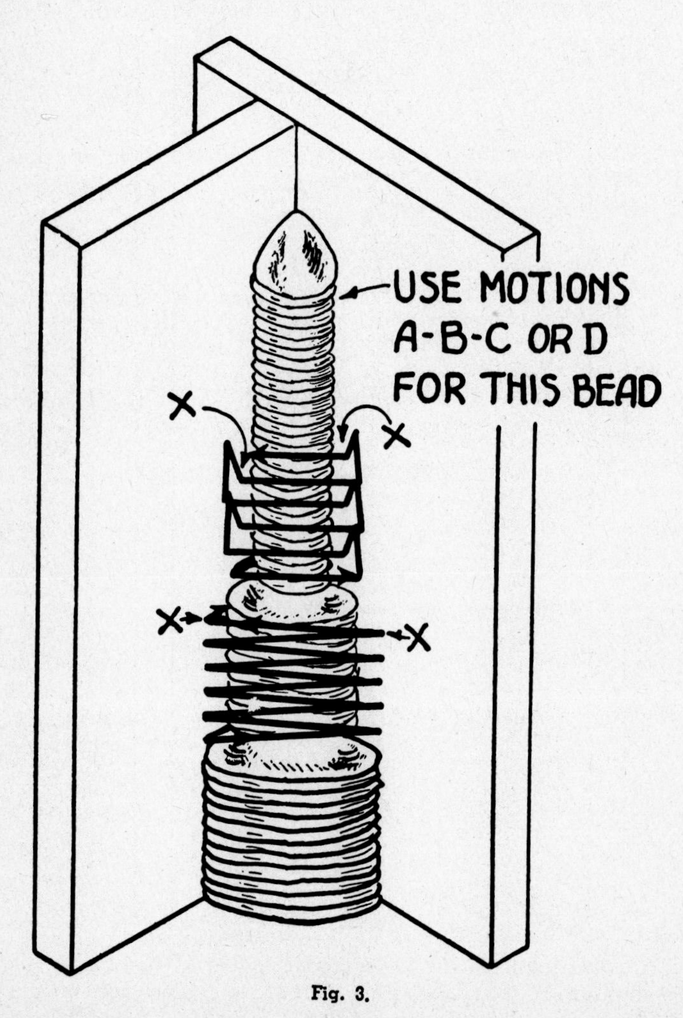

Fig. 3.

4. The second and third beads should be put in using the weaves shown in Fig. 3.
5. Additional beads can be applied, using weaves similar to that shown. A slight pause in the motion should be made at points X. The pause should be somewhat longer for the weave used for the third and fourth beads than for the second bead. These pauses are for the purpose of filling up undercuts in the base metal.

Review Questions

1. Why are fillet and lap welds 1/4" and over welded vertically up?
2. Why use the whipping motion when welding a fillet vertically up?
3. What are two reasons for using the different weave patterns?

LESSON 1.31

Object:
To make a butt weld in the vertical position welding up.

Equipment:
Lincoln "Idealarc" or DC or AC welder and accessories.

Material:
Mild steel plates 1/4" and 3/8"; 5/32" and 3/16" "Fleetweld 5" (E6010) for DC or "Fleetweld 35" (E6011) for AC.

General Information

Review the types of butt joints (Lesson 1.20). This lesson will cover only the vee type butt joint, because it will be used on 1/4" plate or thicker. Similar instructions apply to the groove or double vee joints. Welding may be done from one side, or both sides when possible.

Job Instructions

Job A: Make a butt weld with 1/4" plate.

1. Set amperage 125 to 140 for 5/32" electrode, 150 to 165 for 3/16" electrode.
2. Prepare edges and set up plate for vee butt weld. Tack weld with root spacing and secure in the vertical position as shown in Fig. 1.
3. Hold electrode pointing upward 5 degrees and directly into the joint.
4. Use a 5/32" electrode for the root pass. Strike the arc and establish a puddle at the bottom on both pieces. Carry the bead up using a whipping motion similar to Fig. 3A.
5. Make filler passes welding up with 3/16" electrode. Use motions B, C and D with sufficient weave to fill the vee. Make cover pass for a slight reinforcement bead on top.

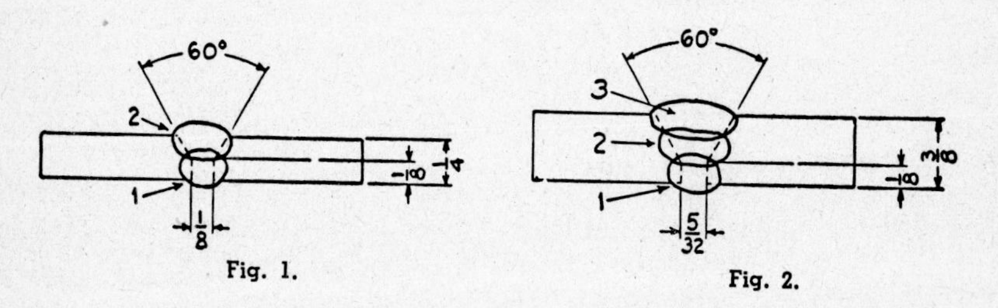

Fig. 1.

Fig. 2.

Job B: Make a butt weld with 3/8" plate.

1. Set amperage as in Job A.
2. Prepare edges, tack weld with spacing and secure plates in vertical position (Fig. 2).
3. Make the root pass with a 5/32" electrode, using motion A in Fig. 3.
4. Make filler passes with 3/16" electrode using the motions shown in Fig. 3.
5. Fig. 4 shows how heavy welds may be made in the vertical position with only two beads, by using a back-up strip. Put in the root pass as before, but make the second bead sufficiently heavy to fill out the vee, using a weaving motion like that of Fig. 3C.

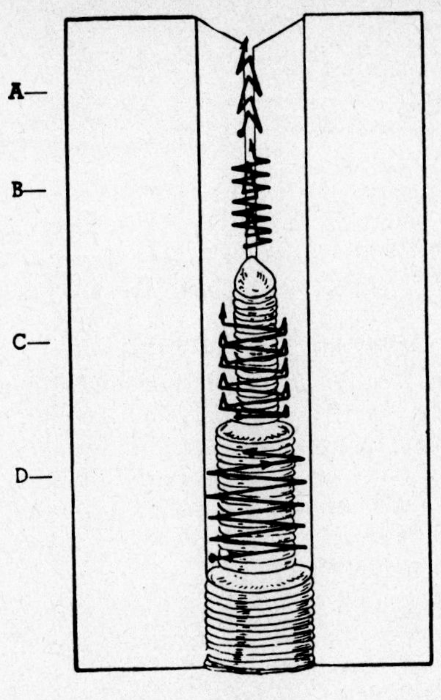

Fig. 3.

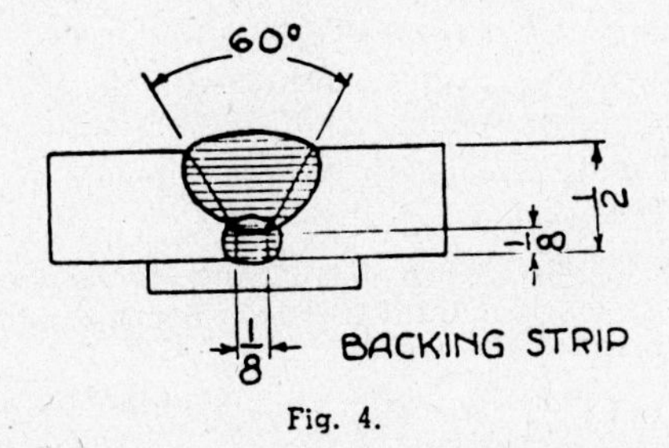

Fig. 4.

Review Questions

1. What type of joint preparation may be used on vertical up butt welds?
2. Can vertical up butt welding be done from both sides of the joint?
3. Why is a 5/32" electrode used for the first bead?

LESSON 1.32

Object:
To run a bead in the overhead position.

Equipment:
Lincoln "Idealarc" or DC or AC welder and accessories.

Material:
Mild steel plate ¼" or thicker; 5/32" "Fleetweld 5" (E6010) for DC or "Fleetweld 35" (E6011) for AC.

General Information

Welding in the overhead position is not difficult after the other positions are mastered. Difficulty may be experienced in keeping the electrode holder steady and the bead from sagging. Using both hands on the electrode holder, and resting one arm or elbow against the body or a solid object will help to steady the electrode. Holding a short arc and depositing the bead with a whipping motion will keep the puddle small and avoid sagging. Amperage settings are comparable to vertical up welding.

Shoulders, arms and head should be protected from falling spatter and sparks. Grasp the electrode holder with the back of the hands upward to ward off falling spatter, and stand to one side of the bead.

Job Instructions

1. Set amperage at 125 to 135 for 5/32" electrode.
2. Fasten the plate firmly in a horizontal position by tack welding or clamping, so that the underside may be easily reached with the electrode by assuming a comfortable position for welding.
3. Hold the electrode perpendicular to the plate, inclined 5 to 10 degrees in the direction of travel (Fig. 1).
4. Strike and establish the arc. When striking the arc, hesitate for two or three seconds with a rather long arc until the base metal is molten above the arc.

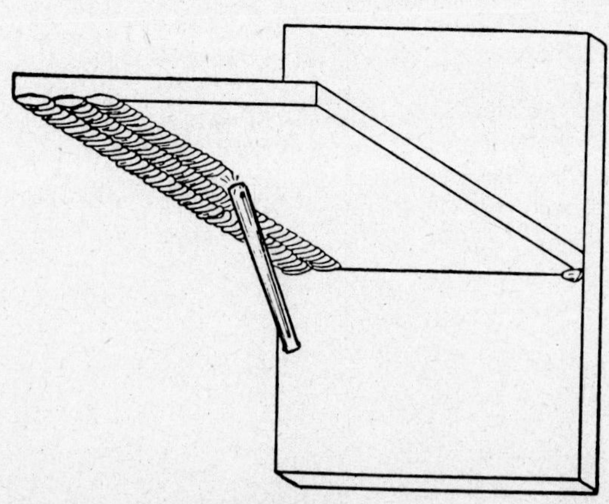

Fig. 1.

5. Move ahead in a straight line holding a very short arc to form a short bead. Try this a number of times until you can easily and quickly strike and maintain an arc overhead.
6. Continue to strike the arc and run longer beads. Move the electrode at a steady rate. The whipping motion will be helpful in controlling bead shape. If there is a tendency for the metal to run down and form drops, a very short arc will stop this. You can melt this drop by using a long arc and, when the drop is molten, quickly shorten the arc.
7. Try a very short arc, a short arc and a long arc. Observe the performance of each type of arc and the appearance of the resultant bead. Slight changes in amperage setting may be helpful in controlling the puddle. Inspect the beads by chipping and cleaning so that the penetration and fusion may be observed. This lesson will require considerable practice to obtain uniform beads.
8. After making satisfactory straight beads, starting and stopping at will, slight weaving motions such as shown in Fig. 2 may be practiced. This will allow you to carry more metal than in the straight bead. The weaving technique has only limited use in the overhead position, and the wide weave is not used.

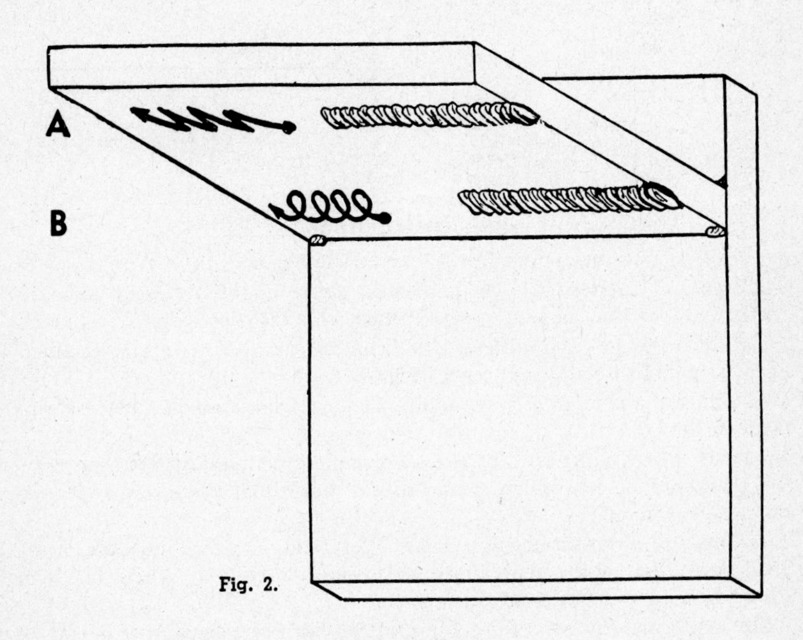

Fig. 2.

Review Questions

1. What are two difficulties of overhead welding?
2. How can you help to steady the electrode when welding overhead?
3. Why is it well to turn the backs of the hands upward during overhead welding?
4. What electrode angle is used for overhead welding?
5. How can you keep the puddle from dropping or sagging?

LESSON

1.33

Object:

To make a fillet and lap weld in the overhead position.

Equipment:

Lincoln "Idealarc" or DC or AC welder and accessories.

Material:

Mild steel plates ¼" and ⅜"; 5/32" and 3/16" "Fleetweld 5" (E6010) for DC or "Fleetweld 35" (E6011) for AC.

General Information

Fillet and lap welds are made in a similar manner in the overhead position. This lesson, therefore, is on fillet welds in detail.

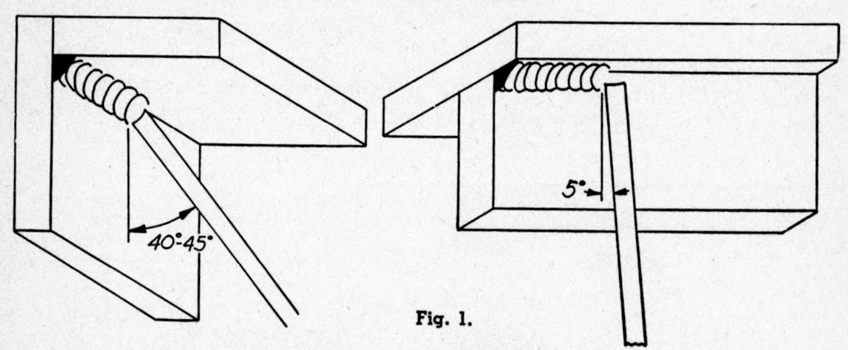

Fig. 1.

Job Instructions

1. Set amperage at 125 to 135 for 5/32" electrode.
2. Tack weld ¼" plates for a tee joint, and secure in the overhead position so that the under side may be easily reached with the electrode.
3. Hold electrode 40 to 45 degrees out from the vertical plate and inclined 5 degrees in the direction of travel (Fig. 1).
4. Strike an arc and establish a puddle evenly on both pieces. Place a single bead in the corner, using the whipping motion.
5. Break the plates apart and inspect for complete penetration into the corner. There should be no undercutting on the horizontal plate, or overlapping on the vertical plate.
6. After making a uniform single bead fillet weld, make a multiple stringer bead weld. Follow the general sequence shown in Fig. 2. These beads may be made with a slight weaving motion.
7. After mastering the procedure for multiple stringer beads, try a two pass weld with weaving. Lay first pass with whipping motion.
8. Lay the second bead using weaving motion (Fig. 3). Use weaving on a two pass weld only. Use stringer if more than two passes are required.
9. Repeat the job using 3/16" electrode on ⅜" or thicker plate.

Review Questions

1. Are the same procedures used for both fillet and lap welds in each of the four basic positions?

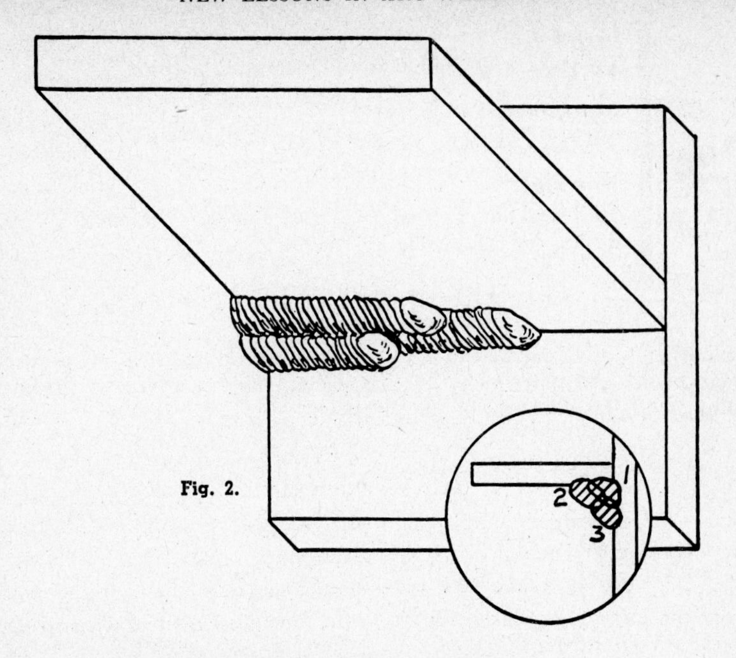

Fig. 2.

2. What two procedures are used for heavy fillets?
3. What is a straight bead sometimes called?
4. What should be the appearance of an overhead fillet bead?
5. How many passes are recommended for a weave technique weld?

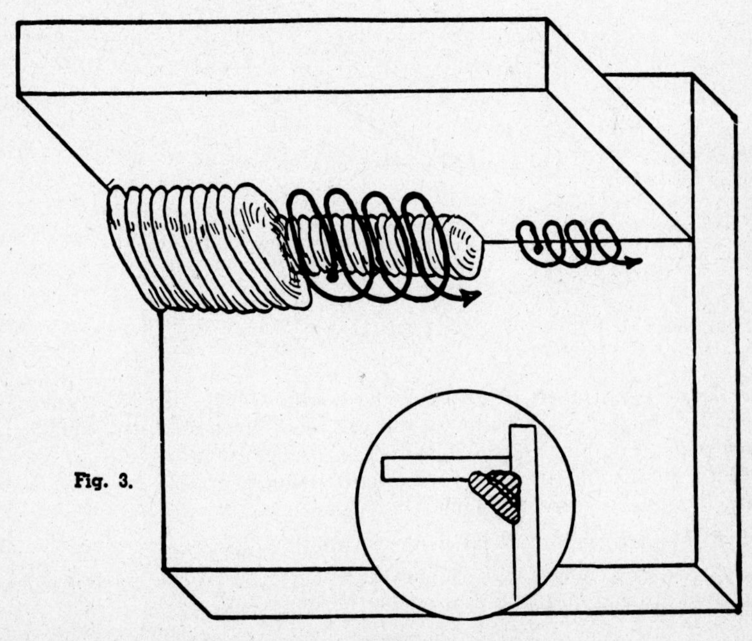

Fig. 3.

LESSON

1.34

Object:
To make a butt weld in the overhead position.

Equipment:
Lincoln "Idealarc" or DC or AC welder and accessories.

Material:
Mild steel plates 1/4" and 3/8"; 5/32" and 3/16" "Fleetweld 5" (E6010) for DC or "Fleetweld 35" (E6011) for AC.

General Information

Review butt joints in Lesson 1.20 for detailed information on preparation and procedure. All types of butt joints are suitable for overhead welding, depending upon the thickness of the metal.

Job Instructions

Job A: Make a square butt weld in the overhead position.

1. Set amperage at 125 to 135 for 5/32" electrode.
2. Tack weld two 1/4" plates with 1/8" gap, and secure in the overhead position.
3. Hold the electrode pointing directly into the joint inclined 5 to 10 degrees in the direction of travel.
4. Strike an arc and run a straight bead along the joint, fusing the two plates together (Fig. 1).

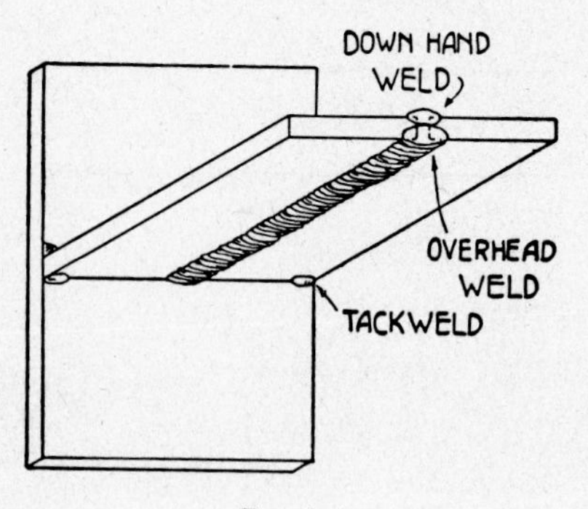

Fig. 1.

5. Break the plates apart to examine the quality of the deposit. Penetration should be slightly over half way through, and the bead fused equally into each plate. For plates up to 1/4" thick, a bead can be run on the opposite side for maximum strength. If the joint is not accessible from the reverse side, it must be veed and welded from one side.

Job B: Make a vee butt weld in the overhead position.

1. Prepare two 3/8" plates with 30° bevel for a vee butt weld. Tack weld with 1/16" root spacing and secure in an overhead position.

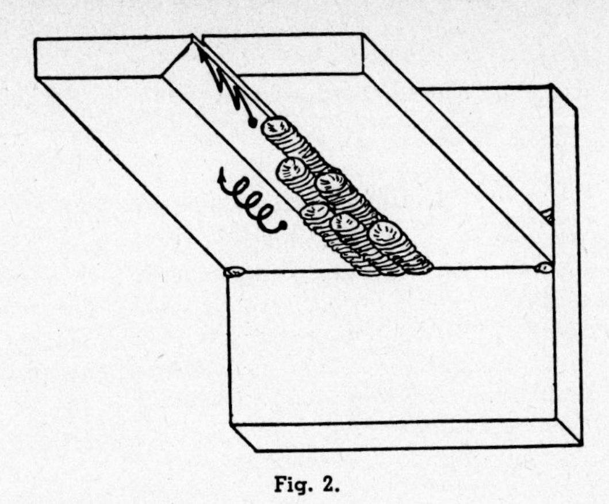

Fig. 2.

2. Following the procedure used in Job A and the sequence of passes shown in Fig. 2, make the weld with a whipping motion.
3. If difficulty is experienced in making the first bead with proper penetration, a back-up strip will be helpful. The back-up strip is often not possible or practical to use on construction jobs, so it is important that this type of weld be practiced without the use of the strip. A larger root spacing may be used with a back-up strip.

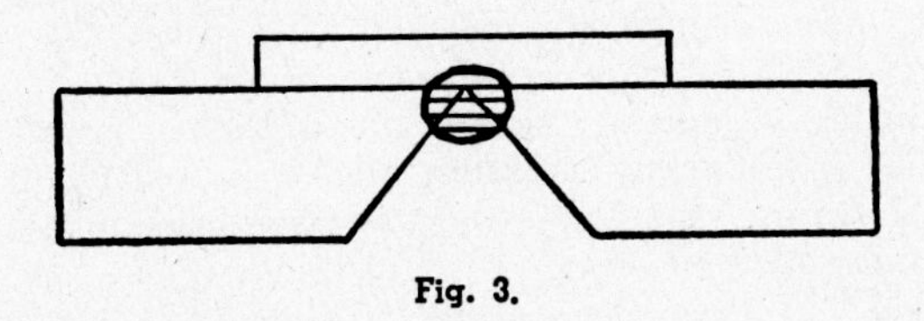

Fig. 3.

4. Welding in the overhead position is much easier using the stringer bead procedure. It is possible to lay succeeding beads in this position using a weave. It is not recommended, however, to make more than a two pass weld by the weaving technique.

Review Questions

1. What arc length is held in overhead butt welding?
2. What happens if travel speed is too slow?
3. Can a back-up strip be used?
4. What type of back-up strip can be easily removed after welding?
5. Why is it necessary to practice welding without a back-up strip?

LESSON

1.35

Object:

To study the standard welding symbols and their applications.

General Information

Welding has become so important and essential as a process of fabrication that it has become necessary to have some standard means of indicating the pertinent information necessary for each welded joint. The American Welding Society set up a method of symbolizing and specifying welding information. Its general adoption by industry for use on shop drawings has made it a standard for designers, draftsmen, and shop men. Its use fits our modern industrial organization of job specialization and standardization of parts inherent in mass production. To the designer and engineer are assigned the job of figuring loads and strength, specifying where the weld is to be made, what size and form shall be used, and any special information needed for its application. To the shop foreman and weldor are assigned the task of seeing that the proper welds are made, and that the correct procedure is used to make the weld specified. It helps to eliminate errors of overwelding, underwelding or the use of the wrong weld form for the stresses involved.

The AWS welding symbols are relatively easy to understand after a person has studied the basic symbols, and their general plan of use. All persons doing or specifying welded fabrication should be familiar with them. From these symbols it is possible to determine the following:

1. Location of each weld.
2. Size of weld (throat, length, spacing)
3. Type of weld (weld form, plate preparation and spacing)
4. Weld application information relative to individual specifications.

This information is easily and accurately read from a standard symbol placed on the drawing. It is no longer necessary to draw in each weld or letter descriptive notes, such as "weld all joints".

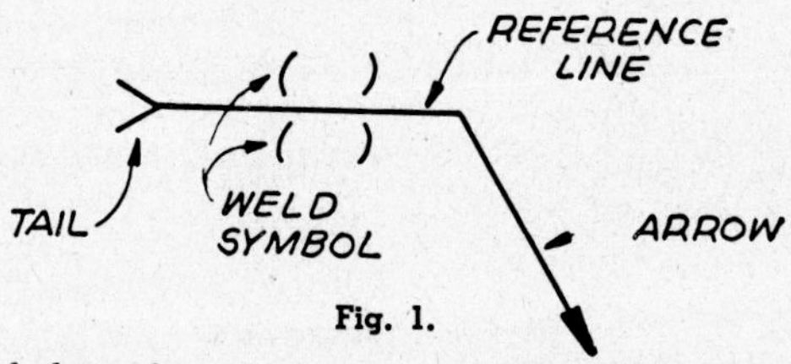

Fig. 1.

Welding symbols have four distinct parts. (See figure 1.) The *arrow* points to the joint, the *reference line* is a base for locating the small *weld symbols,*

TYPE OF WELD							
BEAD	FILLET	PLUG OR SLOT	GROOVE				
			SQUARE	V	BEVEL	U	J
⌓	◺	⏢	||	∨	⁄	Y	⊬

Fig. 2.

and the *tail* is a place to put special procedures or notes. The weld symbols define the type of weld which is required. (See figure 2.) The perpendicular leg of the weld symbol is always to the left. All types of welds and joint preparations may be shown. Additional information concerning size, length, finish, etc., is placed around the reference line.

Each joint has two sides; the "arrow" side to which the tip of the arrow points, and the "other" side. It is possible to designate which side is to be welded. If the small weld symbol is below the reference line the weld is to be made on the "arrow" side; when the symbol is on top of the reference line the weld is to be made on the "other" side.

Sometimes, it is difficult to tell from the drawing just what is the "other" side. Figure 3 illustrates a simple method for properly determining it. If you mentally "look" through the *joint* you can "see" the "other" side. If you try to "look" the wrong way, you will be "looking" at solid metal and cannot "see"

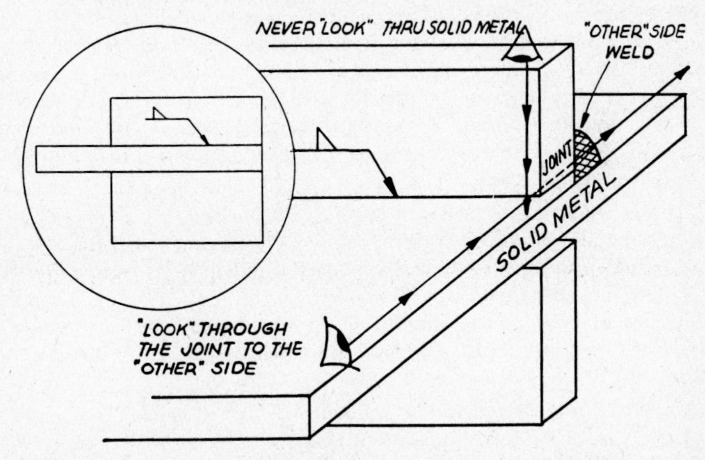

Fig. 3. Mental vision method of determining "other" side of a welded joint as indicated by the welding symbol.

through it. Always look through the joint. Remember that the arrow does not necessarily point to the "other" side. This can be clearly seen in figure 3 where the arrow points through solid metal and does not point to the "other" side. Therefore, don't use the direction of the arrow as a guide to finding the "other" side.

For a detailed discussion of AWS welding symbols and their application see Part 7.4, Section VII.

Review Questions

1. A complete symbol has how many parts?
2. What information is in the tail of the symbol?
3. How many sides are there to each welded joint?
4. A welding symbol placed below the reference line refers to which side of the joint?
5. To determine the "other" side is it ever necessary to "look" through solid metal?
6. Does the tip of the arrow point to the "other" side?

LESSON 1.36

Object:
To study the preparation and welding of cast iron.

General Information

It is necessary to know what cast iron is and how it acts under heat in order to weld it successfully. Cast iron is a mixture of iron with a high percentage of carbon, between 2 and 4 per cent. This carbon is in two forms; part is in solution with the iron (as it is in steel), and part is in a free form deposited as graphite flakes in little pockets throughout the metal.

The high carbon content in cast iron is responsible for the properties causing the difficulty which is encountered during welding; that is, cast iron has high compression strength, but is low in tensile strength and ductility. It becomes hard and brittle by heating and rapid cooling, such as caused by the welding process. This combination of characteristics causes cast iron to be "crack sensitive". If any stress is placed on this metal which tends to pull it apart rather than squeeze it together, its low tensile strength cannot withstand much pull and its low ductility will not allow it to bend or stretch. All it can do, consequently, is break. This may occur as either a complete break of the entire part, or as a localized crack.

The stresses which cause cast iron to crack during welding usually result from uneven heating and cooling. Welding on a piece of cool cast iron involves two violent thermal processes. The first is a very rapid heating caused by the electric arc. The second is a very rapid cooling of the molten metal caused by the small amount of heat in the puddle being rapidly absorbed by the cold cast iron and air surrounding it. Cooling causes shrinkage and shrinkage creates the stresses.

One other factor involved in the tendency of cast iron welds to crack is the effect of the carbon which is picked up from the base metal and mixed with the weld deposit. Though the electrodes used to weld cast iron have a very low carbon content, the high carbon in the melted base metal mixes with it to give a resultant high carbon deposit. High carbon deposits are also crack sensitive. In a good full-thickness bead, this is not much of a problem. In the crater at the end of the bead, however, there is very little thickness and the dilution from the base metal becomes a high percentage of the total deposit. The carbon pickup may make that end of the bead extremely brittle and crack sensitive. It is at this point that the weld is weakest. A crack will practically always form in the crater if the crater is on the cast iron itself. This crack, once started, may proceed through the entire weld.

There are two ways of avoiding trouble from cracking when welding cast iron, usually termed as the "hot" and "cold" methods. The "hot" method means to preheat the entire part slowly and uniformly to prevent building up stresses during this heating operation. The welding operation takes place on an already hot piece of cast iron. When the welding is completed, the entire piece must be slowly and uniformly cooled.

The "cold" method is used when it is impossible or impractical to evenly preheat the entire casting. It is then necessary to insure that the stresses set up by welding are so small that they cannot cause the part to crack. This must be done by a slow back-step method of welding and peening the deposit. The beads are allowed to cool down enough so that the bare hand may be laid on them before starting the next bead.

The work to be welded must be clean. All dirt, oil, grease and paint must be removed. This can be done by wire brushing, grinding, burning, or slowly heating to 400 to 500 degrees.

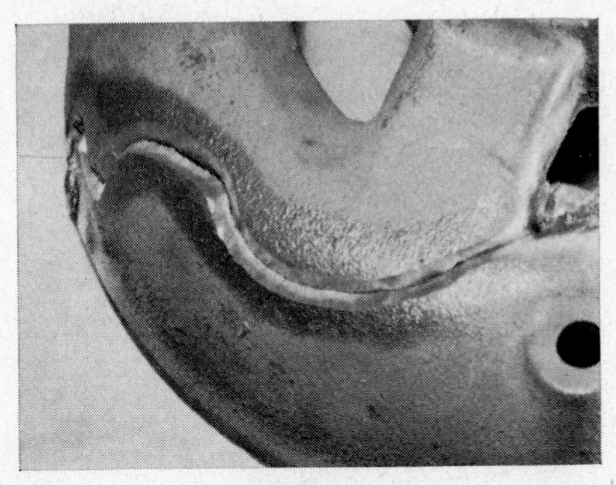

Fig. 1.

Joint preparation in cast iron is important to obtain a strong joint. When the casting is broken in two or more pieces, form a vee (Figure 1) by grinding a bevel off the broken edges, leaving 1/16″ to 1/8″ of the fractured edge to assure proper alignment of the pieces to the original size. This will also insure good fusion throughout the entire thickness of the weld. If the job to be welded is a crack, as is often the case with heavy castings, a hole should be drilled at each end of the crack to prevent its going farther, and to determine the thickness of the casting at that point. The end of the crack can be determined by putting kerosene over the area for a few seconds and wiping it off. Rub some white chalk along the crack. The crack will show up as a dark streak in the white chalk, showing its entire length and any hairline cracks. Vee out all cracks 1/8″ to 3/16″ deep with a grinder or a diamond point cold chisel. If the casting is under 3/16″, vee out half the thickness (Fig. 2).

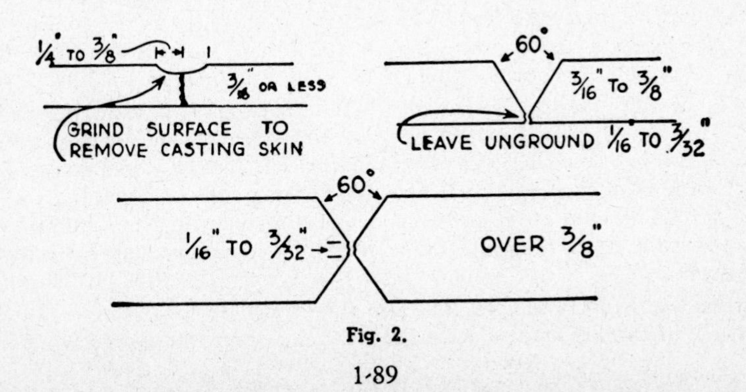

Fig. 2.

Welds in cast iron, if of sufficient thickness, may be strengthened by the mechanical method of studding (Fig. 3). Steel studs 1/4" to 3/8" in diameter should be used. The cast iron should be veed and drilled and tapped along the vee so that the studs may be screwed into the casting. The studs should be long enough to be screwed into the casting to a depth of at least the diameter of the studs, and project 3/16" to 1/4" above the surface. If the casting is of sufficient thickness, it may be veed and studded from both sides.

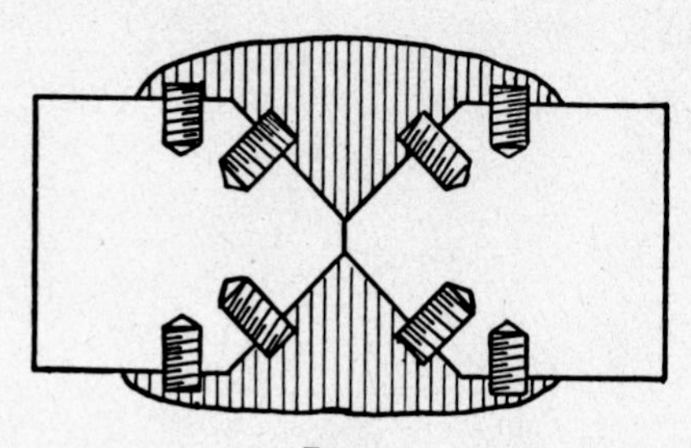

Fig. 3.

The cross-sectional area of the studs should be about 25% to 35% of the area of the weld surface. In such cases, the strength of the weld may be safely and conservatively taken as the strength of the studs. It is considered good practice to first weld one or two beads around each stud, making sure that fusion is obtained both with the stud and the base metal. Straight lines of weld deposit should be avoided insofar as possible. Welds should be deposited intermittently, and each bead peened before cooling.

In all cast iron welding, use the lowest possible current which gives adequate fusion with the base metal. This will keep the carbon pickup to a minimum.

The choice of electrode for welding cast iron is mainly dependent upon cost and machinability. Both steel and nickel electrodes do an excellent job and one is about as easy to use as the other. Nickel electrode (Softweld, Lesson 4.9) is considerably more expensive than the mild steel electrode (Ferroweld, Lesson 4.8). The nickel electrode is used primarily where a soft, machinable deposit is necessary. If there is no need to drill, tap, or machine a weld, the steel electrode should be used. It may be ground to shape, although it is not easily worked with cutting bits. If extensive build-up is necessary, as when filling the groove in a heavy casting, and the top must be machined to shape, the steel electrode may be used for filling. The last two or three layers may be run with the nickel electrode, and a readily-machinable surface will be obtained. Either electrode is usable in both the "hot" and "cold" methods of welding.

When welding using the preheating method, very few difficulties are encountered. This is by far the best method providing adequate means of preheating and slow cooling are available. The welding job can be done as rapidly as desired without taking any particular precautions. It is important that the entire casting be kept at a minimum of 500 degrees (bright metal will turn purple) during the welding operation. Higher temperatures up to cherry red are even better.

If a furnace is not available, it is sometimes possible to make one for small objects. This furnace can be made out of firebrick or asbestos sheets. Preheat may be furnished with an arc torch, gas torch or blow torch. When cooling under these conditions, the entire part should be insulated immediately. This can be done by covering with asbestos or burying in sand, lime or some other insulating substance.

If it is impossible to preheat the part uniformly, no attempt should be made to preheat a small section. In this case, welding must be done in such a manner as to keep the part cool at all times. It should be kept cool enough that the bare hand may be put on the weld area immediately after welding. The method of welding in this case is to make short, intermittent beads with a waiting time between. This may seem slow, but don't try to rush it, as it may cause uneven expansion and contraction with resultant cracks. The beads should be ½" to 1" long. A back-step technique should be used so that each crater lies on top of the previous bead.

The top of the previous bead has a low carbon content, so there is little carbon pickup in the crater and thus little tendency to crack. Immediately after each short bead is made, the bead should be peened. Peening, while the bead is cooling, will relieve stress and avoid building up an accumulation of internal stresses in the part as more beads are applied.

It must be remembered that any amount of heat, if uniform, is helpful when welding cast iron. Room temperature is better than freezing, and 200 degrees is better than 100 degrees.

If it is necessary to have the casting liquid tight after welding, another precaution may be observed. Apply stick sulphur, which may be obtained at any drug store, to the welded joint while it is still warm. The sulphur will melt and seal any cracks or pin holes. Do not apply sulphur until all welding has been completed. Severe porosity will result from welding over sulphur.

Review Questions

1. What happens when cast iron is heated and cooled rapidly?
2. Does cast iron have high compression strength?
3. Does cast iron have high tensile strength?
4. What are the two methods used to weld cast iron?
5. Why does the crater of a cast iron bead crack after the weld is run?
6. What can be done to avoid cracking through the crater?
7. What is the best method of welding cast iron?
8. Is back-step welding used when welding cast iron?
9. What does peening do to the beads?
10. To what temperature must cast iron be preheated for welding?
11. Are vee joints used on cast iron?
12. What does studding do to a cast iron weld?
13. What amperages should be used for cast iron welding?

LESSON 1.37

Object:
To study the welding of higher carbon and low alloy steels.

General Information

Medium carbon steels, high carbon steels and low alloy steels present different problems to a weldor than mild steel. These steels are often given the general name, "tool steel", because of their ability to heat treat to varying degrees of hardness for use as tools. They may be visually identified by spark test and by checking their hardness with a chisel or file (Part 7.1, Section VII). Additional carbon content as well as alloying metals cause these steels to be harder, stronger, and more brittle than mild steel. This results in welding difficulty, because the steels are effected more by heating and cooling. They are apt to break or have hard spots. Similar precautions must be taken as when welding cast iron.

It is possible to weld these steels with a mild steel electrode if the base metal is preheated. This produces a weld with a lower carbon or alloy content and usually of lower tensile strength and impact resistance than the base metal, but is satisfactory for some repairs.

Jetweld electrodes operate satisfactorily on medium carbon steels. Their low penetration reduces admixture and high strength reduces cracking.

The low hydrogen electrodes (Lesson 1.38) are the most satisfactory for welding these types of steels. They may be used without preheating the base metal. Stainless steel electrodes and austenitic hardsurfacing electrodes may also be used. Stainless steel electrodes are especially good for making high strength welds in high carbon, low alloy and wear-resistant steels. They may also be used for joining these metals to mild steel. However, when using stainless remember that it cannot be cut with oxy-acetylene torch; future repairs may be difficult.

When using electrodes other than the low hydrogen type, that is, stainless steel, austenitic and mild steel, it is necessary to preheat. The base metal must be taken up to at least 400 degrees F. (bright metal will turn purple) and it may be taken up to a dull red heat. This heat must be held during the welding operation and the metal allowed to cool off slowly and uniformly in sand, lime or some other insulating material after welding. This reduces the possibility of the base metal cracking along the edges of the bead or through the bead, and eliminates hard spots caused by uneven heating and cooling. Most medium carbon steels can be welded without preheating, but to be on the safe side, it is well to give them this same treatment. When welding with the mild steel electrode use lower amperage setting and the same procedure as when welding mild steel. If low hydrogen, stainless steel, or high carbon steel electrodes are used, close attention should be given the manufacturer's recommended application procedure.

Joint design and edge preparation for fabricating these metals is the same as for mild steel. On repair jobs, certain modifications may have to be made. The process of welding will remove hardness from the carbon steels and many of the alloy steels. When the preheat-slow cooling treatment is used, the hardness will be drawn from the entire piece. Even if the low hydrogen electrode is used without preheating, the temper in the weld deposit, fusion zone, and adjacent area will be altered. This must be taken into consideration when welding a tool, spring, brace or other part which may have been heat treated.

If using austenitic electrodes, the weld deposit may be as easily heat treated as the original part. Stainless steel will not respond to heat treating, but it will to work hardening. Its high tensile strength and impact resistance in the as-deposited state, however, may come close to matching the original part when it is heat treated after welding. The weld deposits of mild steel and low hydrogen electrodes will probably not respond to heat treatment. The low hydrogen deposit, however, has higher strength and impact resistance qualities than the mild steel deposit on similar steels.

Lest the above seem to make welding of high carbon or low alloy steels impractical, remember that it's being done every day on both production and repair applications. The weldors who are doing this work have had a chance to practice with these kinds of steels and electrodes and know what to expect of them. In the shop you may run into different kinds of steel you may not have had a chance to practice on before. Therefore, you must have a broad knowledge of how to weld these steels.

Review Questions

1. Why do the "tool steels" cause more welding problems than mild steel?
2. Can these steels be welded with mild steel electrode?
3. Is the weld produced as strong as the base metal?
4. What electrode is best for most higher carbon and low alloy applications?
5. What other electrode may be used?
6. What added precautions must be taken when using all except the low hydrogen electrode?
7. What electrode may be used for joining higher carbon steels to mild steel?
8. What preheat temperature should be used on the base metals?
9. What type of joint and edge preparation should be used when welding these steels?
10. Will welding draw the hardness out of tempered base metal?
11. What type of electrode may respond to heat treatment?

LESSON 1.38

Object:
To study low-hydrogen electrodes.

General Information

The average weldor owes a great deal to the men who work behind the scenes of any company producing welding materials. It is these research men, metallurgists, chemists and engineers, who have made the process of arc welding much easier and more dependable for the weldor in the field. Each new machine, electrode and coating has been developed, tested and retested as a result of the combined research of these men. They have tried to anticipate many problems encountered by the weldor, and have solved them before they became problems in the field. On the other hand, the weldor can help, and has, by reporting his special problems. In either case, when each problem is solved, another step has been taken in the progress of arc welding.

This was how the low-hydrogen electrode was born.

Certain steels of the medium and high carbon, high sulphur and low alloy groups are crack-sensitive. They can be welded by special methods, using preheating, postheating, or changing welding speed or technique. In most cases, however, the maximum rate of deposition is seriously reduced. Metallurgists found that these steels, while they are in the molten and semi-molten state are effected by the presence of hydrogen. This is contrary to the theory of welding mild steel, as hydrogen is one of the shielding agents for mild steel. In other words, these crack-sensitive metals are allergic to hydrogen. Its presence causes a chemical change resulting in areas of embrittlement which cause stresses and cracks in the weld. Sometimes the crack shows up in the bead or at the edge of it. Other times it occurs in the fusion zone under the bead, and is termed as an "underbead crack".

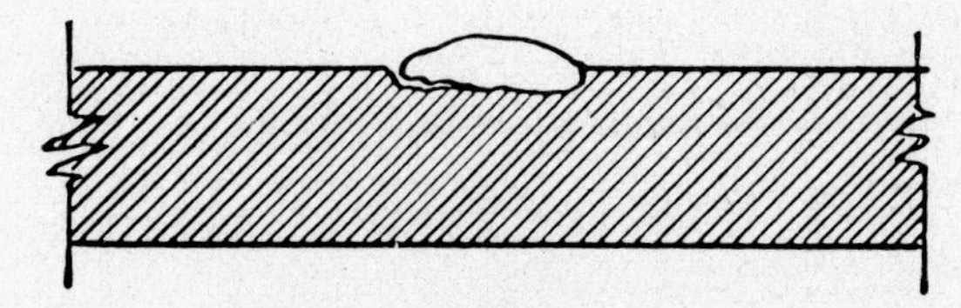

Fig. 1. Underbead cracking.

Through experimentation it was found that a certain type of austenitic electrode would eliminate this trouble. The flux used on this electrode, carbonate of soda and lime, was tried on high tensile and mild steel electrodes. The results were the same, cracking was eliminated or reduced. This produced the low-hydrogen electrode as it was called, because the flux was very low in hydrogen content. This electrode, classified as E6015 became very popular for DC welding. The E6016 electrode soon followed as a similar type for use on AC or DC welders. Further design improvements developed the E-6018 electrode, a modified E-6016 containing powdered iron in its coating.

Low hydrogen electrodes are suitable for all-position welding. Like other iron-powder electrodes, the E-6018 design is easier to apply and exhibits the advantages of a higher deposition rate (up to 30% higher), smoother bead appearance and easy slag removal. For these reasons, powdered iron E-6018 electrodes are rapidly replacing the E-6015 and E-6016 types on production applications that demand low hydrogen electrodes. Typical applications include medium and high carbon steels, many types of low alloy steels, and high sulphur steels. For high tensile steels, the high tensile electrodes of the EXX16 and EXX18 groups are used to raise the strength of the weld deposit.

The coating on low-hydrogen electrodes is slightly thicker than for the same diameter on other types of electrodes. A short arc must be maintained at all times, making the electrode slightly harder to use. The deposited bead lies flat or slightly convex and is easy to clean. In the as-welded condition, mechanical and impact properties are slightly superior to the E6010 and E6011 electrodes.

Some state highway department codes are specifying low-hydrogen electrode for use on bridges and other structures, due to the possibility of obtaining varying sulphur content in construction steels.

With the importance of low hydrogen in the coatings understood, it is easy to see that the coatings should absorb no moisture (H_2O), as this would allow hydrogen in the coating. Open electrode containers should be kept in a low temperature oven. Damp electrodes must be thoroughly dried before using.

Review Questions

1. Who develops and tests electrode and equipment before they come out on the market?
2. What can the average weldor do to help welding progress?
3. What is one of the primary causes for cracking in certain crack-sensitive steels?
4. What has proven to be the best way of eliminating cracks?
5. Does hydrogen cause mild steel to crack?
6. What is an "underbead crack"?
7. Why is the electrode called a "low-hydrogen" type?
8. Why are some welding codes specifying low-hydrogen electrodes on construction work?
9. What care must be taken when storing low-hydrogen electrodes?
10. What are the advantages of the iron powder low-hydrogen electrodes?

LESSON 1.39

Object:
To study the uses of the arc torch and carbon arc.

General Information

Most of your arc welding will be done with the metallic arc with which you have been practicing. Some jobs, however, are more easily done with the flame-type heat produced by the arc torch or carbon arc. Both of these methods were important in the early development of arc welding. The research and development of excellent metallic electrodes have all but replaced the manual carbon-electrode arc welding methods.

The arc torch finds many uses in the shop and broadens the range of work possible with an AC transformer type welder (Fig. 1). Metal bars, rods and

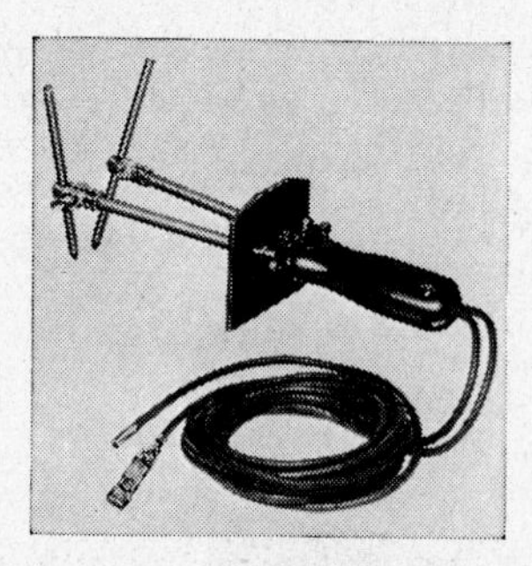

Fig. 1.

straps may be heated and bent quickly and inexpensively with the localized flame-type heat of an arc torch. The arc torch will fusion weld light steels, as well as aluminum, copper and brass. Loosening rusted nuts and bolts, removing paint and scale, brazing and soldering, hard-surfacing, light forging, preheating, and heat treating may also be done with the heat of this torch.

The heat is produced by the arc between two carbon electrodes. The size of the electrodes, amperage setting, and length of arc governs the amount of heat produced. Four sizes of copper-coated, soft center carbons are available for the torch, 1/4", 5/16", 3/8" and 1/2". Fig. 2 shows the suggested amperage range and electrode size for various thicknesses of base metals. The carbons should be

Suggested Ampere Range for Carbon Arc Carbons Used with A. C. Welders

Carbon Diameter	Amperage Range	Size Gauge	Approx. Thickness
1/4"	20–50	28–24	1/64 to 1/40"
5/16"	30–70	22–20	1/32 to 1/20"
3/8"	40–90	16–3	1/16 to 1/4"
1/2"	75–140		over 1/4"

Fig. 2.

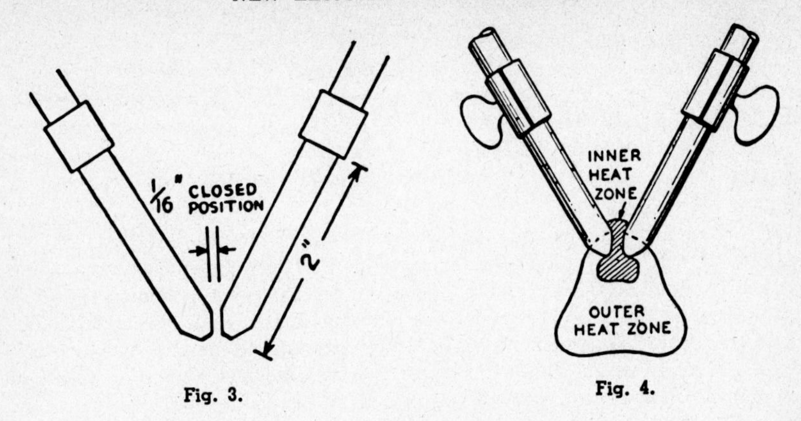

Fig. 3. Fig. 4.

adjusted so that 2 to 3 inches extends below the clamps and the carbons should have a 1/16" gap in the closed position (Fig. 3). The torch is easiest to use on an AC transformer welder. It may, however, be used on a DC welder if the carbon on the positive pole is one size larger than the negative carbon. This reduces the burn-off tendency of the positive carbon.

Connect the torch cables to the welder by placing one terminal in the electrode holder and the other in the ground clamp while the welder is turned "OFF". The work is not grounded. Set the correct amperage on the welder, separate the carbons, turn the welder "ON", lower your shield and start the arc. The arc is started by bringing the carbons nearly together. A good flame will result from the carbons if the gap is set about 3/16" (Fig. 4). To keep from overheating the base metal, the flame may be moved about and the distance between the torch and the base metal varied. The flame is fan-shaped and has no force for penetrating grooves or cavities, but it has good heating qualities. The flame is extinguished by moving the carbons apart, until the voltage will no longer sustain the arc.

The single carbon arc generally utilizes a baked carbon electrode, or in some cases a graphite electrode. The electrode is gripped in the same holder used

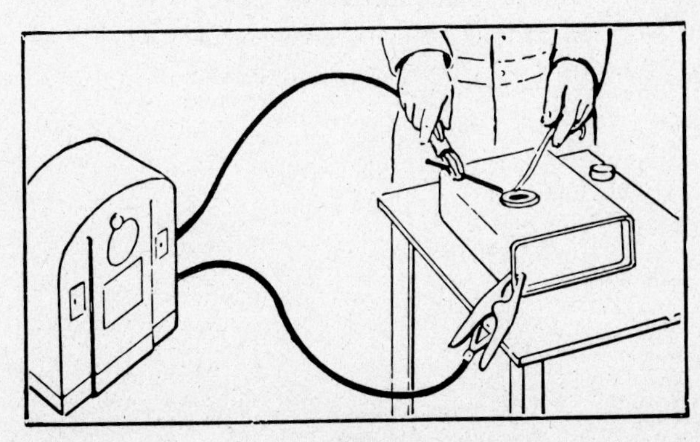

Fig. 5.

for metallic electrodes, or in a special holder designed for use with carbon electrodes. It must be used with the DC welder on negative polarity. A heating type flame may be produced when it is touched and withdrawn from the grounded base metal. The flame is not as versatile as the arc torch, and will extinguish when the electrode is moved too far from the base metal. It has the advantage that it can be pin-pointed and concentrated for work on small or thin material.

The carbon arc may be used for soft soldering with either the AC or DC welder, as shown in Fig. 5. The amperage is set between 20 and 25 amps. The ground is clamped to the work and the carbon electrode is touched to the work and held there. The electrical resistance between carbon and base metal causes heat at that point and melts the solder. No arc should be drawn when soft soldering. The electrode should be removed from thin metal while it is still in the solder, so that an arc will not be formed and burn through the thin base metal. Use a flux to help the solder flow and adhere to the base metal. No shield need be worn during the soldering operation.

When fusion welding with the arc torch and carbon arc, a flange type joint may be set up (Lesson 1.27) and the weld made by fusing the flanges with the arc flame. Other welds will need the addition of filler rod.

Lessons 1.13, 1.27, 1.28, 1.40, 1.41, and 5.2 discuss other uses of the arc torch and carbon arc.

Review Questions

1. What are some of the jobs that can be done with the arc torch?
2. How is the heat produced in an arc torch?
3. What type of carbons are used on the arc torch?
4. What type of welder should an arc torch be used with?
5. What must be done if the torch is used with the other type of welder?
6. Is the work grounded when using the arc torch?
7. What gap is needed to produce a good arc flame?
8. Is the arc torch flame a blowing type flame?
9. What types of carbons are used in carbon arc welding?
10. What is one advantage of the carbon arc flame?
11. Must the work be grounded when welding with the single carbon arc?
12. What is a common use of the carbon arc around the repair shop?

LESSON
1.40

Object:
To cut with the electric arc.

Equipment:
Lincoln "Idealarc" or DC or AC welder and accessories.

Material:
Scrap steel plate 1/4" to 3/4"; 1/4" carbon electrode and 3/16" "Fleetweld 5" (E6010) for DC or "Fleetweld 35" (E6011) for AC.

General Information

Cutting with the arc consists of melting and removing with the force of the arc a cut or kerf in the metal. Most types of metals can be cut with the arc, such as steel, cast iron, alloys and non-ferrous metals. Rivets or bolts may be removed and holes pierced with the cutting action of the electric arc.

Carbons may be the most economical electrode for certain jobs, but can be used only on DC machines. The shielded electrode may be used on either a DC or AC welder, but cuts more efficiently on DC current.

Avoid cutting in confined areas, as the higher amperages used when cutting cause more fumes. Galvanized iron and the non-ferrous metals give off toxic fumes which should not be breathed into the lungs.

Job Instructions

Job A: Cut mild steel using the carbon arc.

1. Taper the carbon point on an abrasive wheel or belt to approximately half its diameter. The taper should be gradual about 5 to 7 times the diameter back from the point.
2. Grip the carbon close to the taper. If a long length of carbon is exposed, excessive heating due to resistance causes the carbon to vaporize and burn, resulting in waste.
3. Set the machine on negative polarity for a carbon electrode.
4. Set amperage at 180 to 200 for 1/4" carbon electrode. Proper amperage depends upon the work to be done. The table will serve as a guide. Amperages given are the maximum which should be used. Lower amperages may be used, depending upon the weight or thickness of the base metal (Fig. 3).
5. Position the plate to be cut so that it projects over the edge of the table.
6. Position a container beneath to catch the molten metal as it drops from the cut. See that the cables are clear of the hot metal.
7. Strike the arc at some point from which the molten metal may readily flow (Fig. 1). Start at a lower corner, then go up the side to the top. Repeat this

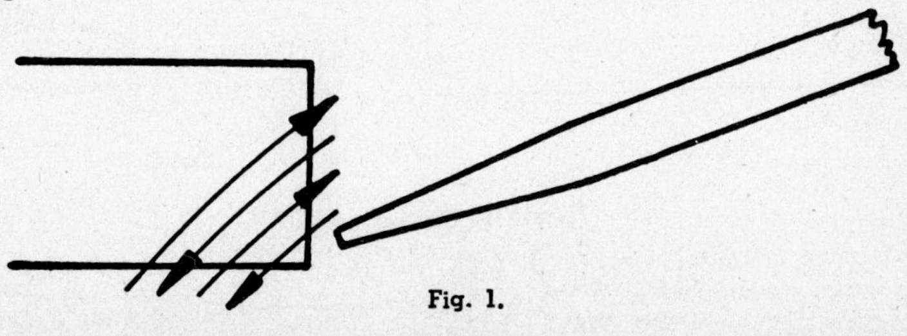

Fig. 1.

action as many times as necessary to cut through and across the plate. This allows the molten metal to flow out of the cut.

Job B: Cut and pierce mild steel with the shielded electrode.

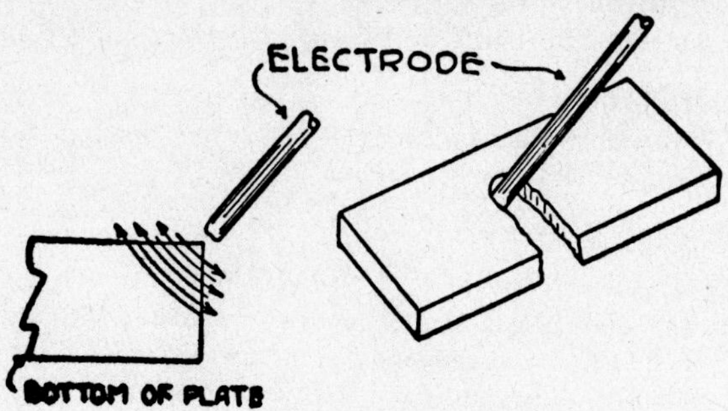

Fig. 2.

1. Put a 3/16″ shielded electrode in the electrode holder.
2. Set amperage at 300 to 350 amps for 3/16″ electrode.
3. Position the plate so that it projects over the edge of the table, as in cutting with the carbon arc. Place a container underneath to catch the molten metal. See that cables are clear of hot metal.
4. Strike the arc at the edge of the plate. Hold a long arc until a large puddle is melted on the edge. Move the electrode downward to shorten the arc and force the molten metal from the plate. Move upward carrying a long arc and move downward with a shorter arc (Fig. 2).
5. Clear the bottom of the plate with each pass to keep "icicles" from forming on the bottom edge of the cut.
6. A 3/16″ electrode will put approximately a 1/2″ hole through a 3/4″ plate.
7. Strike an arc at the point where the hole is to be put. Hold a long arc until a large puddle is formed. Bring the electrode downward and force it through the plate until a hole is pierced. Withdraw the electrode (Fig. 4).

Maximum Currents for Hand Carbon Arc

Size of Carbon Electrode	Maximum Current
5/32″	50
3/16″	100
1/4″	200
5/16″	350
3/8″	450
1/2″	700

Fig. 3.

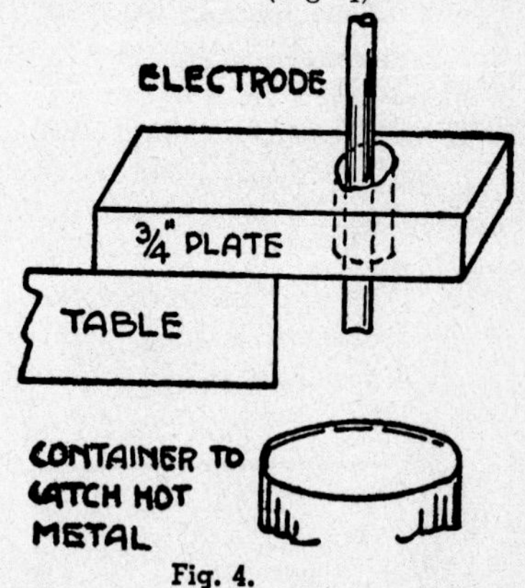

Fig. 4.

Review Questions

1. What type of metal can be cut with the arc?
2. Can holes be pierced with the arc?

LESSON 1.41

Object:

To weld copper with the carbon arc.

Equipment:

Lincoln "Idealarc" or DC welder and accessories.

Material:

Copper plates 1/8"x3"x5"; carbon electrode, copper filler rod, phosphor bronze D filler rod or equal.

General Information

The joint to be welded must be clean. If oxides exist they may be removed by using a warm 10% solution of sulphuric acid. Base metal thickness up to 1/4" need not be beveled for welding. Above 1/4" beveling is recommended, using the same edge preparation as for steel joints. Welding is facilitated when the joint is backed up with carbon or graphite blocks. If this is not practical, elevate the two sheets to be welded about 1/16" so that the bottom of the joint and the work table have a space between them. This permits the penetration of a small amount of excess metal on the underside of the root. When the weld is completed, the underside can be machined flush with the base metal.

The manual carbon arc produces the best results. A high capacity, high efficiency DC welder capable of delivering uniform welding current is recommended. This type of arc welder is necessary to maintain the required 40 volt arc. The carbon arc operates with the electrode negative and work positive.

Composition of the filler rod will vary according to the physical characteristics required of the welded structure. If the weld must have low electrical resistance, the filler metal may be of pure copper or cadmium copper. Where electrical or thermal conductivity are not essential, but where only ductility and physical strength are required, the filler metal may be of Everdur, silicon copper or a suitable grade of phosphor bronze.

Generally there should be a space of about 1/8" between the plate edges. The rod used should be about 20% heavier than the thickness of the metal being welded. It is melted into the plate and fed in by hand as the weld progresses. Best results are obtained at high speed.

Preheating is necessary when light currents are used, due to the high heat conductivity of copper. This can be done with the carbon electrode. Hold a long arc, 1 inch or more in length, and move it rapidly over the surface.

Cold rolled copper may have a strength of 55,000 lbs. per sq. in. The strength of welded copper cannot be higher than about 30,000 lbs. per sq. in., because the welding heat has annealed the weld deposit and the adjacent area. Further mechanical treatment is necessary to raise this strength.

A long arc should be held. This long arc distance between the carbon electrode and the work allows the carbon monoxide produced by the carbon arc to combine with oxygen in the atmosphere to form an oxide, instead of combining with the weld deposit.

The weld should be made all the way through in one pass. This is where the carbon or graphite blocks are helpful in preventing burn-through, or excessive weld deposit on the reverse side.

The results will vary with the quality of copper, probably varying with the oxygen content. Best results will be obtained with deoxidized copper.

Steam or moisture producing fluxes must be kept away from the arc because of the readiness with which molten copper absorbs hydrogen.

There are two characteristics which must be observed. First, welding at high speed produces the best results. Better welds are produced at 20" per minute on 1/4" plate using 600 amperes, than will be obtained at 7" per minute using 200 amperes. Second, the voltage of the carbon arc must be high, that is, a long arc must be used, such as to allow the carbon monoxide produced by the carbon arc to combine with the oxygen in the atmosphere instead of going into combination with the copper. Oxides in the weld deposit cause excessive porosity.

Job Instructions

1. Tack weld two plates together in a square butt joint and position flat. Weld as outlined above.
2. Try the same job on heavier plates. Prepare a vee butt joint and weld by the same procedure.

Review Questions

1. What type of electrode is used to weld copper?
2. What type of edge preparation is used on copper over 1/4" thick?
3. What type of back-up strips are used on copper joints?
4. What arc length is used?
5. What travel speed should be used?
6. What polarity is used?
7. How many passes should be used?
8. What type of copper is best for welding?
9. What type of rod should be used for maximum electrical efficiency?
10. What size rod should be used?
11. What will oxides cause in the weld deposit?

LESSON 1.42

Object:
To study the principles of hardsurfacing with the electric arc.

General Information

Certain surfaces, points or edges on tools, machines and equipment wear away at a faster rate than the rest of the part. Hardsurfacing is the process of building up or overlaying a layer of metal, usually harder or more durable than the base metal. The process has almost unlimited scope and possibilities ranging from building up a worn truck tire chain link to lining a carbon steel tank with a corrosion resistant deposit.

The success of a given hardsurfacing application depends largely on the proper choice of materials. Points to take into consideration when choosing a hardsurfacing material are outlined in the lesson.

1. Service Required

A. Cutting Edge to Be Maintained

1. Wear of the edge effects the efficiency of the part.

Shear blades, punches and metal cutting tools must not only stay sharp, but must hold their original size and shape to operate satisfactorily. Failure of these parts is usually by upsetting, chipping, spalling, or galling, due to the high compressive loads and the flow of metal being cut past the edge of the tool. Impact and abrasion are not excessively high. The "Toolwelds" are used for this application. They give an edge of high strength, are homogeneous and resist galling. "Toolweld 60" maintains high strength during operation at elevated temperatures.

2. Wear of the edge does not impair the operation of the part.

On earth-cutting tools, such as plow shares, blades of rotary drilling bits, scrapers, ensilage knives, and shredder blades, the edge must stay sharp but the tool may wear back. These tools are intentionally surfaced with a layer of wear-resistant material on the leading edge. The backing metal wears away and exposes a sharp edge of hardsurfacing material for fast cutting. Hardsurfacing materials most resistant to abrasion are used for these applications. The "Facewelds" are used on tools when the abrasive conditions are severe. "Abrasoweld" is unsurpassed when the impact is high and a medium sharp edge is satisfactory.

B. Two Surfaces in Contact—Both to Be Protected

This group presents problems of metal-to-metal wear under various combinations of abrasion, impact and corrosion. Machine parts which are operated with lubrication, inadequate lubrication or at elevated temperatures are in this group. Hardsurfacing materials which will wear smooth, have low friction and a minimum tendency to seize or gall are desirable.

Where wear is not complicated by the entrance of abrasive particles between the wearing surfaces, "Wearweld" is used. Overhead crane wheels are a typical application of this type.

If loads are not excessive and abrasive action is mild, such as on journals, "Aerisweld" may be used to deposit a bronze surface.

As abrasive or corrosive material becomes involved it is necessary to go to different materials. "Stainweld", "Manganweld", "Jet-Hard BU-90", "Abrasoweld", and even "Faceweld" may be used, depending on the service conditions.

C. One Surface to Be Protected

This group covers cases where one of the two contacting materials or surfaces is protected while the wear of the other surface is of little or no importance. In some cases it is desirable to have a low friction surface that will polish, such as a plow or scraper moldboard. In other cases a rough, high friction surface is desirable.

Application:	Recommended Material:
Railends	"Wearweld"
Mine car wheels	"Jet-Hard BU-90"
Crusher jaws	"Manganweld" "Abrasoweld"
Sand pump impellers	"Abrasoweld"
Small bucket lips	"Abrasoweld"
Screw conveyers	"Faceweld 12"
Scarifier teeth	"Faceweld 1"
Forging dies	"Toolweld A&O"
Water turbine blades	"Stainweld A"

2. Service Conditions

The various service conditions to which a hardsurfacing deposit may be subjected are classified under several headings.

A. Abrasion—Grinding action due to rubbing, under low or high pressure, against a rough material such as rock, sand, or clay.

B. Friction—Sliding, rolling or rubbing action of one metal part against another at low or high pressures.

C. Impact—Forcible contact with hard or heavy materials in various degrees from light to heavy. It tends to deform the surface, cause cracking or chipping.

D. Corrosion—The action of various chemicals, ordinary water (rusting) and oxidation or scaling at elevated temperatures.

E. Heat deformation—The ability of a metal to resist warping or distortion at elevated temperatures or variations from low to high temperatures. Carbide and austenitic type materials are high in this property.

Before the proper choice of hardsurfacing material can be made for a given application, the service conditions must be known. In most applications more than one of the above factors are at work. It becomes necessary to evaluate the relative importance of each.

3. Composition and Condition of Part to Be Hardsurfaced

There are thousands of alloy compositions. It is obvious that only a few of these can be considered in this discussion. They can be divided into two general groups, regarding their suitability as a base metal for hardsurfacing.

Group A includes those metals or alloys whose physical characteristics are not greatly changed as a result of heating and cooling, and which will withstand sudden localized temperature changes without cracking. This includes plain carbon steel with a maximum of .30 per cent of carbon (mild steel), low carbon low alloy steels, austenitic steels such as the stainless chrome-nickel, and the high manganese steels. Copper and most of its alloys would also be included.

Group B includes those metals or alloys whose physical characteristics are changed considerably, (particularly as to hardness) as a result of the application of heat and subsequent cooling, or which will crack with a sudden localized application of heat. This group includes medium to high-carbon steels, tool steels, medium to high carbon low alloy ferritic steels, cast irons (gray, white, malleable) and semi-steel. In general, all hard metals and alloys are in this group.

Special precautions must be taken when hardsurfacing metals in Group B. In general, if an arc is struck on a metal which is very hard, it will crack due to thermal shock. To avoid or minimize this thermal cracking, we must either reduce the hardness by annealing or reduce the thermal shock by gradual and uniform preheating. Both methods may be used, depending upon the nature of the alloy. Thermal cracking can thus be prevented in most steels. For grey cast iron which is already in its soft condition but is still brittle, and white cast iron which is hard and very brittle, but cannot be annealed easily, there is little that can be done beyond moderate uniform preheating. This will decrease the tendency for thermal cracking. This does not necessarily mean that cast iron should never be hardsurfaced.

In general, preheating to 400 to 600 degrees F. will prevent weld-hardening in medium to high carbon steels. With medium carbon alloy steels, preheating as above plus slow cooling in lime or sand is advisable. With high carbon alloy steels such as tool steels and other special wear resisting alloys, preheating followed by reheating after welding to a temperature in the 800 to 1500 degree range and uniform cooling may be necessary.

Controlled, uniformly slow cooling is always desirable for cast irons to prevent cracking. Their tendency to weld-harden is so great, however, that little can be done to overcome it. When cast iron is hardsurfaced, it is usually a case of building up a hard wear-resistant surface. The part is usually made of white cast iron. The success of this application depends largely upon the choice of hardsurfacing material. The most successful work is done with a deposit which tends to cross-crack upon cooling, such as "Faceweld 12". These cross-cracks tend to relieve the cooling strains and prevent the surface from peeling off in spite of the brittle base metal.

In most cases where the original chilled iron part was not too brittle for the service conditions involved, the hardsurfaced part will also stand up with the added advantage of a superior abrasion resistant surface. In most cases the cross-cracks will not be detrimental. If a strong, tough deposit such as "Abrasoweld" is applied to white cast iron, the shrinkage strains will be relieved by cracking under the deposit with the result that the hardsurfacing will break away.

4. Dimensions

A. Size and Shape of the Part

The heat capacity of the part to be hardsurfaced is largely a function of its size and shape. A part of heavy cross section will heat up slowly during the hardsurfacing operation. It reaches a relatively low maximum temperature, and draws the heat away from the weld area very rapidly. A small part or a thin cross section will heat up rapidly (at least locally). It reaches a higher maximum temperature and draws the heat away from the weld area very slowly. The mass of the metal being surfaced, therefore, determines to some extent the thermal cycle to which it and the deposit are subjected. For this reason, if local welding heat is applied to a large mass of cold metal which is capable of being quench-hardened, the hardening will be drastic. Cracking is likely to occur due to a sharp rise and fall in temperature. If the mass is small, the degree of hardening will be less because of its low heat capacity. Cooling will be more uniform throughout, resulting in less severe thermal stresses. If the hardsurfacing material applied in the above examples happen to be one of the types whose hardness is affected by the thermal cycle ("Jet-Hard" for example), the hardness of the deposit will be much greater in the case of the large mass because of its quench effect.

B. Size and Location of the Area Surfaced

This will determine the heat input and, therefore, the thermal cycle as above. Location or position of hardsurfacing will influence the choice of material to some extent. Very small areas require the use of small electrodes. Small electrodes are not available in all hardsurfacing material, and a second choice must sometimes be made. Hardsurfacing materials are not well adapted for use in vertical and overhead positions. The cost of doing a job will be much lower if the work can be positioned so that welding can be done in approximately a flat position.

C. Thickness of Weld Deposit

The best general rule in terms of service and economy is to avoid thick deposits of the hard alloys. If a large amount of build up is necessary, it should be done with "Fleetweld" or "Jet-Hard BU-90". Thick deposits may also be applied with "Stainweld" or "Manganweld".

5. Finish Required

If the hardsurfacing deposit must be machined, the following can be used in the as-deposited condition: "Fleetweld", "Shield-Arc 85", "Aerisweld", "Stainweld".

The following deposits are not machinable in their as-deposited condition. They may be machined, however, when softened by annealing: "Jet-Hard BU-90", "Toolweld A&O", "Wearweld" and "Abrasoweld".

Grinding must be used to shape "Faceweld" deposits.

6. Types of Hardsurfacing Electrodes

As a matter of convenience in this discussion, the Lincoln line of hardsurfacing materials can be divided into three groups, according to the type of deposit they produce. The mild steel and high tensile low alloy steel deposits will not be discussed, as their properties are well-known. The specific electrodes, their properties and application are discussed in later Lessons.

Ferritic Type

"Fleetweld"
"Shield-Arc"

Martensitic Type (listed in their order of increasing hardness)

"Jet-Hard BU-90"
"Wearweld"
"Toolweld A&O"
"Toolweld 60"

These are heat treatable plain carbon or alloy steels. Deposits of this group show practically their full hardness in the as-deposited state. They show a relatively small increase in hardness upon cold working. Annealing will reduce their hardness, and quenching from above their critical temperatures will harden them. For this reason, the deposit hardness may also vary somewhat because of variable thermal cycles caused by welding.

Austenitic Type

"Stainweld" "Manganweld C" "Mangjet" "Abrasoweld" (semi-austenitic)

Austenitic deposits differ from the martensitic principally in that they possess the property of surface hardening when deformation takes place as in peening, cold working, or through the impact of normal work. The rest of the deposit remains relatively soft and tough.

Carbide Type

"Faceweld 1" "Faceweld 12"

These are primarily abrasion resisting materials, high in carbides of such metals as tungsten, chromium, vanadium and molybdenum. They are inherently hard and respond little or not at all to heat treatment.

7. Hardsurfacing Guide

The "Lincoln Hardsurfacing Guide" is presented on the following page. This chart is intended to give a comparison between types of hardsurfacing materials, to help the user choose the material best suited to his particular job.

The chart lists the relative characteristics of the complete line of manual hardsurfacing materials. It shows in the various columns the ability of each of the materials to resist various service conditions. It also gives the relative hardness, ductility and cost of depositing the material as well as the physical limitations of weld size in applying each. This may serve as a guide to help select the hardsurfacing electrode best suited for a job not previously hardsurfaced, as well as to select a more suitable hardsurfacing electrode for a job where present materials have not produced the desired results.

Where failures occur due to cracking or spalling, it usually indicates that a material higher in impact or ductility rating should be used; where normal wear alone seems too rapid, a material higher in the abrasion rating is indicated.

Many times hardsurfacing failures due to cracking or spalling may be caused by improper welding procedures rather than improper choice of hardsurfacing material. Before changing to a different hardsurfacing material, serious consideration should be given to the question of whether or not the material has been properly applied. For almost any hardsurfacing application, good results can be obtained if the following precautions are observed.

Job Instructions

1. Do not apply hardsurfacing material over cracks or porous areas. Remove any defective areas down to sound base metal.
2. Preheat. Preheating to 400 to 600 degrees F. improves the resistance to cracking and spalling. This minimum temperature should be maintained until welding is completed. The exception to this rule is 11-14% manganese steel which should be kept cool.
3. Cool slowly. If possible, allow the finished part with its hardsurfacing to cool under an insulating material such as lime, asbestos or sand.
4. Do not apply more than the recommended number of layers. When more

than normal build-up is required, apply intermediate layers of "Stainweld". This will provide a good bond to the base metal and will eliminate excessively thick layers of hardsurfacing material which might otherwise spall off.

Review Questions

1. What is hardsurfacing?
2. Where could hardsurfacing be useful?
3. Will one type of hardsurfacing material take care of all jobs?
4. What conditions might hardsurfacing have to withstand?
5. How is hardsurfacing applied to a cutting edge to help it maintain a reasonably sharp edge even though it is wearing back?
6. What electrode will produce a bronze bearing surface?
7. Why is it necessary to know the characteristics of the base metal before applying hardsurfacing?
8. Can cast iron be hardsurfaced?
9. What two major precautions must be taken with metal which is hardenable by heating and cooling?
10. Does thickness or bulk of a part affect the method of applying hardsurfacing?
11. What position is recommended for hardsurfacing applications?
12. Should thick build-up be used to increase the surface hardness of hard alloys?
13. What hardsurfacing materials cannot be machined by cutting tools under any conditions?
14. What are the three most important groups of hardsurfacing materials?
15. What type of hardsurfacing materials will harden with working?
16. What **two hardsurfacing materials have the highest** abrasion resistance?

How to Use the Lincoln Hardsurfacing Guide

The Lincoln Hardsurfacing Guide combines into one chart all the information necessary to the selection of a hardsurfacing electrode for any application. The properties which must be considered in making a selection are listed across the top of the Guide. The bars in the columns above indicate the relative merits of each electrode in the property shown at the column heading. The further the bar is positioned to the right of the column, the better its properties. The length of the bar indicates the degree to which welding procedures influence the properties. A long bar indicates that procedures can greatly affect the results while a short bar means that normal welding procedures will have little effect on the properties. The following is a step-by-step procedure describing the use of the Guide.

STUDY THE JOB—Thoroughly examine the job. Determine the type of wear that it will encounter in service. If this is principally abrasion, check for the amount of impact that will be present. See what special properties are necessary, such as machinability or ability to weld on thin sections.

SELECT THE ELECTRODE THAT WILL BEST RESIST THE WEAR—At the top of the Guide find the column corresponding to the type of wear encountered by the part. Go down the column to the bar situated nearest the right border. On the right hand side of the Guide determine the electrode associated with the bar. Other considerations being equal, this electrode will provide the best wear resistance and longest life.

CHECK OTHER NECESSARY PROPERTIES—If there are other property requirements, such as the machinability, check your initial electrode selection to be sure that it meets the added requirements. If it does not, then you will have to change to another electrode which, while it may not have quite as much wear resistance, will meet the other requirements of your job.

APPLY THE ELECTRODE—The following pages describe the electrodes and how to apply them in more detail. Study these pages before actually welding with the electrodes.

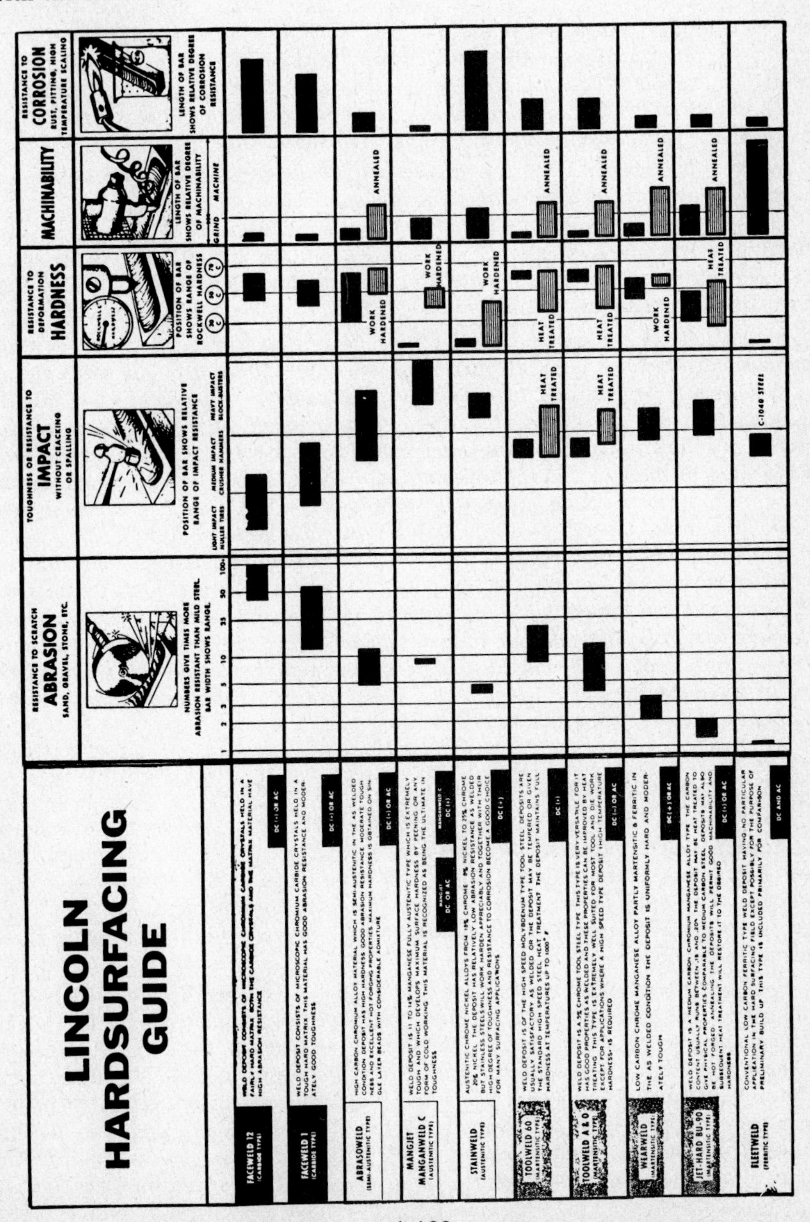

LESSON 1.43

Object:
To study the importance and fundamentals of pipe welding.

General Information

Piping of all shapes and sizes is used in a variety of applications. Probably the most widely publicized is the gas and oil transmission pipe lines, though it is only one of many equally important applications. Refineries, power plants, and chemical plants use miles of pipe each year. Pipe is also used for water lines, industrial heating and ventilating, and structural jobs. Pipes carry liquids, gases, and even solids mixed with liquids. The use of pipe is growing each year.

Welding has greatly aided the growth of pipe usage. It gives a smooth, pressure tight joint which essentially joins any number of separate sections of pipe into one continuous piece. Welding is economical because it is fast, and also because it makes a high quality joint which permits lighter pipe to be used. Welded joints are easier to handle and insulate, and require less maintenance than other methods of joining pipe.

Because of the widespread use of pipe, it is important that every welding operator at least be familiar with pipe welding. Some time in his welding career, he will be required to weld some type of pipe. Pipe welding is usually divided into two categories: the first is transmission line pipe welding; the second is industrial pipe welding and includes all other types of pipe welding. Regardless of the type of pipe welding involved, it is *all* difficult by comparison to other types of welding. Pipe welding requires a maximum of physical dexterity and ability, as well as a lot of patience. Most pipe welding is code work; that is, work which requires that operators and procedures be tested and proven satisfactory before starting on the job. They are also periodically retested. Pipe weldors can expect to be required to periodically pass an operator qualification test, regardless of how long they have been welding. As long as they are welding pipe, they can also expect continual inspection of their work, either by X-ray or other means. This is because great damage to people and property could result from a weld failure.

Transmission Line Pipe Welding

Transmission line pipe welding dates from the early 1920's and has since made tremendous progress. Welded joints have progressed from the old bell-and-spigot joint, through veed butts with a back-up, to the present veed butt without back-up. This progression has improved both welding speed and quality. Improvements both in electrodes and operator skill have made these advances possible.

Welding procedures are based on codes which are generally accepted by the industry. The codes specify the requirements of the completed weld and make certain specific procedural recommendations. Based on these codes, pipeline owners and contractors establish procedures which are used on the lines. Operators take qualification tests using these procedures and *must* pass them before they are permitted to weld on the job. There are no exceptions to this rule. The extent of detail involved in the procedures may vary from one concern to another, but generally are quite specific, even including the current range to be used for each pass. Once the operator has qualified with a particular procedure, he is expected to continue using it on the job. Inspectors are on most jobs to insure that this is done.

Most transmission line pipe welding is welded vertical down, because it is fastest on pipe thickness of less than ½ inch. The joint is a veed butt, approximately 70 degrees, with 1/16th face and 1/16th gap. The first pass is the "stringer" bead and is usually put in with a drag technique. The second pass is called the "hot" pass. Succeeding passes are called "filler" passes, except for the last which is the "cover" pass.

Figure 1 is an illustration of a typical welding spread on a transmission line pipe job. This arrangement is called "stovepipe" welding. In this method, all joints are welded to the pipeline one section at a time. Each joint is position welded. Using this method, it is possible to reduce the size of the crew and the amount of equipment and, thus, keep the operations bunched together under one supervisor. During alignment and tacking, the joint is usually held in place by a line-up clamp "grasshopper". After tacking, two weldors work simultaneously on both sides of the joint making the complete first, or "stringer", bead. A "hot" pass crew then moves in to make the second weld. The welds are finished by the "filler" and "cover" pass crews.

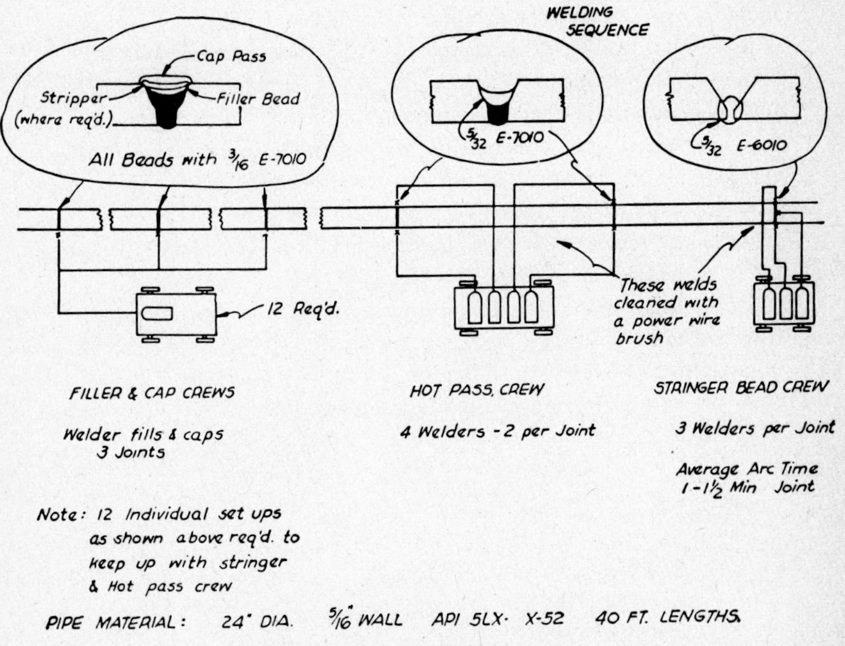

Fig. 1. Typical transmission line pipe welding crew.

Industrial Pipe Welding

Industrial pipe welding is even tougher than transmission line work. Like transmission line weldors, most industrial pipe weldors work under codes and must be qualified and tested regularly. However, because of the greater risk of damage involved, the codes are stricter, and tests are more rigid. In addition to this, a greater variety of alloy steels may be encountered and both vertical and horizontal axis pipe joints must be welded. The final complication, common to the industrial pipe weldor, is having to work in cramped quarters and even using mirrors to see what he is welding.

Contrary to the vertical down welding of the transmission line pipe joints, most industrial pipe is welded vertical up. This calls for an increase in the gap to 1/8 inch to permit full penetration. Otherwise the joint is prepared the same. Welding vertical up requires smaller electrodes and lower currents. Welding speeds are slower. Industrial pipe weldors do an entire joint, instead of splitting into teams as is done on the field work. Industrial pipe weldors encounter a considerable amount of shop work. When it is possible, they use a roll welding technique which permits welding the entire joint in the downhand position.

Review Questions

1. Name four types of applications which could be considered "industrial pipe welding".
2. What are two reasons why welded pipe joints are economical?
3. What are two categories of pipe welding?
4. Should every operator be familiar with pipe welding?
5. Is pipe welding difficult as compared with other types of welding?
6. Is most pipe welding code work?
7. Why is it important that no failures occur in welded pipe joints?
8. Do experienced operators have to take operator qualification tests?
9. On code work, are operators permitted to select their own procedures?
10. How are most transmission line pipe joints welded?
11. What is the second pass on a transmission line pipe joint called?
12. What is the gap on transmission line pipe joints?
13. What are some of the added difficulties involved in industrial pipe welding?
14. How are most industrial pipe joints welded?
15. What is the gap on industrial pipe joints? Why?
16. Is much industrial pipe welding done in the shop?

LESSON 1.44

Object:
To run beads by "roll welding"; to run vertical down drag beads.

Equipment:
Lincoln "Idealarc" or other DC or AC welder and accessories.

Material:
Steel pipe 6 to 10 inches in diameter; 5/32 inch "Fleetweld 5-P" (E-6010) for DC or "Fleetweld 35" (E-6011) for AC.

General Information

Before actually attempting to weld a pipe joint, it is necessary to master certain skills not covered in previous lessons. The exercises in this lesson will introduce you to two of these special skills; welding on the top of a pipe as it rolls under the arc, and running beads vertical down using the drag technique.

Downhand or "roll welding" is the easiest method of pipe welding, when it can be used. The arc is struck on the top of the pipe and held there to deposit the bead as the pipe is steadily revolved under the electrode on dollies or skids.

The usual method of welding pipe will be in the fixed position. Welding must be done in whatever position is necessary to reach the joint. When the pipe is in the horizontal position, the usual method for welding transmission pipelines is by vertical down welding. Thick-walled, high-pressure pipe is welded vertically up.

The fundamental difference between pipe welding and plate welding is the greater amount of travel of the hand compared to the travel of the arc end of the electrode. It is important to maintain a constant electrode angle as the bead is run around the pipe.

The drag technique is used to lay the first pass in the vee of the vertically

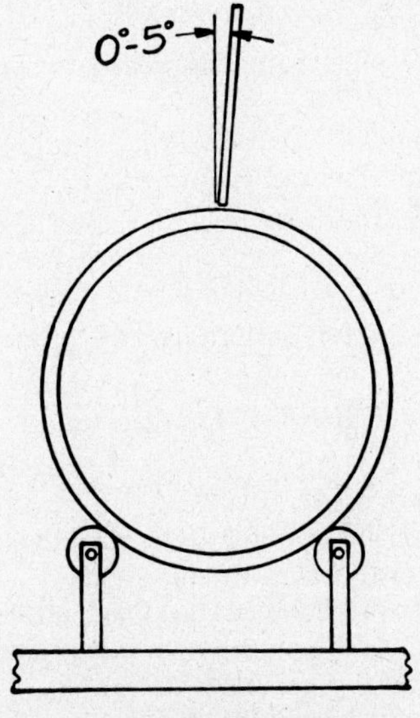

Fig. 1.

down welded pipe joint. This is called a stringer bead, as it is laid with no sidewise motion.

Job Instructions

Job A

1. Clean scale and rust from pipe.
2. Lay pipe in horizontal position on dollies or skids.
3. Set amperage at 130 to 145 for 5/32 inch electrode.
4. Hold the electrode perpendicular to the pipe or inclined slightly in the direction of travel (see Figure 1). The curved surface of the pipe makes it necessary to hold this steep angle. If the electrode is inclined too much the arc force will tend to blow the metal over the far side of the pipe.
5. Strike the arc. As the puddle is established and a bead is formed, roll the pipe ahead, so that the downhand position is maintained. Electrode travel is dependent upon a uniform rate of pipe movement.
6. Clean and examine the bead. It should be smooth and uniform fusing into the pipe on the edges.

Job B

1. Use the same or a similarly cleaned pipe.
2. Position pipe horizontal, from waist to chest high, where it can be reached on top and underneath.
3. Set the amperage by checking with the drag technique on a flat plate. Set the amperage high enough that the arc does not go out when the electrode is dragged along slowly on the surface of the plate inclined 50 to 60 degrees in the direction of travel.
4. Find the most comfortable position for steady welding to lessen fatigue. Right-handed weldors may rest their left arm on the pipe and steady the right hand at the wrist with the left hand in such a way that there is free wrist motion of the right hand. Left-handed weldors would position themselves in the opposite way.
5. Start at the top of the pipe using the nearly perpendicular electrode angle as in Job A.
6. Strike the arc and drag the electrode along steadily with a slight downward pressure, so that a uniform bead is formed.
7. Weld to the bottom of the pipe and repeat on the opposite side. Starting and ending of the pass should be chipped out before starting the next pass. The start and end of the two passes should fuse together.
8. Clean and examine beads for uniformity and proper penetration.

Review Questions

1. What is "roll welding"?
2. What electrode angle is used?
3. Why does the electrode angle differ from working downhand on flat plate?
4. What constitutes electrode travel in this weld?
5. Will most pipe welding be done in the fixed position?
6. What is the fundamental difference between pipe welding and plate welding?
7. What pass is the drag technique used for?
8. Should the ends of the first bead be chipped before starting the second?

LESSON

1.45

Object:

To make butt welds on pipe with the axis horizontal, welding down.

Equipment:

Lincoln "Idealarc" or DC or AC welder and accessories.

Material:

Beveled steel pipe pieces 6 to 10 inches in diameter; 5⁄32" "Fleetweld 5P" (E6010) and 5⁄32" and 3⁄16" "Fleetweld 5" (E6010) for DC or "Fleetweld 35" (E6011) for AC.
(See Lesson 3.2 for properties and application of "Fleetweld 5-P")

General Information

The usual method of welding vertical joints on horizontal transmission piping is from the top down.

The pipe bevel specified by API-AGA is 30 (+5, −0) degrees with a 1⁄16th inch face (Figure 1). In practice this results in an approximate 70 degree vee butt joint. It is extremely important that the face, or land, be properly prepared. Thinner land will result in troublesome burn-through, while heavier land may result in incomplete penetration.

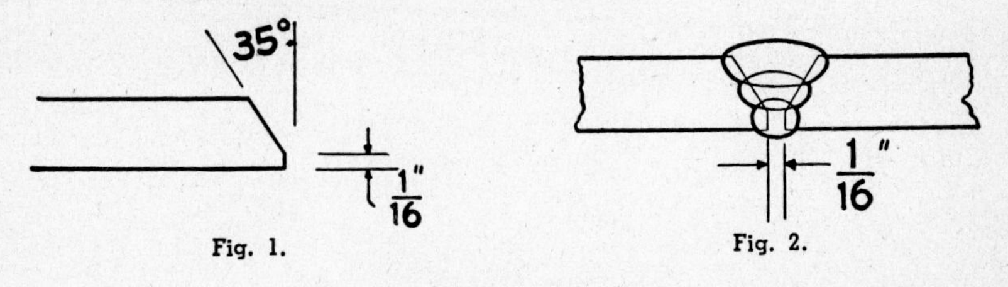

Fig. 1.

Fig. 2.

Job Instructions

1. Check the bevel of the pipe. Insure that it is just as is shown in Figure 1. If necessary the land may be ground to make it even all the way around.
2. Set amperage for drag pass (Lesson 1.44) with 5⁄32" electrode.
3. Tack weld pipe together with 1⁄16" root spacing (Fig. 2) using 4 tacks spaced 90 degrees apart. Use smooth well-fused tacks, so they will fuse in with the stringer bead.
4. Position the pipe about waist high, so it is accessible on top and underneath.
5. Assume a comfortable position for welding.
6. Use "Fleetweld 5 or 5-P" for stringer bead ("Fleetweld 35" for AC). Strike arc at the top of the pipe and carry the stringer bead down to the bottom of the pipe using the drag technique. Drag welding is used only on the root pass.
7. Clean ends of first pass and run stringer bead on opposite side.
8. Clean and examine the entire bead. It should be fused well into each side of the vee with complete penetration through on the inside of the pipe.
9. The second pass may be put in with either 5⁄32" or 3⁄16" electrode. Make necessary amperage adjustment from the drag technique previously used.

10. Do not start or end passes at the same point as the pass underneath.
11. Use a slight weave (A or B, Fig. 3) working the weld deposit into each side of the bevel to form a slightly concave bead. Convex beads in a vee are harder to clean, and there is more danger of having slag inclusions when the next pass is run.
12. Use 3/16" electrode for subsequent filler passes. It will be necessary to use a slightly wider weave as the weld is brought up in the vee. Clean each pass carefully.

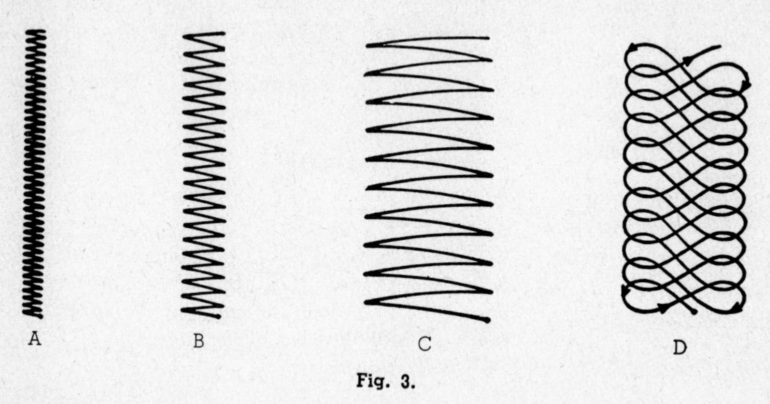

Fig. 3.

13. Run cover pass with a 3/16" electrode using a wide weave or "Figure 8" (Fig. 3C or D).
14. This job may be practiced using "Fleetweld 5-P" for the stringer bead, and filling with "Shield-Arc 85-P" (Lesson 3.2). This electrode is used on alloy pipe.

Review Questions

1. What is the standard bevel for pipe welding?
2. What other two bevels may be encountered in pipe welding?
3. What is the thickness of the nose on the bevel?
4. What root spacing is used on butt welds?
5. How many tacks are put on a pipe?
6. How many passes are welded with the drag technique?
7. Is a weave used when carrying filler passes?
8. What is the correct shape of a deposited filler bead?
9. Must each bead be chipped if subsequent passes are made immediately?
10. Should the ends of subsequent passes be put directly on top of each other?
11. How much penetration should be obtained on the stringer pass?

LESSON 1.46

Object:

To weld joints in horizontal axis pipe vertical up.

Equipment:

Lincoln "Idealarc" or other DC or AC welder and accessories.

Material:

Beveled steel pipe 6 to 10 inches in diameter; 1/8 inch and 5/32 inch "Fleetweld 5" (E-6010) for DC or "Fleetweld 35" (E-6011) for AC.

General Information

On horizontal axis pipe used in industrial applications, the joints are welded vertical up. As in vertical down transmission line pipe welding, the edge preparation and fit-up of the joint are extremely important. A comparison of a joint tacked for vertical down with one tacked for vertical up would show that they are the same, except for the gap. A full 1/8 inch gap is left for vertical up welding to insure full penetration.

Vertical up pipe welding is similar to vertical up plate welding, but considerably more critical. Do not be discouraged if at first these welds seem to be impossible. The guidance of an experienced weldor, and plenty of practice and patience, are required to make a good pipe weldor.

For those who desire to pass operator qualification tests, the "Pipefitter Welder's Review of Metallic Arc Welding for Qualification Under A.S.M.E. Code Rules" is highly recommended. A 36 page pamphlet, it is available from the National Certified Pipe Welding Bureau, Suite 570, 45 Rockefeller Plaza, New York 20, New York, at a nominal charge of $.50.

Job Instructions

1. Insure that pipe is properly beveled (Figure 1, Lesson 1.45).
2. Clean the bevel and a small area of pipe next to it.
3. Set welder to 80-100 amperes for 1/8 inch electrode.
4. Tack weld the pipe together with a 1/8 inch gap. Use the bare end of a 1/8 inch electrode as a spacer. Use four good tack welds equally spaced around the pipe.
5. Clamp or tack weld the pipe at a convenient height and clear of all obstructions. (Figure 1).

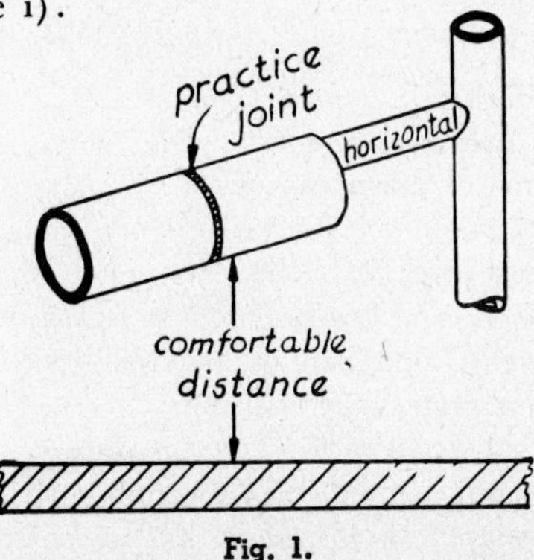

Fig. 1.

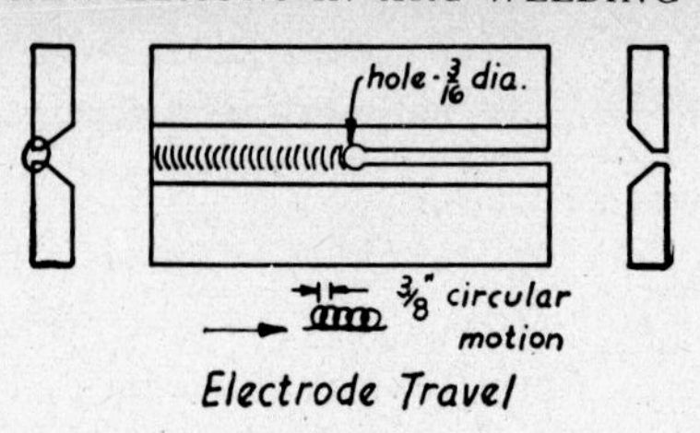

Fig. 2.

6. Using 1/8 inch "Fleetweld 5-P", or "Fleetweld 5" start to weld on the bottom of the pipe. The method of laying in the first pass is shown in Figure 2. A hole slightly larger than the electrode is burned through the joint. The arc is rotated around the hole, so as to burn off the edge in the direction of travel while depositing a bead on the other side. This procedure is continued until the bead is complete to the top of the pipe.
7. Clean the ends of the previous weld and make a similar weld on the other side. Thoroughly clean the entire first pass.
8. Reset the welding machine to 120-150 amperes.
9. Again, starting from the bottom of the pipe (but not at the same spot where the first pass was begun), put in a second pass with 5/32 inch "Fleetweld 5". Use a slight side-to-side weave. Be certain to burn well into each side of previous bead to insure complete fusion with each side of the groove and the previous bead.
10. Run a similar bead on the other side. Clean the entire weld. Figure 3 shows the bead contour you should have.
11. Using a wider side-to-side or a Figure 8 weave with 5/32 inch electrode, put on a third and final pass. This pass should be very neat and should have a contour shown in Figure 4.

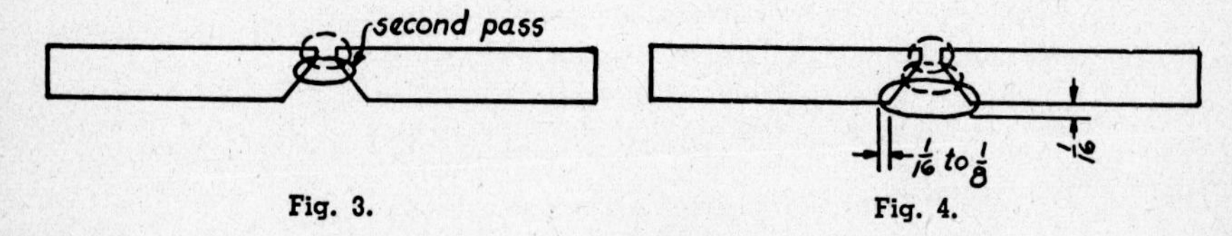

Fig. 3. Fig. 4.

Review Questions

1. How is most horizontal axis pipe welded on industrial applications?
2. What is the difference in joint preparation for vertical up pipe welding, as compared with vertical down pipe welding?
3. What is the angle of the vee for vertical up welded pipe?
4. Is it necessary to use back-up in order to get full penetration?
5. What size electrode is used on the first pass? The second?
6. How many tack welds should be used? How are they located?
7. What is the gap for vertical up welding?
8. How large a hole is burned in the pipe for proper welding on the first pass?
9. What electrode motion is used on the second pass?
10. What is the most important consideration in making the second pass?

LESSON 1.47

Object:
To make a horizontal butt weld on pipe with the axis vertical.

Equipment:
Lincoln "Idealarc" or DC or AC welder and accessories.

Material:
Beveled steel pipe pieces 6 to 10 inches in diameter; 5/32" "Fleetweld 5-P" (E6010) and 5/32" and 3/16" "Fleetweld 5" (E6010) for DC or "Fleetweld 35" (E6011) for AC.

General Information

Wherever there are pipe installations, there will be vertical pipe requiring horizontal welds for fabrication. The bevel will usually be of the standard type given in Lesson 1.45.

Persons interested in passing operator qualification tests are again referred to the "Pipefitter Welder's Review" (see Lesson 1.46).

Job Instructions

1. Clean rust and scale from bevel and adjacent area.
2. Either a drag technique or a short arc may be held when running the root pass. Set amperage according to the technique used.
3. Tack weld pipe together with 1/16" root spacing using 4 equally spaced tacks (Fig. 1).

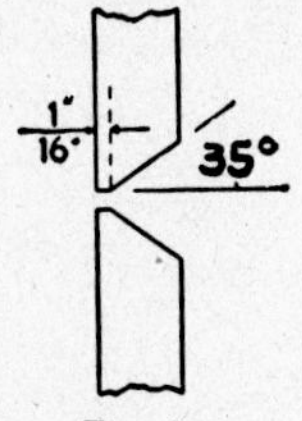

Fig. 1.

Fig. 2.

4. Secure pipe in vertical position with the joint accessible from all sides.
5. Strike the arc at any point on the joint and weld completely around, using either the drag or short arc technique for root pass. Chip out the crater when restarting the bead with a new electrode.
6. Clean and examine root pass for fusion along the sides of the vee, and complete penetration.
7. Use 5/32" or 3/16" electrode for subsequent filler passes. Run straight passes in the sequence shown in Fig. 2.
8. Clean each pass thoroughly before running the next. Do not start and stop beads at the same point as the previous pass. Examine each pass for fusion with the pipe as well as with the previous pass.

Review Questions

1. What type of welding technique is used for root passes?
2. What is root spacing on horizontal butt welds?
3. How many tacks are put on the pipe to hold it while welding?
4. Should craters be chipped out when starting a new electrode?

SECTION II

ARC WELDING MACHINES

LESSON 2.1

Object:
To study the Lincoln "Idealarc TM" AC/DC combination welder.

Equipment:
"Idealarc TM" welder and accessories.

Material:
Scrap steel plate; 3/16" "Fleetweld 35" or "Jetweld 1" electrode.

Fig. 1. AC Transformer welder.

Fig. 2. Combination AC/DC welder.

General Information

"Idealarc TM" welders are available in two models—one is a straight AC transformer welder and the other is a transformer-rectifier welder providing a choice of either AC or DC. Both have continuous mechanical current control, as contrasted to the continuous electrical (saturable reactor) control on "Idealarc TIG" welders. (See Lesson 2.8).

The "Idealarc TM" welders have exceptionally fine arc characteristics and are recommended for all manual welding application except where inert gas welding is used.

In addition to the choice of straight AC or combination AC/DC models, many optional features may be added to the "Idealarc TM" welders to increase their convenience and usefulness. Remote controls, a low voltage contactor and an arc booster are frequently used on these machines.

The remote current control is used on jobs where operators must change electrode sizes frequently and are working some distance from their machines.

The remote current control is a motor inside the case that turns the current control handle. The motor is operated from a toggle switch which the operator carries with him, usually taped to his electrode holder. To change the current, the welder simply pushes the toggle switch in one direction to increase current or in the other to decrease it. Incidentally, the switch is not intended as a means of reducing current for crater filling or to prevent burn-through on thin sections.

Welders may also be equipped with a remote polarity switch that lets the operator flip the selector switch between two preset positions to select the type of current desired. It operates in a manner similar to the remote current control.

The low voltage contactor is frequently used in shipyards and on other jobs where welders work in confined areas surrounded by steel. It is a safety feature. It reduces the voltage between the electrode holder and the work from a relatively high AC voltage to a very low DC voltage when the arc is broken. Thus it is practically impossible for the welder to get a shock should he accidentally put himself between the electrode holder and the work. As soon as the arc starts, the DC voltage drops out and the normal AC output of the welder takes over to provide good welding characteristics. This is automatic and requires no action by the operator.

The arc booster is a feature that aids in arc striking on jobs where the current is set very low for the size electrode being used. The arc booster provides a sudden surge of welding current when the arc is struck. This surg lasts just a few seconds and then cuts out so the current being used is that set on the machine. This surge aids arc striking with hard to strike electrodes. On other applications the arc booster can be switched OFF.

Generally, you will use DC on all 3/16" and smaller electrodes except the iron powder types and AC on all larger electrodes and iron powder types. This is because non-iron-powder electrodes generally operate best on DC except when arc blow is present. As discussed in Lesson 1.14 AC is preferred when the arc blow is present.

To change a combination AC/DC "Idealarc TM" from AC to DC, you simply move the Selector switch to the polarity you desire. Then set the current to the value you want, using the proper scale on the indicator dial. No other changes are necessary.

Note that the "Idealarc TM" welders have considerable welding capacity beyond their rating. As a rule you can expect the machine to have a range of from 10 per cent to 125 per cent of the rating. Thus a 300 ampere machine will have a range of from 30 to 375 amperes. However, when you use higher currents you must reduce duty cycle in order to give the machine adequate time to cool. Even so, it is good to know that you can get plenty of current out of the "Idealarc TM" welders.

Job Instructions

1. Check the "Idealarc TM" welder to see whether it is a straight AC or an AC/DC model. If it has a Selector Switch it is a combination machine; otherwise it is a straight AC welder.
2. Check to see whether there are other optional features.
3. Set current for electrode being used. Run a few beads then adjust the current up and down, each time running a few beads, to see how current adjustment effects welding heat.
4. If the machine is an AC/DC model, run Fleetweld 35 on all three positions of the selector switch and note the effect of polarity on arc action. Be sure to set current to same value for each position.
5. If welder is equipped with other optional features, try each to become familiar with its operation.

LESSON
2.2

Object:
To study the "Shield-Arc" S.A.E. DC welder.

Equipment:
Lincoln "Shield-Arc" S.A.E. welder and accessories.

Material:
Scrap steel; 3/16" "Fleetweld 7" electrodes.

General Information

Dual Continuous Control of the "Shield-Arc" S.A.E. welder gives the operator complete freedom in choice of DC arc characteristics and current. Self-Indicating dials of this control make the proper setting practically automatic. This enables the operator to set the machine for the proper volt-ampere combination to do any kind, type or position of welding.

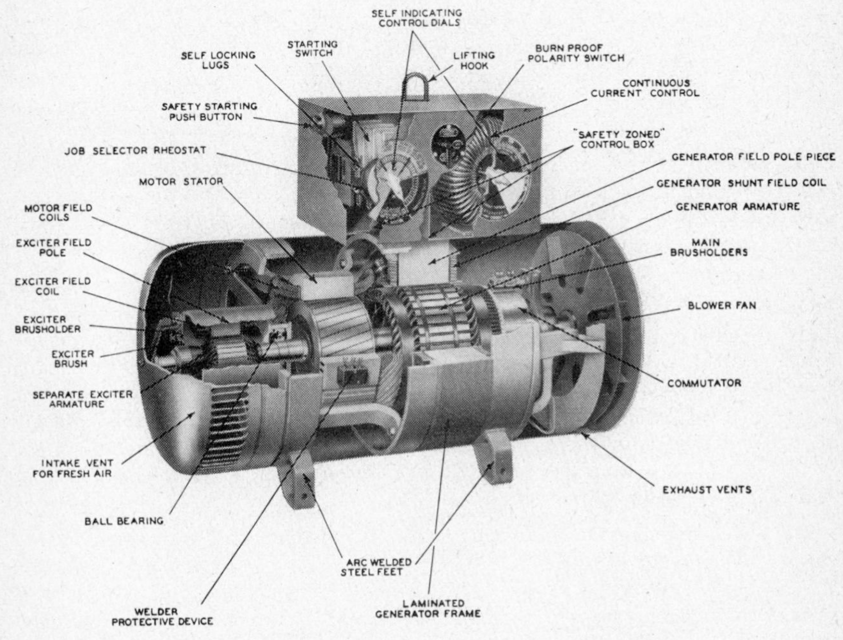

Fig. 1. Cut-away view of "Shield-Arc" S.A.E. welder.

Setting the Self-Indicating dials is as simple as operating the radio. When you tune in a radio station, one dial gives you the station you want while the other varies the volume. In the same manner, weldors can "tune in" the arcs to suit their various welding jobs, (Figure 2). The left-hand dial, the "Job Selector", is for selecting the proper *type* of arc. The other is the current adjustment dial which provides the proper arc *intensity*.

The "Job Selector" or open circuit voltage control is divided into three dif-

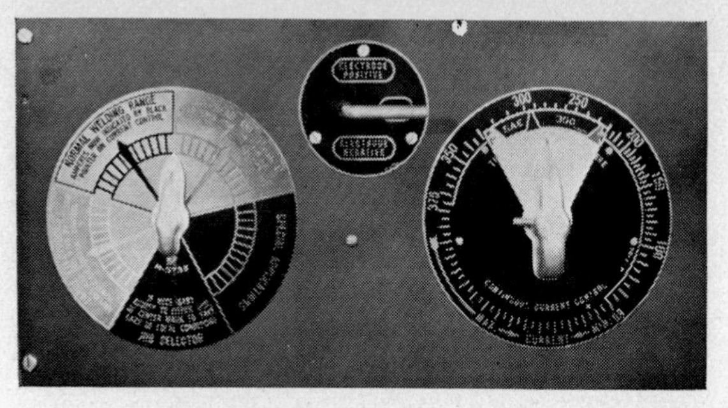

Fig. 2. Control on S.A.E. welder.

ferently-colored segments, labeled "Large Electrodes", "Normal Welding", "Vertical and Overhead", and a fourth segment in black for special low current applications. The setting labeled "Large Electrodes" has the highest voltage, and the others are successively lower in the order named.

The Current Selector has colored arrows corresponding to the segments of the voltage control. In setting the machine, the operator should first choose the voltage control setting required by the work and desired results. Then set the arrow of corresponding color on the current control to the exact current suitable for the work.

To understand the principle of Dual Continuous Control, it must first be realized that there are two types of welding voltage. These are the *open circuit* voltage and the *arc* voltage. The open circuit voltage is the voltage generated by the welding machine when no welding is being done. The arc voltage is the voltage between the electrode and the work during welding.

Open circuit voltages are between 50 and 100; arc voltages are between 18 and 36. The open circuit voltage drops to the arc voltage when the arc is struck and the welding load comes on the machine. Value of the arc voltage is determined largely by the length of the arc and, to some degree, by the type of electrode being used. If the arc is *shortened,* the arc voltage *decreases.* If the arc is *lengthened,* the arc voltage *increases.* The value of open circuit voltage of the welding machine has little effect on the arc voltage, but it does determine the *arc characteristics.*

To understand this, first, recognize that, as the arc length changes the arc voltage, the current also changes. The amount that it changes depends on the volt-ampere curve of the welder. This curve shows what happens to current as voltage changes. Figure 3 shows how this happens. When a short arc is held, the arc voltage is 20 volts and the welding current is 140 amperes. As the arc length is increased, the arc voltage and current move up the curve, until, at 30 volts, the welding current has reduced to 100 amperes.

The volt-ampere curve, shown in Figure 3, is for one particular setting of the "Job Selector" and "Current Control". Changing either of these controls will change the shape of the curve. The Job Selector moves the entire curve up and down voltage axis, as shown in Figure 4; the "Current Control" moves the bottom of the curve back and forth across the current axis, as shown in Figure 5. By operating both of the controls, an infinite number of different curves can be obtained.

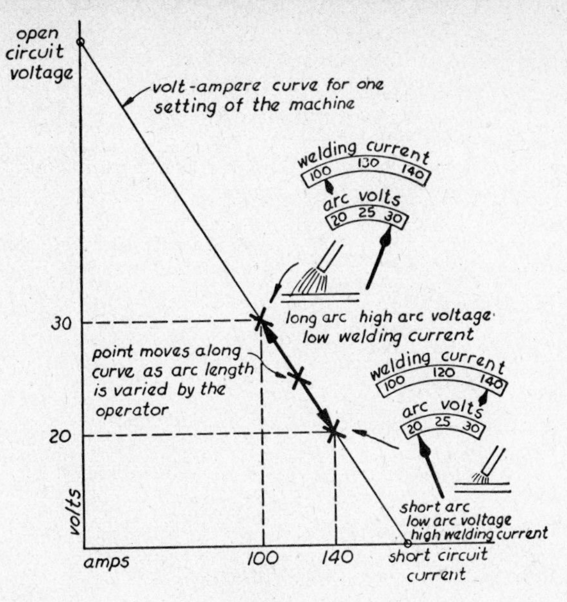

Fig. 3.

The shape of the curve determines the type of arc. Figure 6 shows two different curves obtained for two different settings of the welder. The current and arc voltage at point "X" is the same for both curves. However, note what

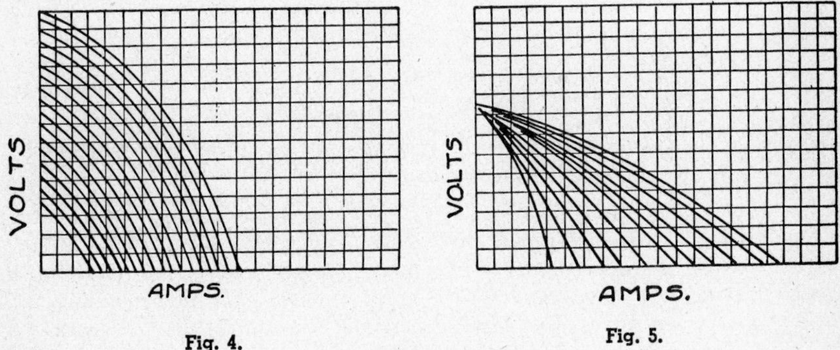

Fig. 4.

Fig. 5.

happens when the arc length is varied. On curve "A", there is very little difference in welding current between the long arc and the short arc; on curve "B", there is a sizeable current change for the same change in arc length. The type of arc resulting from curve "A" is a soft steady arc which is excellent for fast production welding; the type of arc resulting from curve "B" is a sensitive arc which is best for out-of-position welding. With dual continuous control, it is possible to select just exactly the type of arc needed for each job.

Simplifying the above, it can be said that the "Job Selector" controls the type of arc available, or the "arc characteristics". As the dial is moved toward the

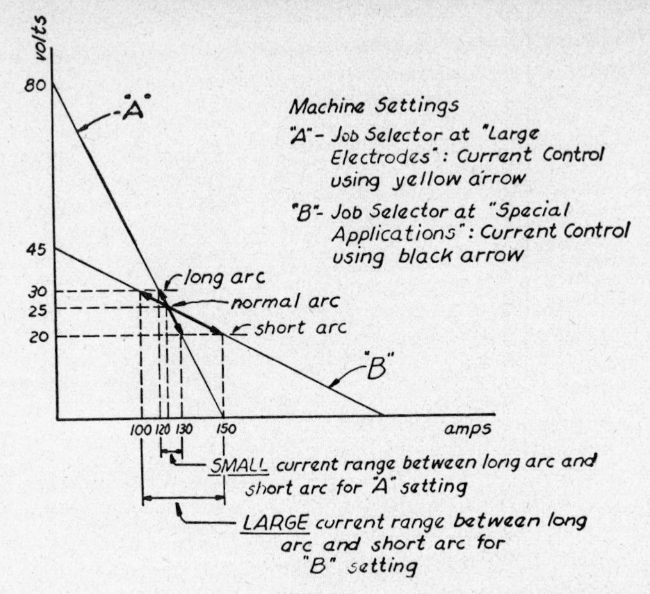

Fig. 6.

"Special Applications" range (low open circuit voltage), a sensitive arc is produced; moving the dial toward the "Large Electrode" range (high open circuit voltage) produces a smooth, steady arc. Further, it can be said that the "Current Control" controls the current. To set the machine, simply set the "Job Selector" to give the type arc desired and then, using the correspondingly colored arrow, set the "Current Control" to the required current.

Job Instructions

1. Connect the welding cables to the output studs on the rear of the welder, being sure that they are connected to the correct stud.
2. Start the welder by pushing the green button on the end of the control box.
3. Set the welder for 200 amperes using the "Large Electrode" setting on the Job Selector. Set the polarity switch to the "Negative" position. Weld with 3⁄16" "Fleetweld 7" electrode. Vary the Job Selector Control either way and note the change in welding heat. Note that this control may be used as a fine current adjustment as well as for control of arc characteristics. Vary the arc length and note that there is little change in welding heat.
4. Set the welder for 200 amperes using the "Overhead and Vertical" range on the Job Selector. Note that, though the current is still the same as for the previous setting, the arc characteristics have changed considerably. Use a short arc length and note the large increase in welding heat; draw the arc out and note that the welding current reduces noticeably.

LESSON
2.3

Object:

To study the "Lincwelder" DC motor-generator welder.

Equipment:

"Lincwelder" DC welder, 180 or 250 ampere size.

Material:

Scrap steel plate; 3/32", 1/8", 3/16" "Fleetweld 37" and "Fleetweld 5" or other mild steel electrodes, both straight and reverse polarity types.

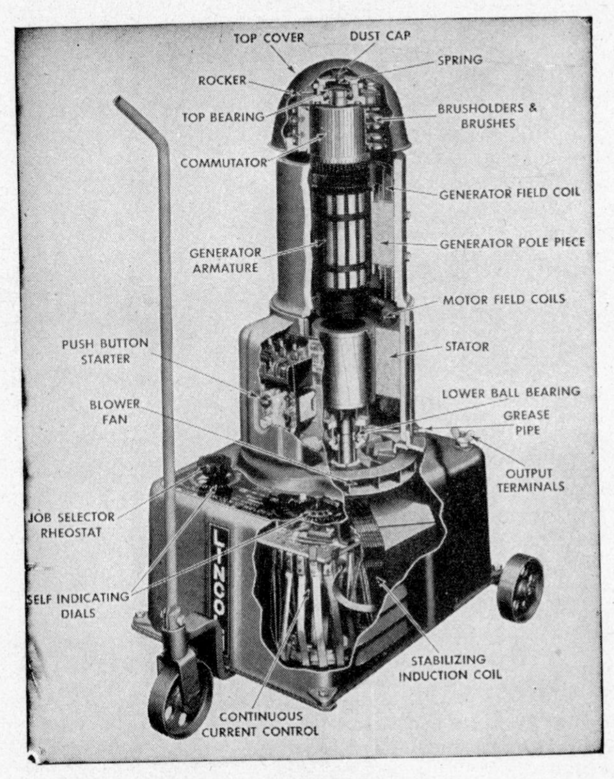

Fig. 1.

General Information

The "Lincwelder" DC welders are designed to do all types of welding with every type of electrode, and are particularly well suited to maintenance, general repair, and sheet metal applications. Similar to the large "SAE" Lincoln welders, they are equipped with dual control to permit selection of desired arc characteristics, as well as welding current.

The fact that these machines occupy a minimum floor space and are very portable make them flexible and desirable where it is frequently necessary to

(R) Fig. 2. (C)

move the welder, such as when they are used for plant maintenance or in repair shops.

Operators find this welder very easy to set and use. It has a steady output, regardless of input voltage fluctuations. The rugged construction of the "Lincwelder" insures long, trouble-free life.

The machine is set to the desired current by first setting the voltage control to the desired value (50-60 is used for most applications; 40-50 is used for sheet metal and out-of-position welding). Then, using the mark on the current control, which corresponds to the voltage setting, select the desired current. For example, to set the machine for 150 amperes for flat welding; set the voltage control to 55; then set the line marked "55" on the current control to 150 amperes. If the arc is found to be too hot, either control may be turned to the left (clockwise) to reduce the current; if it is too cold, either control may be turned to the right.

Facing the welder, the negative terminal is on the right rear of the machine, while the positive terminal is on the left rear. This is indicated on the nameplate. To obtain "straight", or "negative", polarity, connect the electrode cable to the negative stud and the work, or ground, cable to the positive terminal. To use "reversed", or "positive", polarity, simply reverse the cable connections.

Job Instructions

1. Connect the cables for electrode-negative polarity.
2. Set the welder for 100 amperes (50 volts), using the procedure described above.
3. Start the welder.
4. Connect the ground cable to the work and insert a $\frac{1}{8}$ inch "Fleetweld 37" or other straight polarity electrode in the electrode holder.
5. Run a bead.
6. Move the voltage control to 40 volts and run another bead; set to 60 volts and repeat welding. Note how the current varies as the voltage control is changed.
7. Reset the voltage control to 50 and change the current control to 80 amperes. Run a bead. Set to 120 amperes and run another bead. Though both controls will change the current, on most jobs it is best to use a higher voltage setting because this gives a smoother arc.
8. Change the polarity as described above and run several beads with "Fleetweld 5" or other reverse-polarity electrode.
9. Weld with $\frac{3}{32}$ inch and $\frac{3}{16}$ inch electrodes to determine the range of welding current available.

LESSON 2.4

Object:
To study the "Lincwelder" AC welders.

Equipment:
"Lincwelder" AC-180-S or AC-225-S welder and accessories.

Material:
Scrap steel plate; 1/8" and 5/32" "Fleetweld 180" and Fleetweld 37" electrodes.

General Information

The "Lincwelder" AC welder is a transformer welder which has found wide acceptance in a variety of applications. A versatile machine, it is capable of welding, hardsurfacing, soldering, brazing, and metal cutting. It also has a low input voltage switch which permits the welder to be used on low voltage power lines without reduction in output current. The "Lincwelder" can also be used to thaw frozen water pipes.

Fig. 1.

Current control on the "Lincwelder" is extremely simple. A selector switch on the front panel has twelve positions, each corresponding to a different output current. Each position is labeled so the operator knows what current is being delivered. The twelve positions are selected to give finer current control on the lower range of the machine. Operators are able to get just the current they need for any job.

Job Instructions

1. Using "Fleetweld 180" electrodes, run beads at several different current settings. Note that the difference between steps is greater at higher currents than at lower currents.
2. Connect an arc torch to the electrode and ground cables and start the arc. (See Lesson 1.39). Note how the heat of the arc torch is varied when the welding current is changed.
3. Weld first at the minimum position on the current switch and then at the maximum position to determine the total range of welding current available.

LESSON 2.5

Object:
To study the Lincoln "Weldanpower" welder and power unit.

Equipment:
Lincoln "Weldanpower" machine.

Material:
Scrap steel plate; 1/8" and 5/32" "Fleetweld 180" and "Fleetweld 37" or other AC electrodes.

General Information

The "Weldanpower" machine is a combination welder and auxiliary power source. As a welder it supplies up to 200 amperes of alternating (AC) welding current; as a power generator it supplies 5 Kva continuous of either 230 or 115 volt, 60 cycle power. If necessary, it is possible to use the machine both for welding and power at the same time.

The "Weldanpower" unit has a wide range of applications including construction work, farm welding, structural and piping contracting work, and many other applications where it is desirable to have electrical power available on the job.

Fig. 1. "Weldanpower" welder and power generator.

Fig. 2. Belted "Weldanpower."

Operation of the unit is easy. To start the gas engine, first check the gas and oil. Push in the ignition switch and pull out the choke. Crank the engine. Gradually close the choke after the engine has started. Set the throttle control to "Power" if the unit is to be used as a power source, or to "Weld" if it is to be used as a welder. The unit is ready to go!

To use the "Weldanpower" as a power source, set the "Welding Heat Control" to the "Power" position. Then simply plug in the electrical lines. The one 230 volt outlet is in the upper right-hand corner of the control panel of the machine. The four 115 volt outlets are on the left side of the panel.

To use the "Weldanpower" as a welder, first be sure that speed is properly set. Connect the ground cable to the "Work" stud. Plug the electrode cable into the tap marked for the electrode size being used. Set the "Welding Heat Control" to the middle of its range. Weld. If it is necessary to adjust the current setting simply rotate the "Welding Heat Control" in the desired direction.

If it is necessary to use the "Weldanpower" unit as both a welder and a power generator simultaneously, set the speed control to the "Weld" position. Note that less than rated power and welding current must be used. It is not possible to use the full 4 Kva and 200 amperes welding current at the same time.

Fig. 3. Suggested ways of mounting the belted weldanpower on a tractor.

Job Instructions

1. Start the "Weldanpower" as described above. Set throttle to "Power" position.
2. Connect several power tools or light bulbs to the output and operate them. Note the steady operation of both electrical equipment and "Weldanpower".
3. Set the throttle to the "Weld" position.
4. Insert the plug into the 1/8" tap and connect the ground lead.
5. Weld with 1/8" electrode. Vary the "Welding Heat Control" and note the change in welding heat.
6. Leaving the settings as in Step 5 above, connect several light bulbs to the electric power output. Note that they burn slightly more brightly than they did when the speed control was set at the "Power" position. With the light bulbs still burning, weld with the 1/8" electrode. Note that it is possible to use both power and welding current at the same time.

LESSON 2.6

Object:
To study the "Lincwelder" DC engine and belt driven welders.

Equipment:
"Lincwelder" DC-180-BT or DC-225/3-AS welder and accessories.

Material:
Scrap steel plate; 1/8", 5/32", and 3/16" "Fleetweld 5", "Fleetweld 37" or other mild steel electrodes.

General Information

The "Lincwelder" DC engine-driven welders, though designed primarily for farm or general repair welding, have found wide application on a variety of other applications. They are available in 180 and 225 ampere sizes; the 180 size is a belt driven machine. The 225 amp size is driven by an air cooled engine and also supplies auxiliary power to run electric powered hand tools.

Fig. 1. "Lincwelder DC-225/3 AS."

Fig. 2. Belted "Lincwelder DC-180-BT"

These welders are lightweight and compact, and can be easily transported either on a truck or wagon, or on an undercarriage. They provide a wide range of welding current and are easy to use.

The 180 ampere belted welder is simple to operate. Convenient taps on the control panel of the welder provide the correct settings for 5/64, 3/32, 1/8, 5/32 and 3/16 inch diameter electrode. The easily accessible terminals provide a current output ranging from 55 amperes minimum to 205 amperes maximum. To change polarity, switch the welder's ground and electrode cable plugs. The welder is designed to run at 2400 RPM and weighs 203 pounds.

To weld with the 225 ampere size, connect the welding cables to the output studs which are marked "Positive" and "Negative". Set the "Current Range Control" to the number nearest the desired current. Weld. Adjust the "Fine Adjustment" continuous current control to the exact amperage required. The engine is equipped with an automatic idling device that reduces engine speed automatically to a low idle speed when not welding. Current output ranges from 40 amperes minimum to 225 amperes maximum. Welding duty cycle is 50 percent at a 180 ampere welding setting. To change polarity, reverse the cable connections.

Job Instructions

1. Start the welder.
2. Set the machine for the size of electrode being used. Weld.
3. Change the settings to become familiar with the controls.
4. Change polarity and use different electrodes.

LESSON
2.7

Object:
To study the "Shield-Arc" SA-200 gasoline engine driven welder.

Equipment:
"Shield-Arc" SA-200 welder.

Material:
Scrap steel; 3/16" "Fleetweld 5" and "Fleetweld 7" electrodes.

General Information

The "Shield-Arc" DC welder is widely used for all types of field welding such as job repair work, pipeline welding, structural steel erection, etc. It is equipped with dual control to permit selection of the proper amount and type of welding current. This welder is also equipped with a 1 Kw power outlet plug to permit use of 110 volt DC power tools and lights.

The engine of this welder is a water cooled Continental, model F-162. This engine is capable of supplying more than enough power to operate the welder. This reserve power tends to make the arc steadier and settings more stable. Also, the output of the welder will not drop as the engine becomes older and less efficient. An idling device reduces the speed of the welder to a low idle when the operator is not actually welding. The engine speed picks up to normal speed when the operator strikes the arc. The welder may be obtained either with or without a battery start.

To start the engine, first plug the idling device by inserting the pin below

Fig. 1.

Fig. 2.

the frame of the idling device. Push the ignition switch in. Pull the choke out and start the engine either with the starter or by hand. Gradually push the choke back in. After the engine has warmed up a bit, pull the pin out of the idling device and let the engine idle. The machine is ready to weld.

Connect the output cables to the welder. The positive terminal is on the right as you face the end of the welder. The polarity of both terminals is marked on the nameplate. To reverse polarity, it is necessary to reverse the welding cable connections.

To set the machine, first set the Selective Current Control to the value nearest to the desired current. Be sure to set the pointer right on a number and not half way between numbers, because this could cause arcing in the control box. Adjust the Continuous Voltage Control to obtain the desired amount of welding heat. While the Continuous Voltage Control is used principly as a fine current control, it also determines the open circuit voltage of the machine and hence, the arc characteristics. To get a steady arc set the Selective Current Control to a value *less* than the desired current and adjust the Continuous Voltage Control so that it reads a large number. To set the machine for a crisp, sensitive arc, set the Selective Current Control to a value greater than that which is desired and adjust the Continuous Voltage Control so that it is set on a low number. Simply stated, soft steady arc characteristics are obtained with high number (high OCV) Continuous Voltage Control settings, while the crisp, sensitive arc characteristics are obtained with the low number (low OCV) settings.

Job Instructions

1. Start the welder (see instructions above).
2. Connect the welding cables for electrode positive polarity.
3. Tap the electrode to the work, (use a shield) but do not start the arc. Note how the engine speed picks up and then drops back to low idle after a few seconds.
4. Turn the Continuous Voltage Control all the way to the left and set the Selective Current Control on "150". Weld with 3/16" "Fleetweld 5". Without changing the Selective Current Control move the Continuous Voltage Control all the way to the right. Weld with the same size electrode and note the change in welding heat. Repeat for several other settings of each control.
5. Change to electrode negative polarity by interchanging the welding cable connections at the output studs of the welder. Weld with "Fleetweld 7".
6. Connect a DC or universal motor driven power tool to the auxiliary power output plug on the control panel of the welder. Note that it is possible to simultaneously use power and weld.

LESSON 2.8

Object:
To study the Lincoln "Idealarc TIG" AC/DC and inert gas combination welder.

Equipment:
Lincoln "Idealarc TIG" welder and accessories; inert gas, torches and other equipment necessary for tungsten inert gas welding.

Material:
Standard consumable coated electrodes; tungsten electrodes; scrap mild steel, aluminum and thin stainless steel.

General Information

The "Idealarc TIG" will handle any kind of manual welding job that comes along, whether it is done with stick electrodes or with inert gas. Though it is available as a straight AC-only welding machine without other features, most "Idealarc TIG" welders are equipped with high frequency for inert gas welding and many also have rectifiers to provide DC welding power.

For manual welding with coated electrodes, the high frequency part of the machine is turned off. Then the machine operates very similar to the "Idealarc TM" welders described in Lesson 2.1. It has a polarity selector switch that sets the machine's output for AC, DC+ or DC− welding power. There is also a current range selector that divides the machine's current range into three overlapping parts. Finally, a fine tuning control adjusts the current to the value required for the job being done.

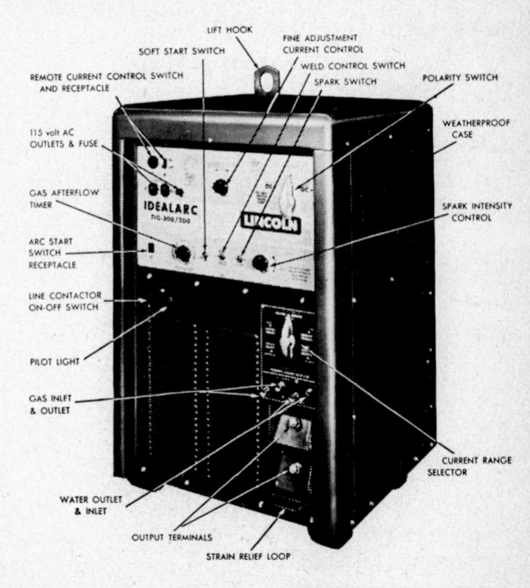

Remote current controls, called "Amptrols" are available in hand or foot operated models. These remote controls provide instantaneous current response and are suitable for tapering off the current in crater filling operations.

When welding thin sections of stainless or welding aluminum or some of the new unusual metals, inert gas welding is preferred. In inert gas welding the arc is maintained between a non-consumable tungsten electrode and the work. The welding torch is connected to a supply of inert gas and feeds the gas around the arc to keep out impurities in the air which would otherwise harm the weld. Filler metal is hand fed into the arc in a manner similar to that used in gas welding.

Some types of metals, such as stainless steel, are best welded with DC. To use "Idealarc TIG" for this type of welding, first set the polarity switch to DC (usually DC—), then turn the high frequency on-off switch ON. This assures control of gas and water flow into the torch.

There is a choice of the type of high frequency spark for DC welding. The spark can be left off, or it can be used just for starting where it automatically cuts off after the arc is established, or it can be left on all the time the arc is established. Usually the spark is used for starting only on DC welding.

The dial on the afterflow timer is keyed to electrode size and should be set for the size of tungsten electrode being used.

Aluminum is usually welded with AC and continuous high frequency. The machine is set as described above except that the spark control is always set to the ON position. Also it is necessary to set the high frequency intensity control. It should be set high enough to assure a smooth steady arc, yet not so high that it makes pock marks on the weld.

The balanced wave filter is an optional feature that improves operation on AC high frequency welding. It consists of three automotive batteries that can be put in or taken out of the welding circuit with a quickly detached connector. It should be in the circuit when welding with AC and high frequency and out at all other times.

Though the "Idealarc TIG" has many switches and controls, it is actually very easy to use. Its versatility in application has made the "Idealarc TIG" machine popular in small maintenance and repair shops that must be equipped to handle any kind of welding job that comes in the shop. These machines are also used in production shops that do nothing but inert gas welding.

Job Instructions

1. Familiarize yourself with all the controls on the machine. Note that all commonly used current and polarity selection controls are on the front of the machine while all the high frequency, water and gas controls are on the side.
2. Set the machine for welding with manual electrodes. Set the current and polarity switches for the electrode being used. Be sure the high frequency is turned OFF and the balanced wave filter is disconnected. For remote control plug in the Amptrol and switch the remote switch ON. Weld with several sizes and types of electrodes to become familiar with the machine's operation and controls.
3. Set the machine for welding stainless. Turn the high frequency switch ON and set the spark control on the OFF or START ONLY position. Make torch, gas and water connections. Adjust current and afterflow timer. Plug in the output contactor remote switch. To start the arc depress the contactor remote switch after the torch is lined up over the work. The arc and gas and water flow will start simultaneously and remain until the arc is broken. Weld several samples to get the feel of the staright DC tungsten arc.
4. Set the machine for welding aluminum. Switch the spark control switch to ON. Adjust the spark intensity to about halfway through its range. Connect the balanced wave filter. Weld to get the feel of an AC with high frequency arc. Adjust spark intensity to minimum and maximum to note its effect on the arc. Use the remote control to see how it reduces arc heat.

SECTION III

MILD STEEL ELECTRODES

LESSON 3.1

Object:
To become familiar with the basic groups of Lincoln mild steel and low alloy electrodes.

General Information

Lincoln mild steel electrodes can be classified and studied in four basic groups according to their operating characteristics: "FAST FREEZE", "FAST FILL", "FILL-FREEZE", which includes "FAST FOLLOW" and a group for welding hard-to-weld steels.

"FAST FREEZE" electrodes include Lincoln electrodes in the A.W.S. classes E-XX10 and E-XX11. These electrodes are basically similar in operation with an outstanding ability for the molten metal and slag to freeze rapidly. Compared to other electrodes, they do not have fast fill or fast follow ability.

The "FAST FILL" group includes Lincoln electrodes in the A.W.S. classes E-XX24 and E-XX27. They are similar in operation, but each electrode has been designed for a particular type of joint. The ability of these electrodes to rapidly fill a joint with weld metal is unsurpassed. They have little fast freeze and limited fast follow.

The "FILL-FREEZE" group includes electrodes in the E-XX12 and E-XX13 AWS classes. Each electrode has, to some degree, both "filling" and "freezing" abilities. However, the proportion of each ability varies considerably from one electrode to another. The differences in operating characteristics of the various electrodes within the group are quite noticeable. Also, some of these electrodes exhibit a special ability termed "FAST FOLLOW".

The fourth group of electrodes is for welding steels high enough in sulfur, carbon or phosphorus to require special consideration in the choice of electrodes to produce crack-free welds. Heavier thicknesses of mild steel also present a problem because of the severe quenching of the welds. This quenching may result in cracks. The first consideration in welding these steels is selecting an electrode which will produce crack-free welds; speed and ease of welding are secondary considerations. Electrodes for these applications are the A.W.S. E-XX15 and E-XX16 classifications.

Lessons 3.2, 3.3, 3.4 and 3.5 present information on each group in the following manner: information which is common to the entire group is presented in the "General Information" portion of the lesson; special characteristics which set apart individual electrodes are given in the sub-lessons (3.2A, 3.2B, etc.)

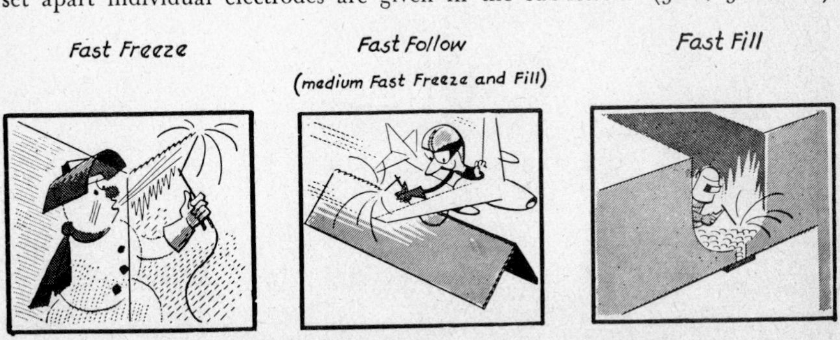

Fast Freeze *Fast Follow (medium Fast Freeze and Fill)* *Fast Fill*

Fig. 1. Basic groups of mild steel electrodes.

along with specific instructions for jobs which will bring out both the group and individual characteristics of the electrode.

Electrode Groups

The terms "fast freeze", "fast fill", "fill-freeze", and "fast follow" describe an electrode's basic characteristic and indicate the type of job for which it is *best* suited. These characteristics correspond to job requirements. Any mild steel production welding job can be described by one or a combination of these terms. Proper electrode selection depends on matching the characteristics of the electrode to the requirements of the job.

It may seem unnecessary to have more than one electrode in each group. It becomes understandable, however, when it is recognized that additional requirements beyond the basic requirements must be considered for maximum efficiency. Operation on AC or DC, bead appearance, penetration, and arc force are some of the other operating characteristics which must be varied and changed to meet different job requirements.

The proper approach to electrode selection, then, is to first study the application and determine what basic characteristics are required. Then, compare the electrodes within the appropriate group to determine which one best meets the special considerations of the job. It's like shopping for household goods. If one needs a chair, he first goes to a furniture store which sells chairs; then, he selects a chair which meets the special considerations of style, color, etc., which are necessary to best compliment the room in which it is to be used.

How to Determine Basic Requirements

An understanding of the terms will help determine basic requirements. "FAST FREEZE" means ability to deposit a weld which solidifies rapidly. This consideration is of importance *only* when there is some danger of the molten metal and slag spilling or running out of the joint. Such conditions are found when welding in a vertical, horizontal or overhead position or on a circular part. Metal and slag must stay in the joint until it has solidified. This consideration overrides all others, including the natural desire to deposit metal faster so that greater welding speeds are attained.

Fig. 2. A FAST FREEZE application.

The other extreme is where fast freezing metal is no concern, but rather the main concern is to pour a given amount of metal into the joint as fast as possible. Electrodes designed expressly to deposit a large amount of weld metal rapidly are termed "FAST FILL" electrodes. These electrodes are limited to the flat, or very near flat, position where gravity helps to hold the deposit in the joint, rather than attempting to pull it out. The more the work is moved out-of-position, the more fast fill ability must be sacrificed, until, eventually, a point is reached where "fast freeze" becomes more important than "fast fill".

Although many jobs clearly fall into either "fast fill" or "fast freeze" classifications, even more applications overlap these groups and require electrodes with a combination of characteristics. Electrodes of this type are called "FILL-

Fig. 3. FAST FOLLOW Applications.

Fig. 4. A FAST FILL Application.

FREEZE". Though belonging to the same group, they have distinctive characteristics. For example, Improved Fleetweld 47 has just enough "fast freeze" ability to weld slightly downhill joints. On the other hand, Fleetweld 7 has considerable "fast freeze" ability and can weld steep downhill joints.

Some "fill-freeze" electrodes also have a distinct characeristic called "fast follow"—the ability to move along a joint at high speeds and make a continuous weld without skips or misses. These welds require that the molten pool closely follow the tip of the fast moving electrode; hence, the name "fast follow".

The "fast follow" characteristic of this group is utilized to best advantage on fillet and lap joint in light gauge metals, 10 gauge down to 20 gauge, for production jobs welded at speeds in excess of 20 inches per minute. For this type of joint, actually, no deposited metal is required. Any deposited metal is in excess of what is actually required to make the joint. This application is the extreme of "fast follow", as the only practical limit to the speed of electrode travel is the rate at which the molten pool is able to follow the electrodes tip, so that a continuous bead will be made.

Determining the application of electrodes FOR HARD-TO-WELD STEELS is easy. Whenever it is known that the material to be fabricated contains high carbon or sulfur or other crack-producing alloys, these electrodes should be used. If while welding a steel of unknown analysis, cracks or porosity are encountered, then these electrodes should be used—regardless of the joint or position. They will produce sound welds and reduce or eliminate preheat.

The next four lessons are arranged by the above groups. For the beginning weldor, the best approach would be to carefully study the "General Information" section of each lesson to first learn the basic differences between the groups. Then, return to the sub-lessons (3.2A, 3.2B, etc.) to pick up the differences between individual electrodes within the group. Study these lessons carefully, for they contain the key to proper electrode selection.

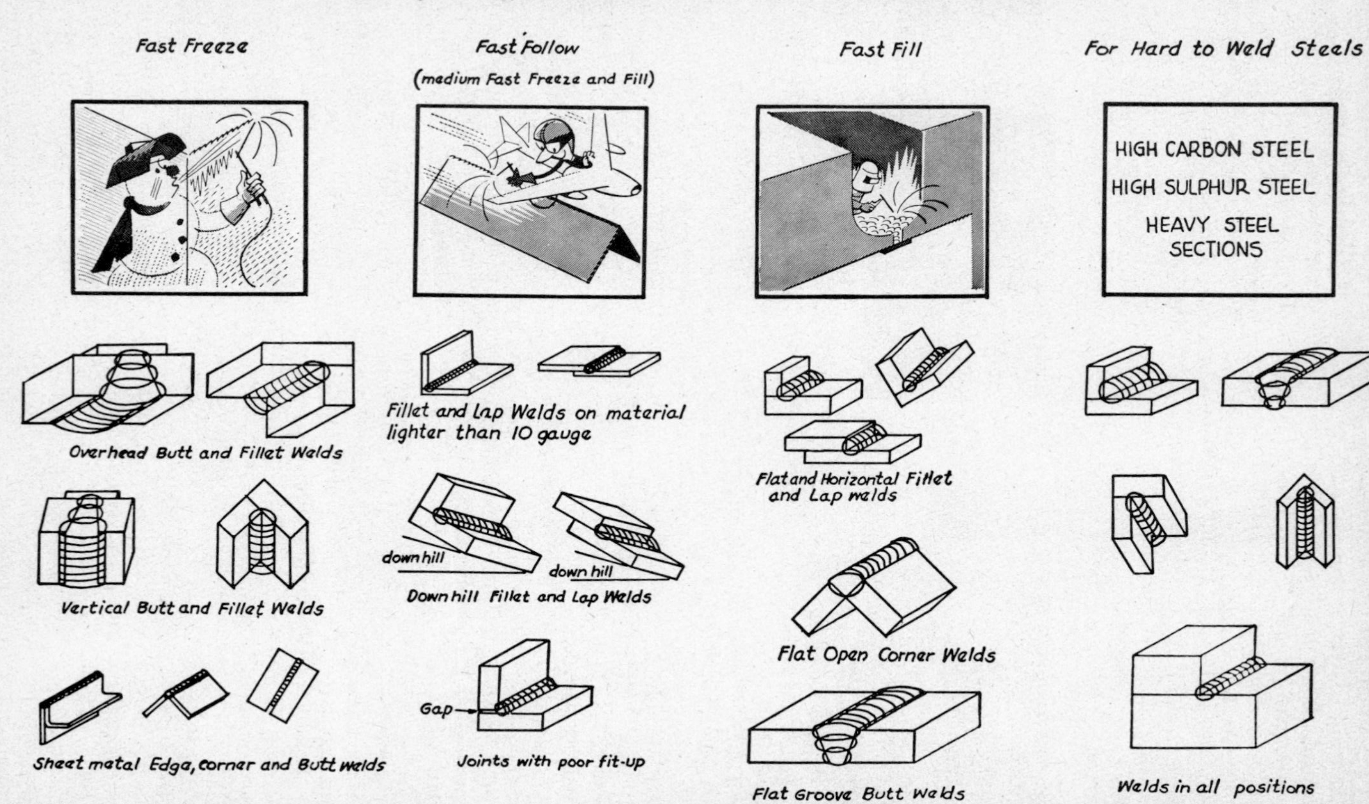

Fig. 5. Typical application of each group of electrodes. These joints illustrate the features of each group; not all types of joints are shown.

LESSON 3.2

Object:
To study the use and applications of Lincoln "FAST FREEZE" electrodes.

General Information

These are versatile all-position, all-purpose Lincoln electrodes, and are *best* for most vertical and overhead welding and some sheet metal applications.

Included in the "FAST FREEZE" group are the following: Fleetweld 5, Fleetweld 5P, Fleetweld 35, Fleetweld 180, and the low alloy, high tensile electrodes, Shield-Arc 85 and Shield-Arc 85P.

Operating Characteristics

Deposition: FAST FREEZE—best of all Lincoln electrodes; when used on DC, electrode negative, also has good "fast follow" ability.

Arc: FORCEFUL—deep penetration with maximum admixture.

Slag: LIGHT—little slag interference and incomplete slag coverage.

Bead: FLAT—with distinct ripples.

Current: DC—normally used on DC, electrode positive (AC with Fleetweld 35 and Fleetweld 180) to give the arc characteristics described above. May also be used on DC, electrode negative, to give a spray-type arc.

Positions: ALL—truly "all-position"; exceptionally good for vertical and overhead work.

Application Information

The versatility, all purpose character, of these electrodes has resulted in their use on many different types of applications. They have excellent physical properties and produce X-ray quality welds. DC, electrode positive, (or AC with Fleetweld 35 and Fleetweld 180) is used on all applications except sheet metal (No. "6" below).

1. ALL VERTICAL AND OVERHEAD WELDING is best done with these electrodes, except for high-speed vertical down lap and fillet welds on sheet metal and light plate. "Fast freeze" permits fastest out-of-position welding on applications such as structural steel erection and pipe welding.
2. GENERAL PURPOSE MAINTENANCE WELDING is best done with these electrodes. Frequently welding is out-of-position and work may be dirty, rusty, or greasy and cannot be cleaned. Deep penetration and light slag produce best results possible under these adverse conditions.
3. X-RAY QUALITY JOBS welded out-of-position require these electrodes. These electrodes produce X-ray quality welds in all positions for all types of joints.
4. JOINTS REQUIRING DEEP PENETRATION, such as square-edge butt welds on 3/8 inch plate, are welded in the flat position with the larger sizes. By taking advantage of the deep penetration, welds can be made faster and plate beveling is eliminated.
5. GALVANIZED STEEL is best welded with these electrodes because the forceful arc with light slag bites through the galvanizing with little bubbling or porosity to produce sound, quality welds.

6. SHEET METAL EDGE AND BUTT WELDS on 10 to 22 gauge sheet are welded with DC, electrode negative. The spray-type arc and fast follow ability obtained with this polarity produce the fastest possible welds.

Fig. 1. Vertical welding, repair work, and X-Ray inspected pipe welding are applications of "Fast Freeze" electrodes.

Procedures

DC, electrode positive (AC for Fleetweld 35 and Fleetweld 180), is used on all applications except the sheet metal.

VERTICAL—For best peneration, welding is usually done vertical up, though vertical down is used for most cross-country pipeline welding because it is faster for the particular joint involved. Use a stringer bead for the first pass. This bead is usually applied with a whipping technique for fillet welds and a circular motion for Vee-butt joints. Succeeding passes are applied with a weave, pausing slightly at the edges to insure penetration and to wash-in without undercut. Use currents in the lower portion of the range.

OVERHEAD—These welds are best made with a series of stringer beads, using a technique similar to that described for vertical welding.

HORIZONTAL BUTT—Use a procedure similar to that employed for overhead welding.

FLAT—Permit the tip of the electrode to lightly touch the work or hold a very short arc. Tip the electrode forward in the direction of travel and move fast enough to stay ahead of the molten pool. Use currents in the middle and higher portion of the range.

SHEET METAL, EDGE AND BUTT—Use DC, electrode negative; hold a comparatively long arc. Lean the electrode forward in the direction of travel and move as rapidly as possible while maintaining good fusion. Welding is fastest when done 45 degrees downhill. Use currents in the middle of the range.

LESSON 3.2A

Object:
To study the special characteristics of Lincoln Fleetweld 5 electrodes.

Equipment:
Lincoln "Idealarc" or other DC welder and accessories.

Material:
Scrap 1/4 inch plate, 14 gauge sheet metal, and beveled pipe (see Lesson 1.45): 1/8", 5/32", and 5/16" Fleetweld 5 electrode.

An E-6010 electrode, Fleetweld 5 was the first shielded-arc electrode. The most widely used electrode of this group, it has been industry's standard for electrodes of this type for many years. It is used on all of the applications discussed above. The easiest way to study electrodes of this group is to first become familiar with Fleetweld 5 and then compare the other Lincoln electrodes of this group with it. The following "Job Instructions" bring out the typical characteristics of this entire group of electrodes; "Job Instructions" for other electrodes of this group bring out characteristics peculiar to the individual electrode.

PHYSICAL PROPERTIES	AS WELDED	STRESS RELIEVED
Tensile Strength	62,000 - 72,000 psi	60,000- -72,000 psi
Yield Point	50,000 - 60,000 psi	48,000 - 56,000 psi
Elongation	22- -30%	29 - 37%
Impact Resistance (Charpy V Notch at Room Temperature)	78 Foot pounds	
Transition Temperature	-75 to -100°F	

Job Instructions

ALL POSITIONS		
Rod Size		Amperage Range
Dia.	Length	
1/8"	14"	75–130
5/32"	14"	90–175
3/16"	14"	140–225
7/32"	14"	160–275
1/4"	18"	190–375
5/16"	18"	250–475

1. With 1/8" Fleetweld 5, make fillet welds in the flat, vertical, and overhead positions. Use 1/4 inch plates. Also, select a particularly dirty plate and weld without prior cleaning. Note the following: (1) the easy operation of Fleetweld 5 in all positions; (2) the forceful arc and deep penetration which is characteristic of this type electrode; (3) the very light slag which does not interfere or tend to spill when welding out-of-position; (4) the ability to weld dirty plate when it is not possible to properly clean it.
2. Using 1/8 and 5/32" Fleetweld 5 and the techniques presented in earlier lessons on pipe welding, make both vertical down and vertical up pipe welds. This is one of the largest applications for Fleetweld 5.
3. Make square edge butt welds on the 1/4" plate in the flat position using the 5/16" Fleetweld 5. Note that it is possible to make a 100% penetration joint with a pass from either side without bevelling.
4. With DC, electrode negative, polarity make edge and butt welds on 14 gauge sheet metal. Note the arc characteristics. Weld in several positions and note that welding is fastest when the work is positioned 45 degrees downhill.

LESSON 3.2B

Object:
To study the special characteristics of Lincoln Fleetweld 5P electrodes.

Equipment:
Lincoln "Idealarc" or other DC welder and accessories.

Material:
Scrap bevelled pipe and plate; 1/8" and 5/32" Fleetweld 5P.

Also an E-6010 electrode, Fleetweld 5P is very similar to Fleetweld 5, but has the outstanding characteristic of having an unusually stable arc at very low currents. It has less slag and better wash-in with a flatter bead. The arc path is truer and will not waver from side to side. It was designed primarily for stringer bead applications, particularly on pipe. Field experience has broadened its usage to other critical applications where arc control is most important.

PHYSICAL PROPERTIES

	AS WELDED	STRESS RELIEVED
Tensile Strength	62,000 - 72,000 psi	60,000 - 72,000 psi
Yield Point	50,000 - 60,000 psi	48,000 - 56,000 psi
Elongation	22 - 30%	29 - 37%
Impact Resistance (Charpy V Notch at Room Temperature)	78 Foot pounds	
Transition Temperature	-75 to -100°F	

Job Instructions

1. Using 1/8" Fleetweld 5P, make pipe welds with particular attention to the stringer beads. Use very low currents for easy control of the arc. Note that the arc is more easily directed than with Fleetweld 5. The arc points straight out from the end of the electrode without wavering from side to side. There is less slag with Fleetweld 5P than with Fleetweld 5.

PIPE—VERTICAL UP, VERTICAL DOWN—STRINGER BEAD

Rod Size		Amperage Range
Dia.	Length	
1/8"	14"	75–130
5/32"	14"	90–180
3/16"	14"	140–225

2. Use Fleetweld 5P on other applications. Note that it is not only very good for stringer beads, but easily controlled on weave passes too.

LESSON 3.2C

Object:
To study the special characteristics of Lincoln Fleetweld 35 electrodes.

Equipment:
Lincoln "Idealarc" or other AC welder and accessories.

Material:
Scrap 1/4 inch plate, 14 gauge sheet metal, and bevelled pipe; 1/8 inch, 5/32 inch and 1/4 inch Fleetweld 35 electrodes.

Fleetweld 35, an E-6011 electrode, is an AC version of Fleetweld 5 and is used on applications identical to those described for Fleetweld 5, except that an industrial AC welder is used instead of a DC welder. It can, of course, also be used on DC. On sheet metal applications where DC, electrode negative is used, Fleetweld 35 is actually preferred to Fleetweld 5 because it has a slightly finer spray which makes welding easier. On other applications, Fleetweld 5 or 5P is normally preferred, if DC welders are available.

PHYSICAL PROPERTIES

	AS WELDED	STRESS RELIEVED
Tensile Strength	62,000 - 72,000 psi	60,000 - 70,000 psi
Yield Point	50,000 - 60,000 psi	48,000 - 55,000 psi
Elongation	22 - 30%	30 - 37%
Impact Resistance (Charpy V Notch at Room Temperature)	78 Foot pounds	
Transition Temperature	-75 to -100°F	

Job Instructions

1. Repeat the jobs done for Fleetweld 5 using an industrial AC welder and Fleetweld 35. In performing these jobs, compare Fleetweld 5 and Fleetweld 35. Note that, with either electrode, DC is preferred over AC for out-of-position welding.

ALL POSITIONS		
Rod Size		Current Range
Dia.	Length	
3/32″	12″	20– 75
1/8″	14″	60–120
5/32″	14″	90–160
3/16″	14″	120–200
7/32″	18″	150–260
1/4″	18″	190–300

LESSON

3.2D

Object:

To study the special characteristics of Lincoln Fleetweld 180 electrodes.

Equipment:

Lincoln "Lincwelder" or other AC limited input welder and accessories.

Material:

Scrap steel plate and angles; 1/8" and 5/32" Fleetweld 180 electrodes.

Fleetweld 180, an E-6011 mild steel electrode, is designed for operation with limited input AC welders for all-purpose welding. It will operate on DC, though other electrodes of this group are usually preferred. The deep penetration, light slag and easy out-of-position welding make Fleetweld 180 the best electrode for maintenance and repair welding when a limited input welder is being used. It is also used for welding thin sections to heavy members where minimum slag inclusion is desired.

PHYSICAL PROPERTIES

	AS WELDED
Tensile Strength	62,000- 77,000 psi
Yield Point	50,000 - 65,000 psi
Elongation	22 - 28%

Job Instructions

1. To become familiar with Fleetweld 180, weld vertical and overhead with the electrode; particularly weld on dirty plate.
2. Use Fleetweld 180 and the procedure described in Lesson 1.40 to cut steel plate.

ALL POSITIONS		
Rod Size		Current Range
Dia.	Length	
3/32"	12"	50–90
1/8"	14"	80–110
5/32"	14"	115–150

Fig. 1. Repairs Cracked Rim on farm implement wheel with single pass weld.

Fig. 2. Adds Reinforcing Strap on rear axle housing of truck to carry heavier loads on rough roads.

LESSON 3.2E

Object:
To study the special characteristics of Lincoln Shield Arc 85 electrodes.

Equipment:
Lincoln "Idealarc" DC welder.

Material:
Scrap 1/4" plate and bevelled pipe; 1/8", 5/32", and 1/4" Shield-Arc 85 electrode.

An E-7010 electrode, Shield-Arc 85 operates very nearly like Fleetweld 5. Its principal difference is in physical properties; Shield-Arc 85 has a higher tensile strength. It is used on all applications similar to those of Fleetweld 5, except for sheet metal where its extra strength is not required.

PHYSICAL PROPERTIES	AS WELDED	STRESS RELIEVED
Tensile Strength	77,000 - 82,000 psi	70,000 - 75,000 psi
Yield Point	62,000 - 66,000 psi	57,000 - 60,000 psi
Elongation	18 - 25%	22 - 25%
Impact Resistance (Charpy V Notch at Room Temperature)	70 Foot pounds	68 Foot pounds
Transition Temperature		-75 to -100°F

Job Instructions

1. To become familiar with Shield-Arc 85, repeat the "Job Instructions" for Fleetweld 5, excluding No. "4". Note that its operation is nearly identical to that of Fleetweld 5 and that it is used on similar applications where the base metal is 70,000 psi tensile strength steel.

ALL POSITIONS		
Rod Size		Amperage Range
Dia.	Length	
1/8"	14"	75–130
5/32"	14"	90–175
3/16"	14"	140–225
1/4"	18"	190–325

Fig. 1. 40 miles of 16" gas line welded with "Shield Arc 85", an excellent electrode for welding high pressure cross country pipeline.

LESSON 3.2F

Object:

To study the special characteristics of Lincoln Shield Arc 85P electrodes.

Equipment:

Lincoln "Idealarc" DC welder.

Material:

Scrap bevelled pipe; 3/16" Shield-Arc 85P electrodes.

Shield-Arc 85P is similar to Shield-Arc 85, but is especially designed and used for cover pass welds on high tensile pipe. It is an E-7010 electrode. Its feature on this work is an ability to produce welds which are free of pinholes. It is important in using this electrode to hold a fairly long arc and weave a narrow bead. It has greater arc force and less slag interference than Shield Arc 85. This accounts for fewer pinholes.

PHYSICAL PROPERTIES

	AS WELDED	STRESS RELIEVED
Tensile Strength	77,000 - 82,000 psi	70,000 - 75,000 psi
Yield Point	62,000 - 66,000 psi	57,000 - 60,000 psi
Elongation	18 - 25%	22 - 25%
Impact Resistance (Charpy V Notch at Room Temperature)	65 - 70 Foot pounds	68 Foot pounds
Transition Temperature	Below -75°F	-75 to -100°F

Job Instructions

1. Practice applying Shield-Arc 85P on cover passes on pipe welds. (See Lesson 1.45.)

ALL POSITIONS		
Rod Size		Amperage Range
Dia.	Length	
3/16"	14"	140–225

Fig. 1. ON THE LINE welding pipe with "Shield-Arc 85-P" welds are produced fast, efficiently . . . are free of surface pin holes.

LESSON 3.3

Object:
To study the use and applications of Lincoln "FILL-FREEZE" electrodes.

General Information

Lincoln "FILL-FREEZE" electrodes are best for joints with poor fitup and for most applications which are downhill, but not actually vertical. Those electrodes in the group that also have "FAST FOLLOW" ability are the best electrodes for high-speed, light-gauge fillet and lap welds.

Included in the "FILL-FREEZE" group are Fleetweld 7, Fleetweld 37, Improved Fleetweld 47, Fleetweld 72, Planeweld 1, and Fleetweld 7MP.

Operating Characteristics

Deposition: FILL-FREEZE, or a combination of both "fast fill" and "fast freeze". Electrodes within the group differ considerably. Some electrodes also have outstanding "FAST FOLLOW" ability.

Arc: QUIET–medium peneration with slight spatter.

Bead: CONVEX–smooth with even ripples in the downhand positions; more distinct ripples on vertical up and overhead joints.

Slag: MEDIUM–gives good slag coverage.

Current: DC or AC–all electrodes will operate on either; AC is preferred for Fleetweld 37 and Improved Fleetweld 47; DC, electrode negative, is best for the other electrodes of this group, unless arc blow is a problem.

Positions: ALL–most widely used in the flat or downhill positions, but has some limited application in vertical and overhead positions.

Application Information

Where costs are important, these electrodes are used because of their distinctive operating characteristics. Of the following applications, the first depends on "fast follow" ability, while the others depend on "fill-freeze" characteristics. They are also sometimes used on other applications because of their ease of operation.

1. HIGH-SPEED LAP AND FILLET WELDS ON 1/4 INCH TO 20 GAUGE MATERIAL, where the welding speed is determined only by fast follow ability and bead size is unimportant, are the specialty of these electrodes. Fast follow ability becomes important when welding speeds in excess of 20 inches per minute are employed. The "Fast Follow" electrodes excel on applications of this type.
2. DOWNHILL LAP AND FILLET WELDS, which are too steep for the "Fast Fill" electrodes, but not steep enough to require the "Fast Freeze" electrodes, are best made with electrodes of this group. This application is the major use for this type of electrode. The application varies widely because of the combinations of angle of incline and the size of the weld. Each electrode in the group has individual characteristics which recommend it for

particular applications. For example, Fleetweld 7 will make a 3/16 inch weld on a 60 degree incline, while Improved Fleetweld 47 is limited to a 20 degree slope for the same size weld. Each job must be considered separately and the various electrodes compared for speed and ability.

3. POOR FIT-UP JOINTS, where the "Fast Fill" electrodes would spill through are welded with this group. The moderate fast freeze eliminates spilling, yet gives reasonably fast deposition rates. Fleetweld 7 is the best electrode of the group for these applications.

4. GENERAL PURPOSE WELDING, where most of the work is of the type described in "2" and "3" above, is a common application for electrodes of this group.

Procedures

DOWNHAND—use stringer beads for first pass and either additional stringer beads or a weave for succeeding passes. Where poor fit-up is encountered, it may be necessary to use a slight weave on the first pass. Hold a very short arc or lightly touch the electrode on the work. Tip the electrode forward into the direction of travel and move as fast as possible consistent with the desired bead size. On light material where bead size is unimportant, travel as fast as possible while maintaining a molten pool. Light material is best welded 45-90 degrees vertical down. Use currents in the middle to higher portions of the recommended range.

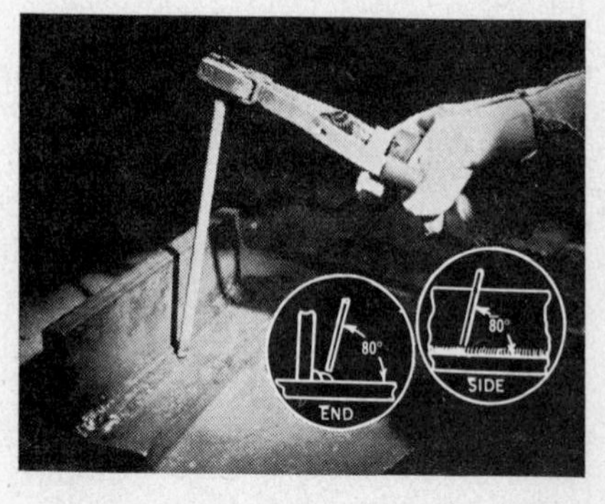

Fig. 1. Angle of electrode for beads against vertical plate—multiple pass fillet.

VERTICAL DOWN—use stringer beads or a very slight weave. Do not try to deposit too much metal with one pass. Drag the electrode on the joint or hold a very short arc. Tip the electrode well into the direction of travel so that the arc force tends to push the molten metal back up the joint. Move fast enough to stay ahead of the molten pool. Use currents in the higher portion of the range.

VERTICAL UP—use a wide triangular weave. Build a shelf of weld metal and, with the weave, deposit layer upon layer of metal as the weld progresses up the joint. Do not use a whip technique or take the electrode out of the molten pool. Point the electrode directly into the joint and slightly upward to permit the arc force to assist in controlling the puddle. Travel slow enough to maintain the shelf without causing the metal to spill. Use currents in the lower portion of the range.

OVERHEAD—use a whipping technique with a slight circular motion in the crater to produce stringer beads without undercut. Do not use a weave. Point the electrode directly into the joint and tip slightly forward into the direction of travel. Use a fairly short arc. Travel fast enough to avoid spilling. Use currents in the lower portion of the range.

LESSON 3.3A

Object:
To study the special characteristics of Lincoln Fleetweld 7 electrodes.

Equipment:
Lincoln "Idealarc" AC and DC welder.

Material:
Scrap plate and 14 gauge sheet metal; 1/8", 3/16" and 1/4" Fleetweld 7 electrodes.

Fleetweld 7 is the original and most widely used electrode of the group. It is an E-6012 electrode and will operate on either DC or industrial AC welders. It is the best of the group for general purpose welding because, of those in the group, it operates best in the vertical and overhead positions as well as downhand. It is fastest for 90 degree vertical down welding of sheet metal lap and fillet welds. On production jobs, it is widely used on jobs with very poor fit-up. It operates better on AC than Fleetweld 72 and is faster than Fleetweld 37. For steep downhill work, it is better than Improved Fleetweld 47.

PHYSICAL PROPERTIES

	AS WELDED	STRESS RELIEVED
Tensile Strength	67,000 - 80,000 psi	66,000 - 81,000 psi
Yield Point	55,000 - 65,000 psi	49,000 - 62,000 psi
Elongation	17 - 29%	24 - 28%
Impact Resistance (Charpy V Notch at Room Temperature)	62 Foot pounds	
Transition Temperature	-25 to -50°F	

Job Instructions

ALL POSITIONS		
Rod Size		Amperage Range
Dia.	Length	
3/32"	12"	25– 90
1/8"	14"	55–140
5/32"	14"	90–200
3/16"	14"	120–275
7/32"	18"	140–325
1/4"	18"	175–500
5/16"	18"	240–625

1. Make fillet and lap welds on 14 gauge sheet metal, both in the downhand and 60 degree downhill positions, using 1/8" electrodes. Note the speed of welding and the outstanding fast follow ability of this electrode. Fleetweld 7 is the fastest electrode for fast follow applications which are welded downhill at an angle of greater than 60 degrees.
2. Repeat the above on heavier stock with 3/16 inch electrode. Vary the angle of incline and, for each position, make as large a weld as possible. Note that the steeper the angle, the smaller the bead must be in order to stay ahead of the molten pool.
3. Intentionally prepare a fillet joint with poor fit-up. Weld this joint with Fleetweld 7. Note that it is possible to make reasonably fast welds without the metal spilling through the joint. If the fit-up is poor enough, it may be

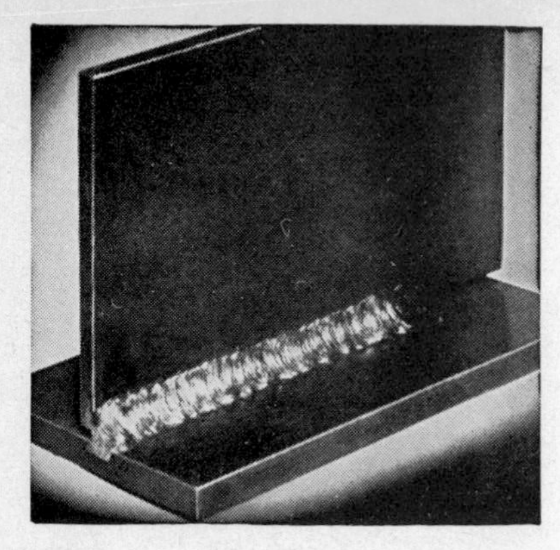

Fig. 1. Downhand fillet weld with poor fit-up (gap of ⅛-in.) in ¼-in. plates welded with "Fleetweld 7."

necessary to use a slight weave on the first pass. Usually it is preferable to use a stringer bead to be sure that adequate penetration is obtained. The ability to weld poor fit-up joints with Fleetweld 7 is based on its ability to deposit a fairly fast freezing weld.

4. Use ¼ inch Fleetweld 7 with high currents on heavier plate with AC to eleminate arc blow. On AC, Fleetweld 7 operates considerably better than Fleetweld 72. Become familiar with the operation of Fleetweld 7 on AC because its operation will be compared with that of other electrodes in the following lessons.

5. Using ⅛ inch Fleetweld 7, make vertical down, vertical up, and overhead welds with the procedures described above. Note that it operates best vertical down, but that vertical up and overhead welds may be made with reasonable ease. Fleetweld 7 is the most versatile out-of-position electrode of the "Fill-Freeze" group, but lacks the complete flexibility of the "Fast Freeze" electrodes on applications of this type.

Fig. 2. Fleetweld 7 is widely used for structural fabrication.

LESSON 3.3B

Object:
To study the special characteristics of Lincoln Fleetweld 77 electrodes.

Equipment:
Lincoln "Idealarc" AC and DC welder.

Material:
Scrap plate and 14 gauge sheet metal; 1/8",3/16" and 1/4" Fleetweld 77 electrodes.

Fleetweld 77 performs like the other E6012 electrodes but has a softer, quieter arc characteristic and therefore greater "operator appeal." Bead shape and appearance are tops for this type of electrode. Fleetweld 77 has better than normal slag control. This plus good fast-follow ability makes the electrode suitable for downhill lap and fillet welds on light gauge materials up to 90° vertical down. Either AC or DC current can be used, but Fleetweld 77 will not match the AC performance of E6013 types.

PHYSICAL PROPERTIES

	AS WELDED
Tensile Strength	67,000 - 78,000 psi
Yield Point	55,000 - 68,000 psi
Elongation	17 - 22%

Job Instructions

1. Repeat the job instructions for Fleetweld 7 and note the following:
   (a) Superior appearance and shape of the weld bead and better slag control.
   (b) The softer, quieter arc made the electrode easier to handle.
   (c) Less spatter and undercut than Fleetweld 7 at any given current setting.
   (d) Good operation when making downhill welds up to 90° vertical down.

ALL POSITIONS		
Rod Size		Amperage Range
Dia.	Length	
1/8"	14"	85-135
5/32"	14"	110-180
3/16"	14"	155-250
7/32"	18"	225-295
1/4"	18"	245-325

LESSON 3.3C

Object:

To study the special characteristics of Lincoln Fleetweld 37 electrodes.

Equipment:

Lincoln "Lincwelder" low open circuit voltage (ocv), limited input AC welder, or other low ocv AC welder and accessories.

Material:

Scrap plate and 14 gauge sheet metal; 1/8" and 3/16" Fleetweld 37 electrodes.

Fleetweld 37 is an E-6013 electrode with better operation on AC than either of the E-6012 electrodes. It also has a softer arc and smoother bead with slightly more slag and less spatter. Fleetweld 37 is widely used on sheet metal lap and fillet welds where appearance and ease of operation are more important than welding speed. It can be used on low open circuit voltage (ocv), limited input welders where general purpose welding is its principal application.

PHYSICAL PROPERTIES

	AS WELDED	STRESS RELIEVED
Tensile Strength	67,000 - 80,000 psi	66,000 - 81,000 psi
Yield Point	55,000 - 65,000 psi	49,000 - 62,000 psi
Elongation	17 - 29%	24 - 28%
Impact Resistance (Charpy V Notch at Room Temperature)	57 Foot pounds	

Job Instructions

1. Make sheet metal lap and fillet welds with 1/8" Fleetweld 37 similar to those made with Fleetweld 7 and Fleetweld 72. Note that the bead is smoother and the arc is softer with less spatter, but that it is slower than the other electrodes.

ALL POSITIONS		
Rod Size		Current Range
Dia.	Length	
5/64"	9"	40–75
3/32"	12"	65–100
1/8"	14"	90–140
5/32"	14"	120–190
3/16"	14"	150–240

2. Make welds in all positions with Fleetweld 37. Note that, like the E-6012 electrodes, it operates best downhand and vertical down, but can also be used on vertical and overhead joints with reasonable ease. This ability qualifies Fleetweld 37 as a general purpose electrode, particularly where low OCV, limited input welders are used.
3. Compare Fleetweld 37 and Fleetweld 7 or Fleetweld 72 for stability and ease of operation on low OCV welders. Note that Fleetweld 37 is markedly superior.

LESSON 3.3D

Object:
To study the special characteristics of Lincoln Improved Fleetweld 47 electrodes.

Equipment:
Lincoln "Idealarc" or other AC or DC welder and accessories.

Material:
Scrap plate; 1/8", 3/16" and 1/4" Improved Fleetweld 47 electrodes.

Improved Fleetweld 47 is an E-6014 electrode. Iron powder in its coating produces operating characteristics that are very similar to those of the "fast fill" electrodes. However, it retains sufficient "fast freeze" ability to operate in all positions, though it is seldom used in the vertical or overhead positions. On those few applications, where it is used vertical, the vertical up procedure is generally preferred. Its principal use is on downhill production welding applications where the "fast fill" electrodes would be used if the joint were in the flat position. It can be used on joints with moderately poor fitup.

PHYSICAL PROPERTIES

	AS WELDED
Tensile Strength	67,000 - 80,000 psi
Yield Point	55,000 - 70,000 psi
Elongation	17 - 29%
Impact Resistance (Charpy V Notch at Room Temperature)	73 Foot pounds
Transition Temperature	-25 to -50°F

Job Instructions

ALL POSITIONS		
Rod Size		Amperage Range
Dia.	Length	
1/8"	14"	125–150
5/32"	14"	150–210
3/16"	14"	190–270
7/32"	18"	280–350
1/4"	18"	350–420
5/16"	18"	400–500

1. Make downhand fillet and lap welds with Improved Fleetweld 47. Note the outstanding speed as compared with other "Fill-Freeze" electrodes.
2. Make downhill fillet and lap welds with varying degrees of incline. Note that where it can be used it is faster than other electrodes of this group. Note, also, that it cannot be operated on as steep an incline. Many production welding jobs contain some welding which must be done at an angle of 10-20 degrees downhill, and it is on these jobs where speed is important that Improved Fleetweld 47 is widely used.
3. Note on all of the above applications the ease of operation and excellent appearance of the weld. Improved Fleetweld 47 is preferred by many operators for these reasons.
4. Weld vertical up and overhead with Improved Fleetweld 47.

LESSON 3.3E

Object:

To study the special characteristics of Lincoln Fleetweld 7MP electrodes.

Equipment:

Lincoln Idealarc, or other AC or DC welder.

Material:

Scrap plate and 14 gauge sheet metal; 1/8, 5/32 and 3/16 inch Fleetweld 7MP electrode.

Fleetweld 7MP, an E-6012 electrode, maintains the speed and versatility of Fleetweld 7, yet has Fleetweld 37's ease of operation, smooth bead shape and ability to operate on AC. It contains metal powder in the coating and actually welds faster than Fleetweld 7. It is capable of welding in all positions, though it is best employed on joints positioned not more than 60 degrees downhill. It is used for general purpose welding, sheet metal production welding, and production welding on joints which might use improved Fleetweld 47, except for downhill or poor fitup conditions.

PHYSICAL PROPERTIES	AS WELDED
Tensile Strength	67,000 - 75,000 psi
Yield Strength	55,000 - 64,000 psi
Elongation in 2 inches	17 - 25%

Job Instructions

1. Make fillet and lap weld on 14 gauge sheet metal, positioned 90 degrees downhill, 60 degrees downhill, and flat. Use 1/8 inch electrodes. Note that the electrode operates best in the 60 degree downhill and flat positions. Note that the speed and appearance of these welds are better than those obtained with Fleetweld 7.

2. Make fillet and lap welds on heavier material, positioned slightly downhill and fit up with a slight gap. Note that Fleetweld 7MP handles these conditions better than Improved Fleetweld 47.

3. Compare Fleetweld 7MP with Fleetweld 37 for sheet metal welding. Note that the speed of Fleetweld 7MP is substantially above that of Fleetweld 37, while the arc characteristics and bead appearance of Fleetweld 7MP is as smooth as that of Fleetweld 37.

ALL POSITIONS			
Rod Size		Amperage Range	
Dia.	Length	DC—	AC
1/8″	14″	90 - 150	100 - 150
5/32″	14″	120 - 210	130 - 210
3/16″	14″	155 - 290	165 - 290
7/32″	18″	230 - 305	250 - 325
1/4″	18″	250 - 350	275 - 375
5/16″	18″	325 - 445	350 - 470

LESSON 3.3F

Object:
To study the special characteristics of Lincoln Fleeweld 73 electrodes.

Equipment:
Lincoln "Idealarc" AC and DC welder.

Material:
Scrap plate and 14 gauge sheet metal; 3/16" Fleetweld 73 electrodes.

Fleetweld 73 has the best fast-follow ability. It is used on joints which are positioned less than 60 degrees downhill. An E-6012 electrode, it is very similar to Fleetweld 7, but has better fast follow ability with a smaller bead, greater mileage (length of weld per electrode consumed), and a higher optimum current on these applications. It has slightly less arc force and fast freeze ability than Fleetweld 7, which makes it less desirable on vertical and overhead applications. Its operation on DC is notably better than on AC.

PHYSICAL PROPERTIES

	AS WELDED
Tensile Strength	67,000 - 72,000 psi
Yield Point	55,000 - 62,000 psi
Elongation	17 - 27%

Job Instructions

ALL POSITIONS		
Rod Size		Amperage Range
Dia.	Length	
5/32"	14"	175-300
3/16"	14"	200-350
7/32"	18"	250-450

1. Repeat the job instruction for Fleetweld 7 and note the Fleetweld 73 has the following:
   (a) Outstanding fast follow ability with small bead, greater mileage and fast speeds on fast follow applications where DC is used and the work is positioned at an incline of less than 60 degrees.
   (b) Higher optimum currents are used than with other electrodes of this group.
   (c) Less spatter than Fleetweld 7 at any given current.
   (d) Better operation on downhill and downhand applications than on vertical or overhead joints.

LESSON 3.4

Object:

To study the use and applications of Lincoln "FAST FILL" electrodes.

General Information

Lincoln "FAST FILL" electrodes are best suited for production welding of downhand joints where it is desired to deposit a specified amount of metal in the shortest possible time, and for exceptionally easy downhand welding by inexperienced welders.

Included in this group are Jetweld 1, Jetweld 2, Jetweld 2–HT. See Lesson 1.17 for further information on iron-powder type electrodes.

Operating Characteristics

Deposition: FAST FILL—exceptionally high deposition rates are possible with these electrodes.

Arc: SOFT—has little force or penetration and negligible spatter.

Slag: HEAVY—very fluid and covers the weld completely.

Bead: FLAT—or slightly concave with a smooth, glossy appearance.

Current: AC—DC may also be used, but is less desirable.

Positions: DOWNHAND—may also be used slightly downhill.

Application Information

The industrial fast fill electrodes are used because of their very high deposition rates, excellent appearance and ease of operation. The group is limited to operation in the downhand positions.

1. FLAT AND HORIZONTAL FILLETS on 1/4 inch and heavier material are made with Jetweld 1, which produces fast, smooth welds with easy slag removal. Where X-ray requirements must be met, Jetweld 2 is used.
2. DEEP GROOVE WELDS IN THE FLAT POSITION are best made with Jetweld 2, which washes well into the sides of the joint and has an extremely friable slag which is easily removed and will not lock in the joint.
3. MEDIUM CARBON CRACK-SENSITIVE STEELS are readily welded with Jetweld 1. Its light penetration and high physical properties permit fastest welding without cracks.

Procedures

FLAT—Use a drag technique to deposit stringer beads. Hold the electrode perpendicular to the work and tipped forward into the direction of travel approximately 30 degrees. Travel fast enough to stay 1/4 to 3/8 inch ahead of the molten pool. Avoid high currents, especially on X-ray quality work.

HORIZONTAL FILLETS—The procedure is the same as that described for FLAT, except that the electrode is pointed directly into the joint at an angle of 45 degrees touching both legs of the joint instead of being held perpendicular.

LESSON 3.4A

Object:
To study the special characteristics of Lincoln Jetweld 1 electrodes.

Equipment:
Lincoln "Idealarc" AC and DC welder.

Material:
Scrap 1/4" or heavier plate, including some of high carbon content; 3/16" Jetweld 1 electrodes.

An E-6024 electrode, Jetweld 1 is very different from the fast freeze or fast follow electrodes; the arc is very soft and hardly visible, the travel speed is astonishing, and the heavy slag curls up by itself without any chipping to reveal a bead which looks as if it had been made with an automatic submerged arc machine. It operates best on AC, though DC, electrode positive, may also be used. Compared with Jetweld 2, it has different type of slag, operates better on fillet welds, produces a slightly smoother bead, has higher physical properties, and can be used on crack-sensitive steels.

Fig. 1. Horizontal fillet weld. Plate size is 3/8", fillet size 1/4", "Jetweld 1" electrode size, 3/16". Current is 275 amps A.C., arc speed 15 inches per minute.

Fig. 2. "Jetweld 1" bead made on flat plate was bent to show ductility. Notice the smoothness of the weld and its freedom from face cracks.

PHYSICAL PROPERTIES	AS WELDED
Tensile Strength	67,000 - 90,000 psi
Yield Point	55,000 - 86,000 psi
Elongation	17 - 25%
Impact Resistance (Charpy V Notch at Room Temperature)	57 Foot pounds
Transition Temperature	-50 to -100°F

FLAT OR HORIZONTAL POSITIONS					
Rod Size		Amperage Range		Recommended Amps. for Fillets	
Dia.	Length	D.C.+	A.C.	D.C.	A.C.
3/32″	12″	80–110	80–110	90	90
1/8″	14″	130–190	130–190	170	175
5/32″	14″	190–250	190–250	200	225
3/16″	18″	220–280	240–300	250	275
7/32″	18″	290–360	300–380	325	350
1/4″	18″	330–400	360–440	350	380
5/16″	18″	460–550	520–620	500	550

Job Instructions

1. Make downhand fillet and lap welds with Jetweld 1. Note the characteristics described above.

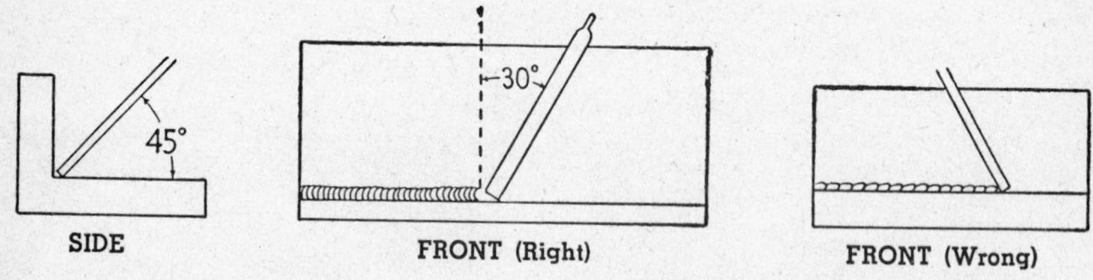

Fig. 3. Electrode position for making fillet welds with Jetweld 1.

2. Repeat the above several times, each time slightly increasing the amount of incline of the joint to determine the limits of this electrode for downhill welding.
3. Make welds on crack-sensitive steels to note that such welds can be made at high speeds with Jetweld 1 without danger of cracking. Jetweld 1's ability to weld crack-sensitive steels approaches that of the low hydrogen electrodes.
4. Note on all the above that it is important to stay ahead of the molten pool in order to maintain proper bead shape. Note also that excessive currents tend to make the slag stick to the weld, and, though it is not visible, may reduce weld quality by causing the weld to have fine, internal porosity.
5. Use Jetweld 1 with both AC and DC, electrode positive. Note that it is easier to maintain proper bead shape with AC, while DC tends to make the bead slightly humped in the middle, or stringy.
6. Use Jetweld 1 for intermittent welds and note the ease of restriking.

LESSON 3.4B

Object:
To study the special characteristics of Lincoln Jetweld 2 electrodes.

Equipment:
Lincoln "Idealarc" AC and DC welder.

Material:
Bevelled scrap 3/8" or heavier plate; 3/16" Jetweld 2 electrodes.

Jetweld 2 is similar to Jetweld 1, but is different in that it washes up better in a deep groove, has a slag which crumbles for easy removal, and produces a weld of X-ray quality. It is an E-6027 electrode and operates best on AC, though DC, electrode negative, may also be used.

PHYSICAL PROPERTIES

	AS WELDED
Tensile Strength	**62,000 - 72,000 psi**
Yield Point	**50,000 - 64,000 psi**
Elongation	**25 - 30%**
Impact Resistance (Charpy V Notch at Room Temperature)	**73 Foot pounds**
Transition Temperature	**-100 to -125° F**

Electrode Size		Overall Range Amps.	Max. Amps. for X-Ray	Recommended Amps. for Fillets
Dia.	Length			
5/32"	14"	200–275	240	220
3/16"	14"	250–350	300	270
7/32"	18"	275–375	350	310
1/4"	18"	350–450	400	350

Job Instructions

1. Make deep groove butt welds with Jetweld 2 and note the characteristics described above.
2. Make fillet and lap welds similar to those made with Jetweld 1. Note that, while Jetweld 1 is preferred, Jetweld 2 is readily applied and can be used on joints of this type where X-ray quality is required.
3. Weld with Jetweld 2 using both AC and DC, electrode negative. Note that the same results are obtained as were noted for Jetweld 1.

LESSON 3.4C

Object:
To study the special characteristics of Lincoln Jetweld 2-HT electrodes.

Equipment:
None.

Material:
None.

Jetweld 2—HT is identical to Jetweld 2 in operation and appearance, and differs only in physical properties. It is an E-7020-A1 electrode for use on low alloy, high tensile steel. It produces welds of X-ray quality and is applied in exactly the same manner as Jetweld 2.

PHYSICAL PROPERTIES

	STRESS RELIEVED
Tensile Strength	70,000 - 79,000 psi
Yield Point	57,000 - 63,000 psi
Elongation	22-27%

Job Instructions

1. If the "Job Instructions" for Jetweld 2 have been accomplished, there is no need to repeat them for Jetweld 2—HT. Simply note that it is identical in application, but is used on higher tensile steels. Complete "Job Instructions" for Jetweld 2 if not previously accomplished.

Electrode Size	Overall Range Amps.	Max. Amps. for X-Ray	Recommended Amps. for Fillets
5/32″x14″	200–275	240	220
3/16″x18″	250–350	300	270
7/32″x18″	275–375	350	310
1/4″x18″	350–450	400	350

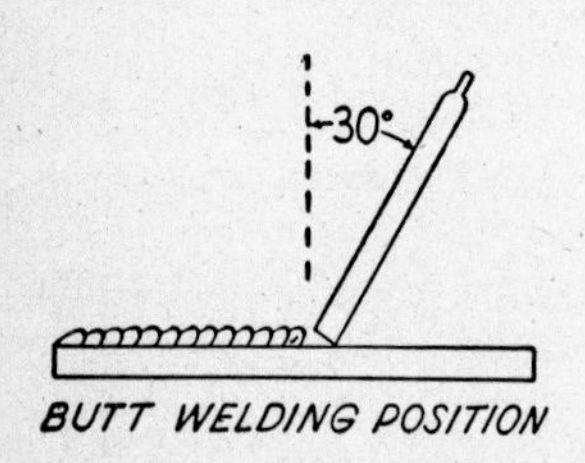

Fig. 1. Electrode position for Jetweld 2 and Jetweld 2—HT.

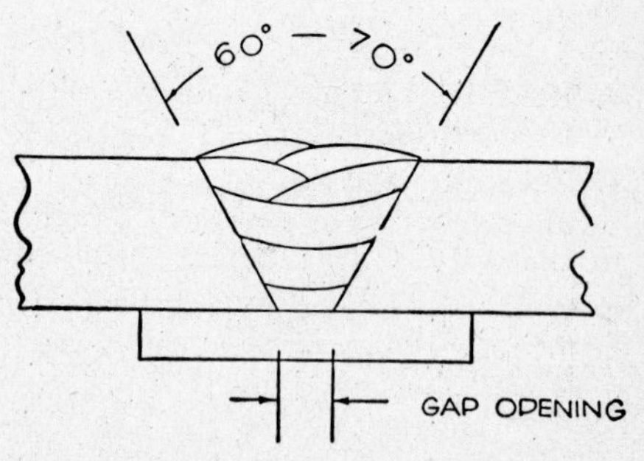

Fig. 2. Joint preparation for deep groove welds.

LESSON 3.5

Object:

To study the use and application of Lincoln low hydrogen electrodes for hard-to-weld steels and alloy steels.

General Information

Certain types of steel are crack sensitive in that the heating and cooling of welding tends to create cracks. These are medium and high carbon steels, high sulphur and low-alloy steels. Certain types of cracks in these steels are directly traceable to hydrogen in the weld deposit. Low hydrogen electrodes produce hydrogen free welds eliminating the cracking. They also prevent the formation of gases, such as hydrogen sulphide, which cause either internal or surface porosity. Consequently, deposits are dense with exceptional ductility. (See Page 1-94 for further information.)

Lincoln low hydrogen electrodes are available in three tensile strengths. The operating characteristics of all three are similar, so that they differ only in physical properties and deposit analysis. Iron powder in their coatings produces higher deposition rates than that of low hydrogen electrodes without iron powder.

Application Information

The ability to satisfactorily weld hard-to-weld steels is the outstanding ability of these electrodes. They produce porosity free and micro-crack free welds and reduce the amount of preheat required on heavier sections.

Low hydrogen coatings substantially improve ductility by depositing dense, sound welds. This is particularly valuable in the higher strength steels.

Low hydrogen electrodes generally reduce the amount of preheat required by 300 degrees F. Also, they weld on phosphorus or sulphur bearing steels without "hot-short" difficulties or surface holes.

Applications of these electrodes utilize the above characteristics of the low hydrogen family. Which particular electrode is used on any job depends principally on the tensile strength of the base metal. Jetweld LH-70 and LH-3800 are use on mild steels and steels with 70,000 psi tensile stength, while the other electrodes are used on higher tensile steels. Typical application include:

1. ALL TYPES OF JOINTS ON HARD-TO-WELD STEELS, including high tensile steels. The low hydrogen characteristics reduce or eliminate the need for preheat. They produce porosity free welds on high sulphur material and prevent "hot-shortness" in phosphorus bearing steels.
2. LARGE WELDS ON HEAVY PLATE, where shrinkage stresses tend to cause cracks at the root of the weld. Bead shape of the Jetweld LH electrodes, plus high physical properties produce sound welds and reduce the need for stress relief.
3. PRESSURE PIPING and other applications which require the specific deposit analysis available in Jetweld LH-90 or Jetweld LH-110.
4. WELDS WHICH ARE TO BE PORCELAIN ENAMELED, particularly if the weld is not to be annealed before enameling. Low hydrogen properties produce the best possible bonding between the porcelain and the weld metal.

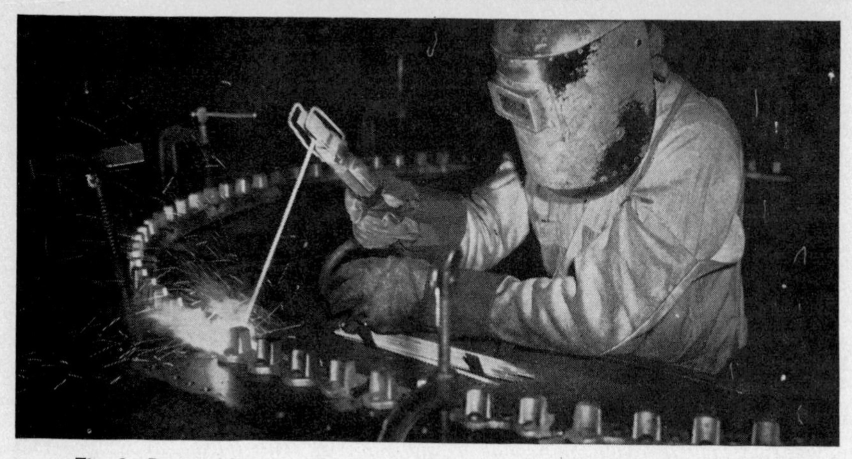

Fig. 1. Butt welding medium carbon steel sprocket forgings with "Jetweld."

Procedure

DOWNHAND—on the first pass, or wherever it is desired to reduce admixture with a base metal of poor weldability, use low currents. On succeeding passes, use currents as high as possible consistent with the plate thickness. Lightly drag the electrode or hold a very short arc. Do *not* use a long arc at any time, since this type of electrode relies principally on molten slag for shielding. Stringer beads or a small weave are preferred to wide weave passes. When starting a new electrode, strike the arc ahead of the crater, move back into the crater and then proceed in the normal direction. Use lower currents with DC than with AC. Electrode should point directly into the joint and tipped forward about 45 degrees into the direction of travel. Travel speed should be governed by the desired bead size.

VERTICAL—Weld vertical up. Use a triangular weave. Build a shelf of weld metal, and, with the weave, deposit layer upon layer of metal as the weld progresses up the joint. Do not use a whip technique or take the electrode out of the molten pool. Point the electrode directly into the joint and slightly upward to permit the arc force to assist in controlling the puddle. Travel slow enough to maintain the shelf without causing the metal to spill. Use currents in the lower portion of the range.

OVERHEAD—Use a slight circular motion in the crater. Maintain a short arc with the whip. Do not use a weave. Point the electrode directly into the joint and tip it slightly forward into the direction of travel. Travel fast to avoid spilling weld metal, but do not be alarmed if the slag spills some. Use currents in the lower portion of the range.

LESSON 3.5A

Object:

To study the special characteristics of Lincoln Jetweld LH-70 electrodes.

Equipment:

Lincoln Idealarc, or other AC or DC welder.

Material:

Scrap high sulphur and high carbon steel; 5/32" Jetweld LH-70 electrodes.

Jetweld LH-70, an E-7018 electrode, is a 70,000 pound tensile strength electrode of the Lincoln low hydrogen electrode family. It is also suitable for 60,000 psi tensile mild steels and meets AWS E-6018 specifications. It is the workhorse of the Lincoln family of low hydrogen electrodes and can be used on all low hydrogen applications except those requiring the physical or chemical properties of the Jetweld LH-90 or LH-110.

PHYSICAL PROPERTIES	AS WELDED	STRESS RELIEVED
Tensile Strength	70,000 - 82,000 psi	65,000 - 77,000 psi
Yield Point	60,000 - 72,000 psi	55,000 - 67,000 psi
Elongation	28 - 33%	30 - 35%
Impact Resistance (Charpy V Notch at Room Temperature)	100 foot pounds	120 foot pounds

Job Instructions

1. Weld on high sulphur, nut stock, scrap steel with both Jetweld LH-70 and Fleetweld 5. Note that Jetweld LH-70 eliminates the porosity encountered with the conventional electrodes. This is because Jetweld LH-70 does not have hydrogen in its coating and so will not form the hydrogen sulphide gas which causes the porosity with conventional electrodes.

2. Weld with both Jetweld LH-70 and Fleetweld 37 on heavy stock, high carbon steel. Note that the Jetweld LH-70 eliminates cracks encountered with Fleetweld 37. This is, again, due to the low hydrogen content of the coating.

ALL POSITIONS			
Rod Size		Amperage Range	
Dia.	Length	DC+	AC
3/32"	12"	70-100	80-120
1/8"	14"	95-155	115-165
5/32"	14"	120-200	145-230
3/16"	14"	160-280	200-310
7/32"	18"	190-310	240-350
1/4"	18"	230-360	290-410
5/16"	18"	375-475	400-500

LESSON 3.5B

Object:

To study the special characteristics of Lincoln Jetweld LH-90 electrodes.

Equipment:

Lincoln Idealarc, or other AC or DC welder.

Material:

1¼% chromium, ½% molybdenum scrap pipe or plate; 5/32 inch Jetweld LH-90 electrode.

Jetweld LH-90, an E-9018 electrode that also meets the requirements of E-8018-G, is a 90,000 psi tensile strength electrode of the Lincoln low hydrogen electrode family. It has a nominal alloy content of 1¼% chromium and ½% molybdenum, and is intended for pressure piping or general fabrication where the joint requires either the specific alloy or the high strength deposits of this analysis. Its operating characteristics are similar to those of Jetweld LH-70.

PHYSICAL PROPERTIES	AS WELDED	STRESS RELIEVED
Tensile Strength	97,700	95,000
Yield Strength	85,800	82,000
Elongation in 2 inches	24%	24%
Impact Resistance (Charpy V-Notch at Room Temperature)	65 foot pounds	

Job Instructions

1. Use Jetweld LH-90 on pressure piping to become familiar with the characteristics of this electrode when welding pipe. Generally follow the procedures for vertical-up welding as outlined in Lesson 3.5 with modifications to compensate for the pipe joint.

2. Weld downhand, vertical and overhead on 90,000 psi tensile scrap plate to become familiar with the operating characteristics of the electrode.

ROD SIZE		CURRENT RANGE	
Dia.	Length	DC+	AC
1/8″	14″	90-150	110-160
5/32″	14″	110-200	140-230
3/16″	14″	160-280	200-310

LESSON
3.5C

Object:
To study the special characteristics of Lincoln Jetweld LH-110 electrode.

Equipment:
Lincoln Idealarc, or other AC or DC welder.

Material:
Scrap high tensile steel; 5/32 Jetweld LH-110 electrode.

Jetweld LH-110 is the 110,000 psi tensile strength electrode of the Lincoln low hydrogen electrode family. It is classified E-11018-G. It is particularly intended for welding the new high tensile steels as well as any other application where high strength weld metal is necessary. Its operating characteristics are similar to those of Jetweld LH-70.

PHYSICAL PROPERTIES	AS WELDED	STRESS RELIEVED
Tensile Strength	128,000 psi	119,000 psi
Yield Strength	117,000 psi	103,000 psi
Elongation in 2 inches	15-18%	15-18%
Impact Resistance (Charpy V-Notch at Room Temperature)	53 foot pounds	

Job Instructions

1. Weld with Jetweld LH-110 on all types of joints in scrap HY-80 or T-1 steels to become familiar with its operation. Note that its operating characteristics are very similar to those of Jetweld LH-70 and Jetweld LH-90, but that it is used on applications where its high tensile strength is required.

ROD SIZE		CURRENT RANGE	
Dia.	Length	DC+	AC
1/8″	14″	90-150	110-160
5/32″	14″	110-200	140-230
3/16″	14″	160-180	200-310
7/32″	18″	190-310	240-350
1/4″	18″	230-360	290-410

LESSON 3.5D

Object:
To study the special characteristics of Lincoln Jetweld LH-3800.

Equipment:
Lincoln "Idealarc" AC and DC welder.

Material:
Scrap high sulphur and high carbon steel plate; 3/16" Jetweld LH-3800 electrodes.

Jetweld LH-3800 is an E-7028 iron powdered low hydrogen electrode. It is particularly suited to making groove and fillet welds in the flat position and other production welding applications where the high speed and high deposition rate can be effectively used to lower welding costs. AC current is preferred with DC (+) and 10 percent lower current the second choice. Jetweld LH-3800 is limited to making welds in the flat and horizontal positions only.

PHYSICAL PROPERTIES	AS WELDED	STRESS RELIEVED
Tensile Strength	85,000-95,000 psi	84,000-89,000 psi
Yield Point	75,000-79,000 psi	73,000-77,000 psi
Elongation	24-26%	25-27%
Impact Resistance (Charpy V Notch at Room Temperature)	75-80 foot pounds	80-85 foot pounds

Job Instructions

ROD SIZE		CURRENT RANGE	
Dia.	Length	DC (+)	AC
5/32"	18"	170-240	180-270
3/16"	18"	210-300	240-330
7/32"	18"	260-380	275-410
1/4"	18"		360-520

1. Weld on high sulphur and high carbon scrap stock as you did with Jetweld LH-70. Note that Jetweld LH-3800 duplicates the performance advantages previously observed when performing the job instructions 1. and 2. for Jetweld LH-70.
2. Make a flat position weld with Jetweld LH-3800 and then Jetweld LH-70. Note the difference in deposition rate and welding speed. Also observe that Jetweld LH-3800 developes a larger weld crater and produces more slag. These characteristics limit its use to flat and horizontal position welds.

SECTION IV

STAINLESS STEEL ELECTRODES

LESSON 4.1

Object:
To study the use and application of Lincoln lime coated stainless steel electrodes.

Equipment:
Lincoln "Idealarc" or other DC welder and accessories.

Material:
Scrap stainless steel plate, manganese steel casting, and mild steel plate; ⅛" and 3/16" Stainweld 308-15 electrodes and 3/16" Stainweld 310-15 electrodes.

General Information

Lincoln lime coated electrodes are made in a variety of analyses for all position use on stainless steel applications.

Characteristics

The operating characteristics of all Lincoln lime coated stainless steel electrodes are quite similar. A fairly forceful arc with deep penetration and light slag produce a FAST-FREEZE ability which permits good operation on out-of-position joints.

Type:	E308-15—Stainweld 308-15	– 19-9 unstabilized
	E347-15—Stainweld 347-15	– 19-9 Columbium stabilized
	E310-15—Stainweld 310-15	– 25-20 unstabilized
Coating:	LIME TYPE	
Current:	DC—normally used on DC, electrode positive, though DC, electrode negative is sometimes used on sheet metal applications.	
Positions:	ALL—versatile all position electrodes.	

Application Information

Within the group of lime coated stainless electrodes the selection of a particular electrode for a given application depends entirely upon the alloy of the stainless base metal. The stainless steels for which each electrode is suited are given above. The stabilized electrodes may be used on either stabilized or

Fig. 1. Welding stainless steel liner in a pulp-mill smelter blow nozzle is representative of a typical stainless steel electrode application.

Fig. 2. Corner weld in 18-8 stainless steel welded with "Stainweld 308-15". Left-hand portion is ground. Note smoothness.

unstabilized base metal, while the unstabilized electrodes are used only on unstabilized base metal.

When welding a stainless steel to a mild steel it is best to use a stainless electrode with higher alloy than the material being welded. This additional alloy will compensate for the loss in alloy caused by admixture with the mild steel.

Out of position welding is a specialty of the lime coated electrodes where their fast-freeze ability is most desirable.

Strength welds in manganese steel are best made with the higher alloy electrodes Stainweld 310-15 is most widely used.

Hardsurfacing buildup is usually done with Stainweld 308-15. On such applications the stainless may be used as the only surfacing material to prevent wear or corrosion, or it may be used as a base between the original base metal and more abrasion resistant material such as one of the Faceweld electrodes. In the latter case, the ductile stainless acts as a cushion to absorb the shock of impact loads and prevent the harder surfacing material from spalling off.

These electrodes may be used on the following AISI types of stainless steels:

Stainweld 308-15 — Types 301, 302, 304, and 308 unstabilized stainless steels.

Stainweld 347-15 — Types 301, 302, 304, and 308 unstabilized stainless steels, plus types 321 and 347 stabilized stainless steels and the Extra Low Carbon (ELC) 18-8 stainless steels.

Stainweld 310-15 — Type 310.

Procedure

All of these electrodes operate on DC, electrode positive (Reverse polarity).

Three characteristics peculiar to stainless steel make welding with stainless steel electrodes different than welding with mild steel; higher electrical resistance, lower thermal conduction, and higher thermal expansion. This means that the electrode gets hotter, the molten pool cools more slowly, and

FLAT VERTICAL & OVERHEAD		
Rod Size		Amperage Range
Dia.	Length	
5/64"	9"	20– 55
3/32"	9"	30– 70
1/8"	14"	50–100
5/32"	14"	75–130
3/16"	14"	95–165
1/4"	14"	150–225

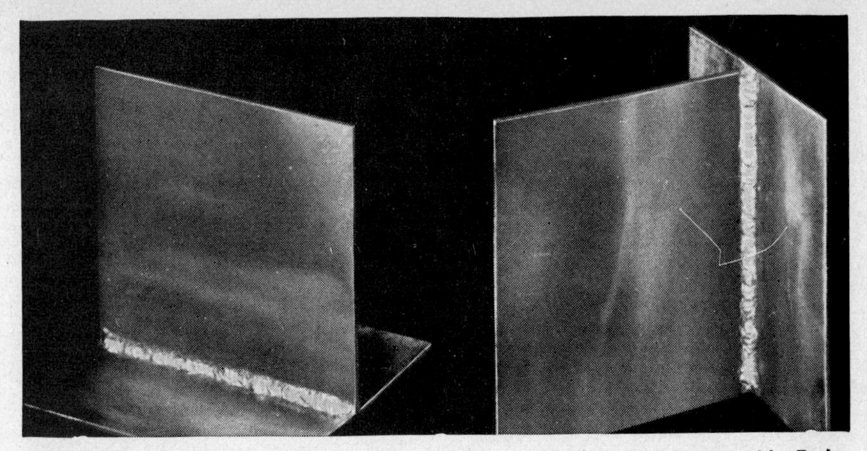

Fig. 3. Fillet welds in 16 gauge 25-12 stainless steel. Left: Downhand weld. Right: Vertical weld, welded upward.

distortion is a greater problem. To counteract these characteristics less current is used, greater care must be taken in cleaning and fitting the joint, and provisions must be made to take the heat away on light sheet metal by such means as copper chill bars placed under the joint.

FLAT–Use as short an arc as possible without choking or sticking. Use stringer beads for all passes and avoid weaving. Use currents as low as possible consistent with good arc action and proper fusion.

VERTICAL AND OVERHEAD–Use only stringer beads made with a slight whip and a circular motion in the crater. Use currents as low as possible.

PROPERTIES

"STAINWELD 347-15":

Average chemical analysis of all-weld metal deposit. "Stainweld 347-15", .065% carbon, 19.7% chrome, 9.5% nickel, .80% columbium.

As Welded:

Tensile strength—95,000 lbs. per sq. in.
Elongation in 2 inches—37%.

"STAINWELD 308-15":

Average chemical analysis of all-weld metal deposit. "Stainweld 308-15", .060% carbon, 19.7% chrome, 9.5% nickel.

As Welded:

Tensile strength—87,000 lbs. per sq. in.
Elongation in 2 inches—40%.

"STAINWELD 310-15":

Average chemical analysis of all-weld metal deposit: .13% carbon, 25.8% chrome, 20.4% nickel.

As Welded:

Tensile strength—87,000 lbs. per sq. in.
Elongation in 2 inches—40%.

Job Instructions

1. Make welds in the flat position. Note how the electrode has a tendency to overheat and become red. Reduce the current until the arc becomes sticky. The proper setting for the electrode is one which will just maintain the arc smoothly without overheating the electrode.
2. Make welds in the vertical and overhead position to become familiar with the procedure.

3. Weld on sheet metal. Note that the best results are obtained if a copper chill bar is placed behind the joint.
4. Join two pieces of manganese steel with Stainweld 310-15 electrodes. When welding on manganese steel it is important to keep the base metal cool; each bead should be allowed to cool enough so that it can be touched with the bare hand before another bead is applied. If the manganese is allowed to stay at a high temperature for very long it will become very hard and may crack either during welding or in service.
5. Build a pad of surfacing with Stainweld 308-15. Note that this deposit is very tough and will stand severe impact. A layer of hardsurfacing material on top of the stainless pad will increase the abrasion resistance of the surface without materially reducing the impact resistance.

Fig. 4. Milk cooler of 25-12 stainless steel in process of construction.

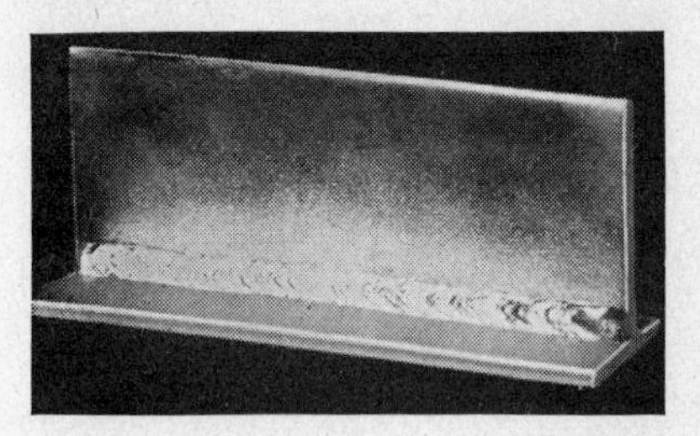

Fig. 5. Fillet weld on 3/16-inch 18-8 SMO stainless steel plate.

LESSON 4.2

Object:
To study the use and application of Lincoln titania-coated stainless steel electrodes.

Equipment:
Lincoln "Idealarc" AC or DC welder and accessories.

Material:
Light and heavy gauge scrap stainless steel; 3/32 inch and 5/32 inch Stainweld 308-16 electrode.

General Information

Lincoln titania-coated stainless steel electrodes are for use with AC on 18-8 stainless steels.

Characteristics

The titania-coated electrodes have a less forceful arc than the lime-coated stainless electrodes with medium penetration and a smoother bead shape.

Type:	E308-16—Stainweld 308-16—19-9 unstabilized. E347-16—Stainweld 347-16—19-9 Columbium stabilized. E308L-16—Stainweld 308L-16—19-9 Extra Low Carbon. E316L-16—Stainweld 316L-16—18-12Mo Extra Low Carbon. E310-16—Stainweld 310-16—25-20 unstabilized.
Coating:	TITANIA TYPE.
Current:	AC & DC—Electrode negative polarity is used with DC. On AC, industrial type welders are preferred.
Positions:	ALL—Is capable of operation in all positions, though downhand welds are preferred.

Application Information

The titania-coated stainless steel electrodes are principally used where AC is the power source. Because of their softer arc and smoother bead, they are sometimes preferred even when DC is available.

GENERAL PURPOSE WELDING on stainless steels, particularly where exceptionally smooth appearance is desired.

SURFACING APPLICATIONS either where it is desired to obtain a particularly tough and ductile or corrosion-resistant surface, or for intermediate layers between base metal and a hardsurfacing deposit.

Fig. 1. Corner joint in stainless steel edge band of gas range welded with "Stainweld 308-16." Grinding and buffing make weld indistinguishable.

These electrodes may be used on the following AISI types of stainless steels:

STAINWELD 308-16

Types 301, 302, 304, and 308 unstabilized.

STAINWELD 347-16

Types 301, 302, 304, and 308 unstabilized plus types 321 and 347 stabilized stainless steels.

STAINWELD 310-16

Types 310, stainless clad carbon steels **and** dissimilar alloy steel combinations.

PROCEDURE

The procedures used with the titania-coated electrodes are identical to those used with the lime-coated electodes (See Lesson 4.1)

FLAT, VERTICAL & OVERHEAD	
Electrode Sizes	Amperage Range
5/64″x 9″	20-45
3/32″x 9″	30-60
1/8″ x14″	55-95
5/32″x14″	80-135
3/16″x14″	115-185
1/4″ x14″	200-275

''STAINWELD 308-16'':

Average chemical analysis of ''Stainweld A7'' all weld metal deposit: .06% carbon, 19.7% chrome, 9.5% nickel, no columbium.

As Welded:

Tensile strength—85,000 to 95,000 lbs. per sq. in.
Elongation in 2 inches—35% to 50%.

''STAINWELD 347-16'':

Average chemical analysis of ''Stainweld A7-Cb'': .065% carbon, 19.7% chrome, 9.5% nickel, .80% columbium.

As Welded:

Tensile strength—90,000 to 100,000 lbs. per sq. inch.
Elongation in 2 inches—35% to 50%.

''STAINWELD 310-16'':

Average chemical analysis of ''Stainweld D7'' all weld metal deposit: .14% carbon, 26.8% chrome, 21% nickel.

As Welded:

Tensile strength——87,000 lbs. per sq. inch.
Elongation in 2 inches—37%.

Job Instructions

1. Weld with Stainweld 308-16 using both AC and DC welders, and compare the operating characteristics of the electrode on both types of current.
2. Repeat job instructions No. 1 through 4 for Lesson 4.1 and compare the operation of Stainweld 308-16 with that of the lime-coated electrodes.

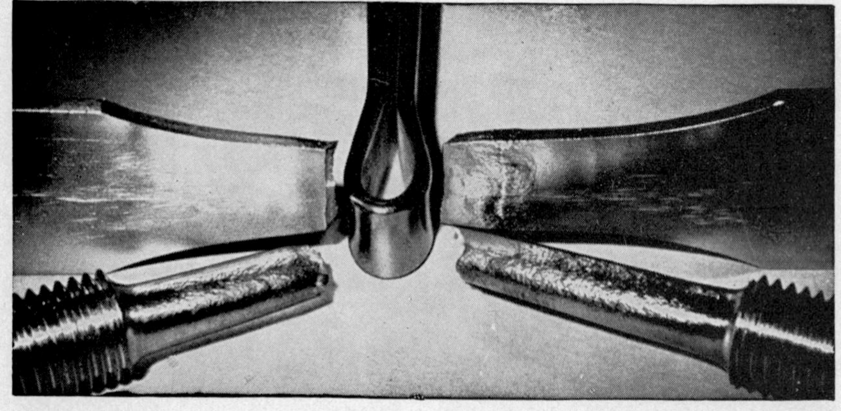

Fig. 2. Specimens of 18-8 stainless steel welded with "Stainweld 308-16." As welded: weld metal tensile strength 88,800 lbs./sq. in.; free bend elongation 53% in outer fibres; tensile pull sample failed in plate at 88,600 lbs. per sq. inch.

SECTION V

HARDSURFACING ELECTRODES

LESSON 5.1

Object:
To study the use and application of Lincoln Hardsurfacing electrodes.

General Information

The principles of hardsurfacing are discussed in Lesson 1.42. This lesson should be studied and thoroughly understood before studying the following lessons in Section V.

As with the mild steel electrodes discussed in Section III, the important thing in hardsurfacing electrode selection is to properly match the requirements of the application with the properties of the electrode. Also like mild steel electrodes, the characteristics of applications and electrodes are not clear-cut but are relative to each other. For any application there is one electrode which has the best combination of properties, but frequently it is necessary to actually test several electrodes before the best one is found.

A careful study of the application is the starting point for electrode selection. Consider the following: Must two surfaces be protected, or only one? Is a cut-

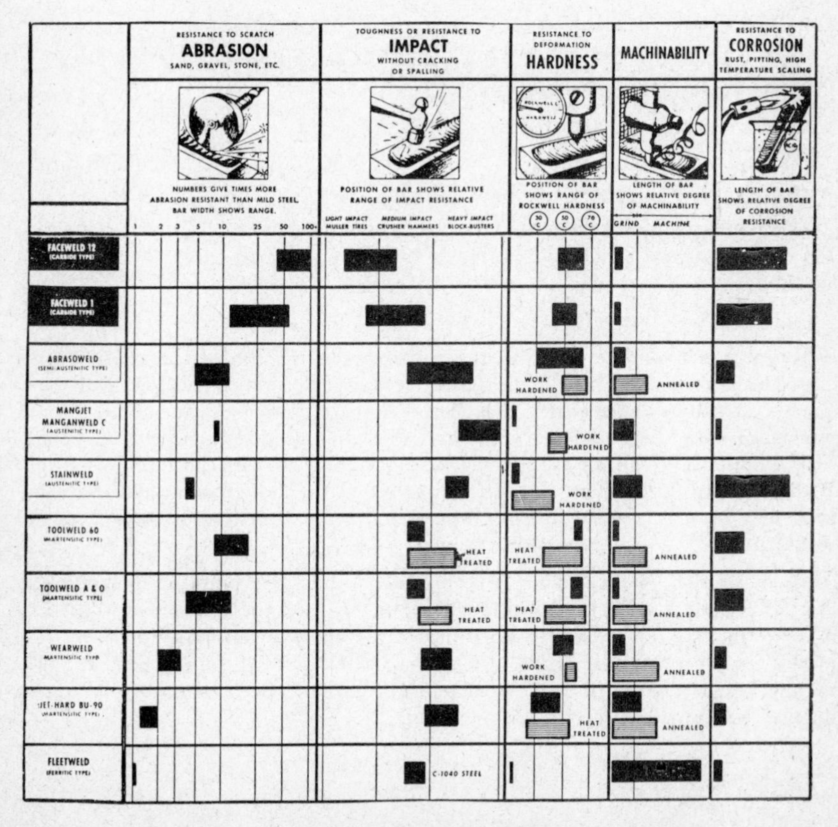

ting edge required? What is the relative importance of *abrasion, impact,* and *corrosion* on the wear of the part? How large is the part? What finish is required? and what is the composition of the part? The cost and available sizes of the electrodes may also have a bearing on the decision.

The following lessons are arranged to assist in matching the properties of the electrodes with the requirements of the job as determined by the careful study discussed above. In the "Deposit Properties" portion of the lesson is a chart, similar to the one shown on the previous page, which compares some of the basic characteristics of the electrodes. Other pertinent characteristics are discussed in the text of the "Application Information" portion.

The procedure used to deposit the hardsurfacing electrode also affects the electrode selection. Most of the properties will vary with the procedure used. The amount of admixture of weld metal and base metal will affect the alloy content of the resulting deposit. Welding current is usually the controlling factor because it controls the amount of peneration. Welding current may also affect the cooling rate, though the use of preheat and slow cooling techniques is usually more important. Cooling rate affects the structure of the deposit, which in turn affects the properties of the deposit. Information on the effect of procedures on each electrode is found in the "Procedure" portion of the lesson.

Study the following examples:

APPLICATION: Dragline bucket tooth working in sandy gravel with some good-sized rocks.

Only one surface needs protection.

No cutting edge is required.

Abrasion is the most important source of wear on the part, though impact must also be considered.

The part is large enough to take a heavy bead and cooling will be rapid. The surface of the bead as-deposited is sufficient; no machining or grinding is required.

The teeth are manganese steel.

As abrasion is most important, the electrodes at the top of the chart receive first consideration. Faceweld 12 has excellent abrasion resistance, but little impact resistance. Faceweld 1 has better impact resistance and with proper procedure may be satisfactory. Abrasoweld has sufficient impact resistance with some sacrifice in abrasion resistance as compared with Faceweld 1. Either of these electrodes may be used on manganese steel, though a layer of Manganweld may be necessary to insure proper bonding to the work hardened surface of the part.

APPLICATION: Same as above except that the bucket is working in sandy soil.

The fact that the bucket is working in sandy soil changes the above study only in that there is no impact to be concerned with. Now it possible to use the most abrasion resistant material which is Faceweld 12. Had the soil been changed to one which contained large rocks it would have been necessary to use a deposit made up entirely of Manganweld.

LESSON 5.2

Object:
To study the use and application of the Faceweld electrodes.

Equipment:
Lincoln "Idealarc" AC and DC welder and accessories.

Material:
Scrap steel, including a large part; Surfaceweld 3/16" Faceweld electrodes.

General Information

Lincoln Faceweld 1 and Faceweld 12 electrodes are used for hardsurfacing new and worn parts to resist *severe* abrasion and *moderate* impact.

Deposit Properties

The deposit consists of microscopic chromium carbide crystals held in a tough, hard metal bond. Both the carbides and the bonding metal have high abrasion resistance and moderate toughness.

ELECTRODE	DEPOSIT	ABRASION	IMPACT	HARDNESS (30 C / 50 C / 70 C)	FRICTION	MACHINABILITY (GRIND — MACHINE)
FACEWELD 12	CHROMIUM CARBIDE					
FACEWELD 1	CHROMIUM CARBIDE					
Abrasoweld	Semi-Austenitic					
Manganweld	Austenitic					
Stainweld	Austenitic					
Jet-Hard BU-90	Martensitic					
Fleetweld	Ferritic					

x - WORK HARDENED ▲ HEAT TREATED ⊙ ANNEALED

Fig. 1. Dredge pumps worn by sand abrasion, are reclaimed by building up the worn areas with "Fleetweld" and then hard-surfacing with "Faceweld."

Application Information

The Faceweld electrodes produce the *most* abrasion resistant deposits and are used wherever parts are worn away by the action of a "gritty" or sand-like material but are not subjected to a severe pounding by hard, rock-like particles. The choice between the two Faceweld electrodes depends on the amount of pounding or impact involved on the application. If there is no impact Faceweld 12 is best; if moderate impact is present, Faceweld 1 should be used. When it is possible to experiment on a job, first try Faceweld 12; if this deposit spalls off in service, then use Faceweld 1.

TYPICAL APPLICATIONS	
Screw conveyors	Conveyor sleeves
Scarifier teeth	Gyratory crusher mantles
Grader blades	Shovel bucket lips
Cement mill parts	Pulverizer jaws
Coke machinery parts	Crusher mill plates
Brick plant parts	Crusher rolls
Bradley and Griffin rings	Harrow discs
Pug mill paddles	Cultivator shovels
Plow shares	Bean knives
Wire feed rolls	

High temperatures and corrosive atmospheres have little effect on the deposits of these electrodes. Multilayer deposits will maintain hardnesses in the neighborhood of 30 Rockwell C at temperatures of 1500°F.

Procedures

Faceweld electrodes should be used in the flat position. Either DC, electrode positive (Reverse polarity), or AC may be used.

No more than three layers of Faceweld 12 should be used; four layers of Faceweld 1 are permissible. High preheat must be used if a greater number of passes are required. Frequently it

FLAT		
Rod Size		Amperage Range
Dia.	Lgth.	
3/16"	14"	60-150
5/16"	14"	145-350

Fig. 2. Aggregate, falling five feet on to this chute, wears the metal down due to severe abrasion. The plate is hard-faced with "Faceweld 12."

Fig. 3. Crusher roll must withstand severe abrasion. Outside diameter is hard-surfaced with multiple layers "Faceweld 12."

is possible to build up with several layers of Jet-Hard BU-90 (or Manganweld on manganese steel) and then finish off with one or two passes of Faceweld. Not only is this more economical than applying many layers of Faceweld, but it also reduces the possibility of spalling.

Beads frequently have hairline cross checks across the bead. These checks are a method of stress relief and normally are not detrimental to the quality of the deposit. If a particular application is such that they *must* be eliminated, preheat the work and cool slowly after welding.

Hold a fairly short arc when using the Facewelds. Weave beads the full width of the deposit if possible.

On single layer deposit use lower currents to reduce admixture. When welding on manganese steel keep the part cool. Let each bead cool down to the point where it can be touched with the bare hand before starting the next bead.

Job Instructions

1. Weld a pad of Faceweld hardsurfacing on scrap plate to become familiar with the technique required.
2. On a heavy piece of scrap deposit a multilayer pad. Note the cross checks which occur when the pad cools. Preheat the part and repeat the welding. Cool the part slowly. Note that the cross checks have been eliminated.
3. By grinding a sample of each, compare the abrasion resistance of Faceweld deposits with mild steel and other hardsurfacing deposits. Note how difficult it is to grind away the Faceweld deposits.

Fig. 4. Worn cutting edge of bulldozer blade is built up with impact and abrasion resisting "Faceweld 1."

Fig. 5. Die ring worn by moderate impact and severe abrasion is hardsurfaced with "Faceweld 1."

LESSON 5.3

Object:
To study the use and application of Abrasoweld electrode.

Equipment:
Lincoln "Idealarc" AC and DC welder and accessories.

Material:
Scrap steel, 3/16" inch Abrasoweld electrodes.

General Information

Abrasoweld electrodes are used for hardsurfacing new and worn parts to resist wear from both abrasion and impact.

Deposit Properties

The deposit is a high carbon, chromium alloy material which is semi-austenitic in the as-welded condition. It has an unusual combination of good abrasion resistance with moderate toughness and excellent hot forging properties. Maximum hardness is obtained with single layer deposits; best abrasion resistance is obtained with multiple layer deposits. The deposit work-hardens. This means that as the surface is pounded the thin surface layer becomes still harder while the material under the surface retains its relative softness.

ELECTRODE	DEPOSIT	ABRASION	IMPACT	HARDNESS (30 C, 50 C, 70 C)	FRICTION	MACHINABILITY (GRIND / MACHINE)
Faceweld 12	Chromium Carbide					
Faceweld 1	Chromium Carbide					
ABRASOWELD	SEMI-AUSTENITIC			x		⊙
Manganweld	Austenitic			x		
Stainweld	Austenitic			x		
Jet-Hard BU-90	Martensitic			Δ		⊙
Fleetweld	Ferritic					

X-WORK HARDENED Δ HEAT TREATED ⊙ANNEALED

Fig. 1. Dipper tooth for power shovel. Built-up with "Manganweld" and hardfaced (partially) complete with "Abrasoweld."

Application Information

Abrasoweld has the advantage of very good abrasive resistance in combination with toughness and impact resistance. Similar to the Facewelds, they are used wherever parts are worn away by the action of a "gritty" or sand-like material. Unlike the Facewelds, however, they may be used where the part is subjected to severe pounding by hard, rock-like particles. It is for this reason that the bucket lip of a shovel working in sand is best when hardsurfaced with Faceweld, but the same bucket lip working in a rocky soil should be surfaced with Abrasoweld.

TYPICAL APPLICATIONS

Dipper teeth	Screw flights
Tractor grousers	Plow shares
Shovel tracks	Scarifier teeth
Shovel drive sprockets	Truck chains
Dipper lips	Pump housings
Rock crusher hammers	Rock crushers
Sand pump impellers	Coal mining cutters
Dredge cutter teeth	Conveyor buckets
Scraper blades	Conveyor rolls
Charging rams	Gears
	Crusher mantles
	Pulversizer plows
	Mill hammers

Procedures

Abrasoweld is best applied in the flat position. Either DC, electrode position (Reverse polarity), or AC may be used. No more than three layers of Abrasoweld should be applied. If a thicker deposit is required, several layers of Jet-Hard BU-90 (or Manganweld on manganese steel) should be used under the Abrasoweld. If thicker deposits of Abrasoweld must be made, apply a layer of Stainweld between every two layers of Abrasoweld and preheat each bead while it is hot.

Rod Size		Amperage Range
Dia.	Length	
1/8″	14″	40—250
5/32″	14″	75—200
3/16″	14″	110—270
1/4″	14″	160—400

Abrasoweld deposits are not machinable as welded. Grinding is recommended if shaping is necessary. If machining *must* be done, the deposits should be fully annealed at 1650°. To restore abrasion resistance, reheat the part to 1450°, quench, and draw at 400°F.

When applying Abrasoweld, do not permit the preheat or inter-bead temperature to remain above 600°F. Below this limit, higher temperatures will result in higher hardnesses, though the abrasion resistance is not greatly affected.

Hold a fairly short arc when using Abrasoweld. Weave beads the full width of the deposit if possible. Use at least ¾ inch wide beads.

When welding on manganese steel, keep the part cool. Let each bead cool down to the point where it can be touched with the bare hand before starting the next bead.

Job Instructions

1. Weld a pad of Abrasoweld hardsurfacing on scrap plate to become familiar with the technique required.
2. By grinding a sample of each, compare the abrasion resistance of Abrasoweld with mild steel and other hardsurfacing deposits. Also, note that it is possible to pound on Abrasoweld deposits without causing them to spall off or crack.

LESSON 5.4

Object:
To study the use and applications of Lincoln Mangjet and Manganweld C electrodes.

Equipment:
Lincoln "Idealarc" AC and DC welder and accessories.

Material:
Scrap manganese steel; 3/16 inch Mangjet and Manganweld C electrodes.

General Information

Mangjet and Manganweld C are used for building up high (12-14%) manganese steel to resist *moderate* abrasion and extremely *severe* impact. Mangjet is also used for joining manganese steel, either to another piece of manganese steel or to carbon steel, and for building up a manganese steel deposit on carbon steel.

Deposit Properties

The deposit is a 12-14% manganese steel which is fully austenitic and extremely tough. It develops maximum surface hardening by peening or other cold working.

ELECTRODE	DEPOSIT	ABRASION	IMPACT	HARDNESS 30 C / 50 C / 70 C	FRICTION	MACHINABILITY (GRIND / MACHINE)
Faceweld 12	Chromium Carbide					
Faceweld 1	Chromium Carbide					
Abrasoweld	Semi-Austenitic			x		◎
MANGJET MANGANWELD	AUSTENITIC			x		
Stainweld	Austenitic			x		
Jet-Hard BU-90	Martensitic			x		◎
Fleetweld	Ferritic					

x WORK HARDENED △ HEAT TREATED ◎ ANNEALED

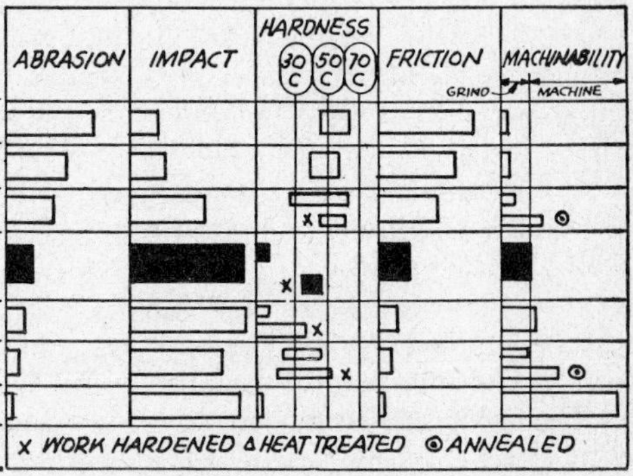

Application Information

Build up—Both electrodes are used for rebuilding worn manganese steel parts. Mangjet, a coated electrode, deposits a smooth, thin bead. Manganweld C, a bare electrode, deposits a thick bead. Depending on the service conditions, other hardsurfacing materials may be put over the manganese deposit to improve abrasion resistance. Mangjet is also used to deposit a manganese steel weld on carbon steel parts.

TYPICAL APPLICATIONS

- Rail cross overs
- Rail frogs and switches
- Dipper teeth
- Dipper lips
- Shovel drive sprockets
- Shovel tracks
- Crusher pads
- Rolling mill parts
- Crusher hammers
- Chain hooks
- Strip mill wobblers
- Manganese buckets
- Crusher rolls
- Crusher screens
- Dragline pins and links
- Cement grinder rings
- Dredge parts

Joining parts—Mangjet is used to weld manganese steel-to-manganese steel and manganese steel-to-carbon steel joints. These welds are as strong as the manganese base metal.

Procedures

Probably the most important thing to remember in welding manganese steel is to keep the work cool. Other important general considerations are outlined in the "Job Instructions" below and should be followed carefully, if the best results are desired.

Mangjet on manganese steel—Mangjet operates with an exceptionally steady arc for electrodes of this type. To weld on manganese steel, use either DC, electrode negative, or AC and the lowest current possible consistent with good fusion. Travel fast enough to stay ahead of the slag. Use either a drag technique, a short arc, stringer beads, or a weave, depending on which best fits the job. Stay about 1/4 inch away from work edges and let the molten metal flow to the edge.

Mangjet on carbon steel—DC, electrode positive, is preferred. Also, as compared with the procedures on manganese, it is best to increase the current, lengthen the arc to about 3/16 inch, and tip the electrode into the direction of travel, so that the arc force washes metal toward the finished bead.

Manganweld C on manganese steel—Use DC, electrode positive and the lowest current possible consistent with good fusion. Deposit beads with a weave 1/2 inch to 1 inch wide and preferably not more than three inches long.

MANGJET				MANGANWELD C		
Size		Amperage Range		Size		Amperage Range
Dia.	Lgth.	DC	AC	Dia.	Lgth.	
5/32"	14	125-210	140-230	3/16"	18	130-170
3/16"	14	150-260	165-285	1/4"	18	170-225
1/4"	14	200-350	220-385			

Job Instructions

1. Deposit beads with each of the three Manganweld electrodes to become familiar with the characteristics of each.
2. Using either of the electrodes, deposit a large area of Manganweld deposit. In so doing, check the following precautions:
   (a) Clean the work and remove any spongy or excessively hard areas of base metal before welding.
   (b) Disperse the beads around the work so that no two beads are laid side-by-side as the area is covered.
   (c) Be sure that the work never gets so hot that you can't touch it with the bare hand.
   (d) Cool the work, not the deposit, if necessary.
   (e) Peen the surface of the hot bead to relieve shrinkage stresses.

LESSON 5.5

Object:

To study the use and application of Lincoln Wearweld and Jet-Hard BU-90 electrodes.

Equipment:

Lincoln "Idealarc" AC and DC welder and accessories.

Material:

Scrap steel: 3⁄16 inch Wearweld and Jet-Hard BU-90 electrodes.

General Information

Jet-Hard BU-90 and Wearweld electrodes are used for hardsurfacing new and worn parts to resist rolling or sliding abrasion, and for fast build-up.

Deposit Properties

Wearweld electrodes are low carbon-chrome-manganese alloy which are partly martensitic and partly ferritic in the as-welded condition. The deposit has uniform hardness and is moderately tough.

Jet-Hard BU-90 is a medium carbon-chromium-manganese alloy which may be heat-treated to give physical properties comparable to medium carbon steel. Deposits may be hot forged. Good machinability can be obtained by annealing, and subsequent heat treatment will restore it to the desired hardness.

ELECTRODE	DEPOSIT
Faceweld 12	Chromium Carbide
Faceweld 1	Chromium Carbide
Abrasoweld	Semi-Austenitic
Manganweld	Austenitic
Stainweld	Austenitic
WEARWELD	MARTENSITIC
JET-HARD BU-90	MARTENSITIC
Fleetweld	Ferritic

ABRASION | IMPACT | HARDNESS 30 C, 50 C, 70 C | FRICTION | MACHINABILITY (GRIND — MACHINE)

X-WORK HARDENED Δ HEAT TREATED ⊙ ANNEALED

Fig. 1. Wearweld is used on rail ends.

Application Information

While similar in nature, the applications for each of these electrodes are fairly distinct.

Wearweld has a spray-type arc which permits the deposition of a thin and relatively smooth layer of surfacing. This is particularly advantageous when building up parts which are to resist rolling or sliding abrasion, such as rail ends. For applications of this type, Wearweld has an excellent combination of hardness and toughness to resist wear.

Jet-Hard BU-90 is a versatile electrode used both for preliminary build-up under higher alloy hardsurfacing materials and also for hardsurfacing without the addition of other electrodes. It is used both to withstand the abrasion of gritty or sand-like material and also rolling or sliding friction. It will withstand severe impact or pounding. When used on preliminary build-up, Jet-Hard BU-90 acts as a cushion under the more abrasion-resistant hardsurfacing deposits.

TYPICAL APPLICATIONS	
BU-90 For Both Build-Up and Finish Passes	BU-90 As Build-up Prior To Other Hardsurfacing Overlay
Trunnions Shafts Crane and mine car wheels Tractor grousers Rolls and idlers Shovel tracks Drive sprockets Gears Churn bit points	Shovel and bucket lips Pump impellers and housings Dredge and bucket teeth Scraper blades Pulverizer plows Mill and crusher hammers
Wearweld	
Rail ends Coal mine Car wheels Crane wheels	Cams Craneways Caterpillar treads, sprockets & track links

Jet-Hard BU-90 deposits can normally be machined in the as-welded condition with the use of high-speed or carbide tools. Preheat and slow cooling will aid machinability.

The hardness and abrasion resistance of Jet-Hard BU-90 can be increased by water quenching the deposit from 1600°F. and drawing at 400 to 800°F. to achieve the desired toughness.

Procedures

These electrodes are best applied in the flat position. Both electrodes will operate on AC; with DC, use electrode negative with Jet-Hard BU-90 and electrode positive for Wearweld.

There is no limit to the number of layers of Jet-Hard BU-90 which may be applied; no more than three layers of Wearweld should be deposited without using very high preheating and slow cooling.

Use a short arc with both electrodes with a narrow weaving motion.

FLAT			
Rod Dia.	Lgth	Amperage Range	
		DC	AC
Jet-Hard BU-90			
5/32″	14″	130-195	145-195
3/16″	14″	165-235	180-260
1/4″	18″	225-365	250-400
Wearweld			
3/16″	18″	110-275	110-275
1/4″	18″	150-400	150-400

When welding on manganese steel, keep the work cool. Let each bead cool down to the point where it can be touched with the bare hand before starting the next bead.

Job Instructions

1. Weld a pad of hardsurfacing with each electrode to become familiar with the technique required. Note that the Wearweld has a fine spray-type arc and deposits a flat, smooth bead. Jet-Hard BU-90, with its iron powder coating, has a very high deposition rate with a very steady arc and very little spatter. This ease of operation of Jet-Hard BU-90 has made it the favorite of welding operator trade.
2. By grinding a sample of each, compare the abrasion resistance of Wearweld and Jet-Hard BU-90 deposits with mild steel and other hardsurfacing deposits. At the same time, note that these deposits are capable of withstanding severe impact.

Fig. 2. Jet-Hard BU-90 is used to combat wearing friction on billet manipulators in steel mills.

LESSON 5.6

Object:

To study the use and application of Lincoln Toolweld electrodes.

Equipment:

Lincoln "Idealarc" or other DC welder and accessories.

Material:

Bar stock or plate which is prepared to resemble die or cutting tools; 1/8 inch Toolweld 60 and Toolweld A & O electrodes.

General Information

Lincoln Toolweld electrodes are used for depositing hardsurfacing on cutting edges of tools and dies.

Deposit Properties

Toolweld A & O has a 5% chromium tool steel deposit which has good as-welded properties and which may be further improved by heat treating. This versatile electrode is well suited for most tool and die work, except where a high-speed (high temperature hardness) deposit is required.

Toolweld 60 produces a deposit of the high-speed molybdenum tool steel type. Deposits may either be used as-welded or may be tempered or given standard high-speed steel heat treatment. The deposit maintains full hardness at temperatures up to 1,000°F.

ELECTRODE	DEPOSIT	ABRASION	IMPACT	HARDNESS 30 C / 50 C / 70 C	FRICTION	MACHINABILITY (GRIND / MACHINE)
Faceweld 12	Chromium Carbide					
Faceweld 1	Chromium Carbide					
Abrasoweld	Semi-Austenitic					
TOOLWELD 60	MARTENSITIC					
TOOLWELD A & O	MARTENSITIC					
Wearweld	Martensitic					
Jet-Hard BU-90	Martensitic					
Fleetweld	Ferritic					

X - WORK HARDENED △ HEAT TREATED ⊙ ANNEALED

Fig. 1. Toolweld 60 for high speed, high temperature applications.

Fig. 2. Toolweld A & O for edge strength on low temperature applications.

Application Information

These electrodes are used on all types of tool and die repair and alteration work. They are frequently used to make composite tools of a mild steel body and a tool steel cutting edge.

They withstand severe metal-to-metal impact, while maintaining a sharp edge.

Toolweld A & O is used on all applications, except those requiring high temperature hardness, in which case, Toolweld 60 is used.

TYPICAL APPLICATIONS	
Toolweld A & O	
Forming dies	Shearing dies
Punching dies	Punches
Paper cutting knives	Punch press dogs
Sewage scraper blades	Slag cleaning tools
Skelp hooks	Dogs on ingot tongs
Clutch parts	Spot weld flash trimmer
Guides	Molds
Shredder knives	Woodworking tools
Chuck jaws	Axes
Toolweld 60	
Lathe tools & centers	Broaches
Shear blades for hot work	Forming cutters

FLAT		
Rod Size		Amperage Range
Dia.	Length	
Toolweld 60		
1/8"	13"	65-100
Toolweld A & O		
3/32"	12"	40- 85
1/8"	14"	65-130

Procedures

DC, electrode positive is used with Toolweld 60, while DC, electrode negative or AC is used with Toolweld A&O.

No more than four layers of Toolweld should be applied on any application. For maximum hardness, two layers are recommended.

When used on large hardened alloy tools and dies, preheat and interpass temperatures of 900-1000° are preferred. If such is not possible, maintain as high a temperature as possible without softening the base metal. This would be the case when the part is re-heat treated. On parts which were not hardened, preheat and interpass temperatures of 400-500°F. is recommended. When the deposit is to be heat treated by tempering only, allow the work to cool to room temperature, so that the weld deposit will transform to martensite before tempering.

Where only one layer is to be used, reduce the current to as low a value as possible, so that admixture will be reduced. Use a medium length arc and weave the bead the entire width of the deposit if possible.

Job Instructions

Weld with both Toolweld 60 and Toolweld A & O to become familiar with the operation of the electrodes. Make small welds on bar stock to become familiar with the operations of the electrodes on small stock. Make welds along the edge of bar stock or light plate to become familiar with the procedure which would be used in hardsurfacing a cutting edge. Note that this edge may then be ground to a sharp point which will hold its edge under severe impact loading.

SECTION VI

CAST IRON, NON-FERROUS, AND WASH-COATED MILD STEEL ELECTRODES

LESSON 6.1

Object:
To study the use of "Ferroweld" and "Softweld" electrodes.

Equipment:
Lincoln "Idealarc" welder and accessories.

Material:
Some broken pieces of cast iron; ⅛" "Ferroweld" and "Softweld" electrodes.

General Information

"Ferroweld" is a mild steel electrode for depositing non-machinable welds on cast iron; "Softweld" is a nickel electrode for depositing machinable welds on cast iron. See Lesson 1.36 for further information on cast iron welding.

Properties

"Ferroweld" deposits are stronger than the cast iron itself. It operates with very low currents to reduce heat input to the base metal and, consequently, to minimize the possibility of weld cracking and hardening along the fusion line. It is very good where studding is used.

"Softweld" deposits are soft and easily machined. They are also ductile and have the least tendency to crack.

Application Information

Both electrodes are used on similar applications: "Ferroweld" is used on most work except where machining is required; "Softweld" is used on applications requiring machining.

TYPICAL APPLICATIONS	
Machine bases	Compressor blocks
Cast boiler sections	Steam radiator sections
Headers	Gear teeth
Motor blocks	Cast iron to steel
Cast iron wheels	Lamp posts
Motor heads	Fire plugs
Diesel water jackets	Pulleys
Tractor transmission cases	Wheels
Tractor differentials	Frames
	Blow holes
	Printing press rolls
	Gate valves

Procedure

POLARITY—"Ferroweld" – Electrode positive, work negative, or AC.
"Softweld" – Electrode negative, work positive, or AC.

ALL POSITIONS	
Rod Size	Amperage Range
"Ferroweld" ⅛"x14"	80-100
"Softweld" ⅛"x14"	75-110
5/32"x14"	90-135

Job Instructions

Set up broken casting, matching edges. Tack together or hold in position. If piece is 3/16" or over in thickness, Vee out pieces 45° on each side, a total of 90°.

Weld with both "Ferroweld" and "Softweld" to become familiar with the operation of each.

LESSON 6.2

Object:
To study the use of "Aerisweld" electrode.

Equipment:
Lincoln Idealarc or DC welder and accessories.

Material:
Copper plates 1/8" x 3" x 5"; 5/32" "Aerisweld."

General Information

"Aerisweld" is a shielded arc electrode for welding bronze, brass and copper in the many applications. In repair work, "Aerisweld" builds up and fills in bronze castings. Many types of bronze which are difficult to braze are easily welded with "Aerisweld." Conforms to AWS Class E CuSn-C.

Properties

Dense, high-strength deposit with characteristics of true phosphor bronze.

Characteristics of the base metal are of great importance in determining the characteristics of the joint, and fusion zone due to possible admixture of base metal in welding.

Application Information

TYPICAL APPLICATIONS	
Contact points	Bus bars
Bearing surfaces	Malleable iron
Cast iron	Bronze ship propellers
Galvanized iron	Bronze check valve discs
Aluminum bronze castings	"Ground" connections
Copper rivet heater blocks	Bronze tubing
Fire hose couplings	Copper vats
Copper piping	Copper clad fabrication
Copper-to-steel	Bushings
Viscose Mixers	Pipe line bonds
Brass pads on steel	Caustic pump blocks
Ornamental work	Copper brew stills
Bonding rails	
Repair brass valves	
Impeller blades	

Procedure

POLARITY DC, Electrode positive.

On ferrous metal or thin copper or bronze, it is generally unnecessary to preheat the metal. As the work progresses and the heat builds up, it may

ALL POSITIONS	
Rod Size	Amperage Range
1/8"	50-125
5/32"	70-170
3/16"	90-220

Fig. 1. Racks for storage of machined parts. Bronze cushions are welded on steel brackets with "Aerisweld."

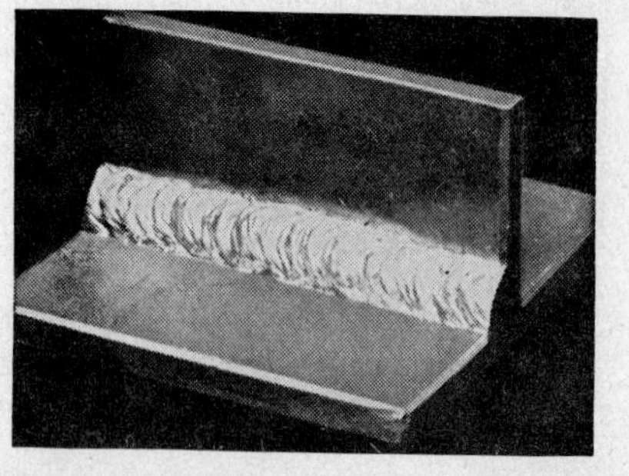

Fig. 2. Fillet in 3/16-in. bronze plate welded with "Aerisweld."

be necessary, in some cases, to reduce the current.

On heavy copper and bronze, preheating may be necessary due to the high heat conductivity of these metals. This preheating can be accomplished by using a carbon electrode with negative polarity and rapidly moving the arc over the area to be welded.

High current, high temperatures, or considerable penetration will cause a great admixture of the base metal and the procedure should take this into account in the case of the first layer. It is therefore advisable to put on as much metal per bead, or layer, as can be conveniently and easily done.

Types of metals which evolve gases in the molten state, at the point of solidification, result in porosity. In some cases the use of higher current, keeping the work hot, will tend to reduce this porosity.

Holding the electrode at an angle so that the flame of the arc is directed back over the work will aid in permitting the gases to bubble through to the surface.

Where the work has to be machined it is, of course, necessary that the original, or base metal, be cut away so that when the deposit is made the line of machining will come through near the top of the deposit and not at the junction zone. The work should be laid out to obtain this result.

Caution: Weld brass and bronze in well ventilated area. Avoid breathing fumes, as they are toxic.

Job Instructions

Tack the two copper plates together for butt weld. Use 5/32" electrode, 100 to 115 amperes, and weld.

Try this with copper and then with bronze plates or bronze castings.

Note the easy flow of metal, and the free flowing, well formed type of bead.

Repeat, making lap and fillet welds.

Fig. 3. A Church bell 116 years old repaired with "Aerisweld." Saved $250 over the cost of recasting a new bell.

Fig. 4. Copper condenser of plate and sheet—typical of many copper or brass containers welded with "Aerisweld."

LESSON 6.3

Object:
To study the use of "Aluminweld" electrode.

Equipment:
Lincoln Idealarc or DC welder and accessories.

Material:
Aluminum plates ⅛" x 3" x 5" and several pieces of aluminum casting; 5/32" "Aluminweld."

General Information

"Aluminweld" is a 5% silicon aluminum alloy shielded arc electrode for welding aluminum in any form—cast, sheet, shapes or extruded forms. Designed for either metallic or carbon arc welding. Conforms to AWS Class A1-43.

Properties

"Aluminweld" is provided with a coating which prevents excessive oxidation and will dissolve any aluminum oxide which might be formed. The coating also assists in giving a very smooth operating arc which is so particularly essential in welding aluminum. The resulting weld is very dense without porosity and possesses high tensile strength. The weld can be polished satisfactorily with practically no discoloration.

Fig. 1. Group of aluminum storage tanks for olive oil. Fabricated with "Aluminweld" Electrode.

Application Information

TYPICAL APPLICATIONS	
Bottle brackets	Tanks
Racks	Crank cases
Cylinder heads	Transport truck tanks
Ornamental work	Cooking utensils
Laundry chutes	Portable drill castings
Viscose plant piping	Appliance parts
Beer barrels	Structural aluminum
Outboard motor gas tanks	Trim
Moulding	Window frames
White metal die castings	Soap kettle liners
Washing machine tubs	

Fig. 2. Left: Butt weld in ⅛-in. aluminum made with "Aluminweld" and ground down at end to show denseness of weld metal. Right: Butt weld in 14-gauge aluminum made with "Aluminweld."

Procedures

POLARITY—DC, Electrode positive.

A short arc should be held, the coating approximately touching the molten pool.

In general, use the highest current possible without melting the edges too far back or burning through. With the high melting rate of aluminum, little heat is dissipated into the plate with consequent chilling of deposits. Hence, it is sometimes desirable to preheat the seam to 600° to 700° F. before welding.

On striking, the best results are obtained by "scratching" the electrode. To restart an electrode, strike the arc in the crater, then move the electrode quickly back along the completed weld for ½", then proceed as usual, making sure the crater is completely remelted.

Even melting off of the flux is facilitated by holding the electrode approximately perpendicular to the work at all times.

Always direct arc so that both edges to be welded are properly and uniformly heated and the electrode advanced along the seam at such a rate that a uniform bead is made.

FLAT	
Rod Size	Amperage Range
3/32"	20- 55
1/8"	45-125
5/32"	60-170
3/16"	85-235
1/4"	125-360
VERTICAL AND OVERHEAD	
Not recommended. However, where vertical or overhead welding is required, follow same Procedure as for flat work, using 3/32", 1/8" or 5/32" electrodes.	

TACKING—Increase recommended currents approximately 50% for tacking. Use short arc and rotary motion.

Remove slag from the weld by light hammering and brushing. Last traces may be removed with warm water and a wire brush or by soaking the weld in 5% nitric acid or 10% warm sulphuric acid followed by a warm water rinse.

Vertical and overhead welding should as far as possible be eliminated. However, where imperative, vertical welding can be done downward or upward with straight beads or by weaving. Overhead welding should be done with a number of straight beads.

"Aluminweld" produces excellent welds used as a filler rod with the carbon arc.

Caution: Avoid breathing fumes from electrode. Work in a ventilated area.

Joint	Beads or Passes	Electrode Size	Current Amps.
Fig. 3 18 ga. sheet	1	3/32"	40
Fig. 3 14 ga. sheet	1	1/8"	65
Fig. 3 1/8" plate	1	5/32"	120
Fig. 4 3/16" plate	1	3/16"	170
Fig. 5 1/4" plate	1	1/4"	250
Fig. 6 3/16" plate	1 2	3/16" 3/16"	170 170
Fig. 6 1/4" plate	1 2	3/16" 3/16"	170 170
Fig. 7 3/8" plate	1 2	1/4" 1/4"	250 250

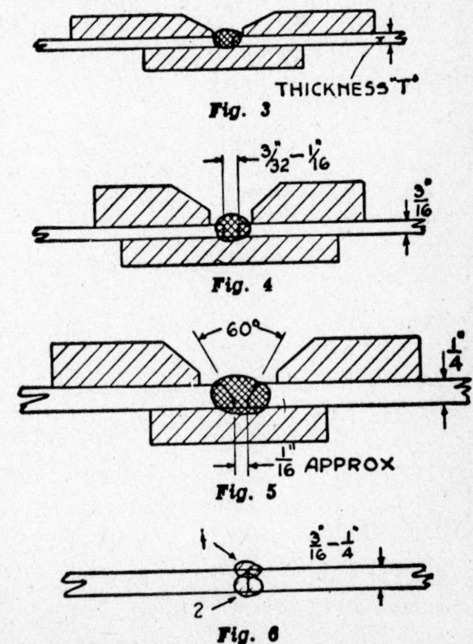

Fig. 3
Fig. 4
Fig. 5
Fig. 6

Butt Welds—The work should be held in position by jigs and backed up by copper, as illustrated in Fig. 3. When butt welding plates ⅛″ and thicker, the copper backing should be slightly grooved beneath the joint to be welded.

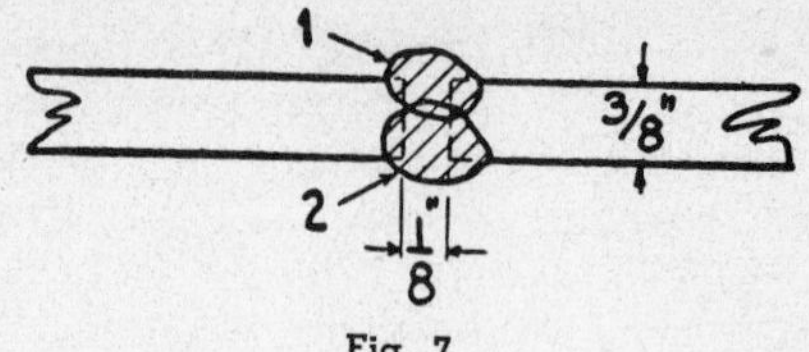

Fig. 7.

Welding of butt joints in 3/16″ plate and heavier should be done with two beads, as indicated in Figures 6 and 7. No backing or clamping is required for welding joints in this manner.

The general procedure as given previously should be followed in making these types of welds.

Fillet Welds—In making fillet welds the electrode should be held in such a position that the angle between the electrode and the horizontal plate is approximately 45°. The electrode should be manipulated with a small rotary motion with the arc being played first on the vertical member and then on the horizontal member of the joint. With the above exceptions the general procedure given previously should be followed in making a fillet weld.

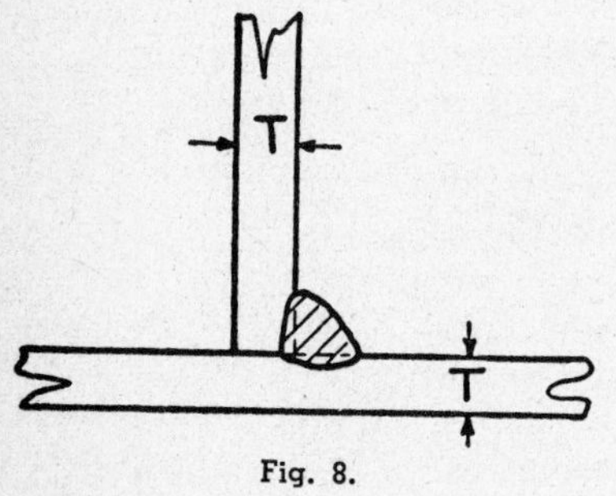

Fig. 8.

Joint	Beads or Passes	Elec-trode Size	Cur-rent Amps.
Fig. 8 1/8″ plate	1	5/32″	120
Fig. 8 3/16″ plate	1	3/16″	170
Fig. 8 1/4″ plate	1	3/16″	170

Corner Welds—This type of weld is made by following the general procedure previously given.

Joint Fig. 9	Beads or Passes	Elec-trode Size	Cur-rent Amps.
1/16″ plate	1	1/8″	65
1/8″ plate	1	5/32″	120
3/16″ plate	1	3/16″	170
1/4″ plate	1	1/4″	250

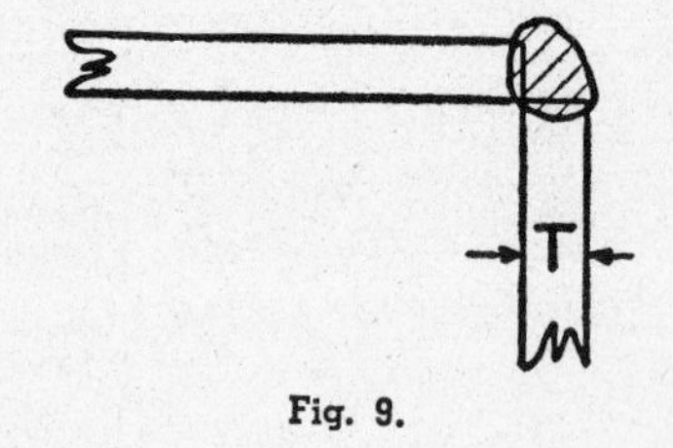

Fig. 9.

Job Instructions

Make butt weld on ⅛″ plate. If material is available repeat with thicker plate. Make lap, fillet and corner welds in light and heavy plate if possible.

Prepare butt weld in casting by veeing. Preheat casting before welding. Remove all traces of flux after welding.

LESSON 6.4

Object:
To study the use of "Stable-Arc" electrode (E4510).

Equipment:
Lincoln "Idealarc" and accessories.

Material:
Mild steel plate 1/4"; 5/32" "Stable-Arc."

General Information

"Stable-Arc" is a wash coated electrode, operating on DC for welding mild steel in all positions where welds with no slag interference and no slag cleaning are desired, and where high ductility and tensile strength are not required. Produces welds with a minimum admixture of plate metal. Gives almost any desired amount of build-up.

Application Information

TYPICAL APPLICATIONS	
Filling holes	Cable way drums
Welding tanks for galvanizing	Building up shafts
Filler rod for large gaps	Ornamental iron
Well screens	High silicon steel parts
Repair steel castings	Castings
Tanks	Welds requiring galvanizing
Build-up work	Stacks
	Hot water tanks

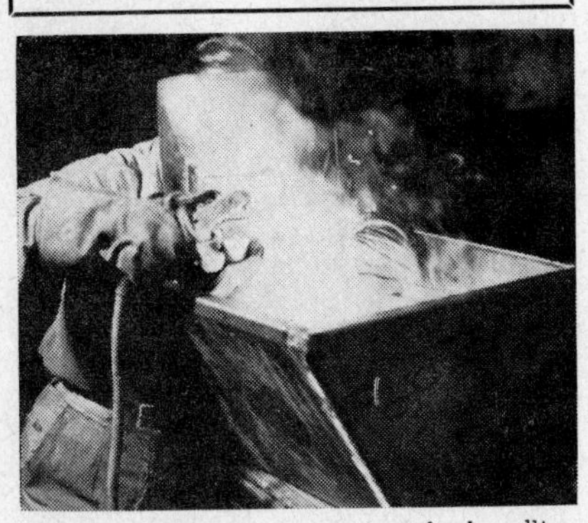

Fig. 1. Fabricating container for handling liquids from 18 gauge metal. To be galvanized later.

Procedure

POLARITY DC Electrode negative.

Hold short arc. Tip of electrode should be just close enough to forward edge of crater to prevent sticking. Good short arc of proper voltage usually indicated by steady snappy crackling sound and steady shower of small sparks. Too long an arc will hiss and sputter.

Where bead cannot be made in a straight run, use weaving. This helps keep metal hot; also aids in releasing gases and eliminating undercutting. Clean each bead with wire brush to remove oxides accumulated as molten metal cools.

ALL POSITIONS		
Rod Size		Amperage Range
Dia.	Length	
3/32"	12"	50— 80
1/8"	14"	75—125
5/32"	14"	110—175
3/16"	14"	150—225
1/4"	18"	210—300

Job Instructions

Position plate downhand. Set amperage at 140 to 150. Run straight beads using different arc lengths to observe the action of the arc and the resultant bead. Clean with a wire brush and observe the difference of the bead from that of the shielded electrode. Run beads using a weaving motion. Standard weld forms may be made if further practice is desired with the bare electrode.

SECTION VII

SUPPLEMENTARY DATA

INTRODUCTION

The additional information on welding in Section VII is included to help weldors advance their careers and help advance the welding industry. To progress in a career as a weldor, it is necessary to continually develop and grow with the industry. Welding is a relatively young industry and new developments and ways of doing things are created every day. The progressive weldor will keep up with these developments and be able to make suggestions on better ways to do any job which comes to him.

The welding industry not only expands through the development of new and better ways of welding, but also has one of its greatest prospects for growth in the use of present welding methods on jobs which are done by other means. Two fields in which the use of welding is rapidly expanding

Fig. 1. Welding is used in constructing modern schools, homes, and skyscraper buildings.

are structural work, where rivets and bolts are presently used, and machinery manufacture, where castings are widely used. In these two fields are many jobs which can be done better and at less cost with welding. Welding will reduce costs because welded design puts exactly the right amount of metal in the exact spot where it is needed for the best efficiency. While the trend towards using welding on these jobs must come from the designer who plans and lays out the work, the weldor can do several things to assist the progress of welding and his own future.

First, the weldoor can, to the best of his ability, always make highest quality welds. On the rare occasions when a poorly-made weld fails in service, it is invariably widely publicized—to the discredit of the entire welding industry. It creates, in the minds of those who are not thoroughly familiar with welding, a doubt as to the reliability of the process. This doubt retards the further application of welding. So, it is important that every weldor do his best to insure that all the work he does is top quality. Further, he can tell others the true facts about the advantages and dependability of welding. It *can* be depended upon to produce quality joints which are actually stronger than the base metal, if such is desired and called for by the designer.

Secondly, the weldor can learn to use the fastest possible welding procedures which are consistent with the quality desired. Overwelding is to be avoided as useless and expensive. No one, including the weldor, benefits by a slow procedure. Fast procedures lead those who are designing to appreciate the full cost advantages of welding and encourages them to make greater use of the process. When this occurs, the entire welding industry benefits. Part 7.2 includes considerable data which can help the weldor to produce the fastest quality welds.

The weldor can be observant and recommend changes which convert castings to weldments and reduce costs. While this is not a design book, the following is included to assist the weldor to recognize the possible applications of welding on parts which are presently cast.

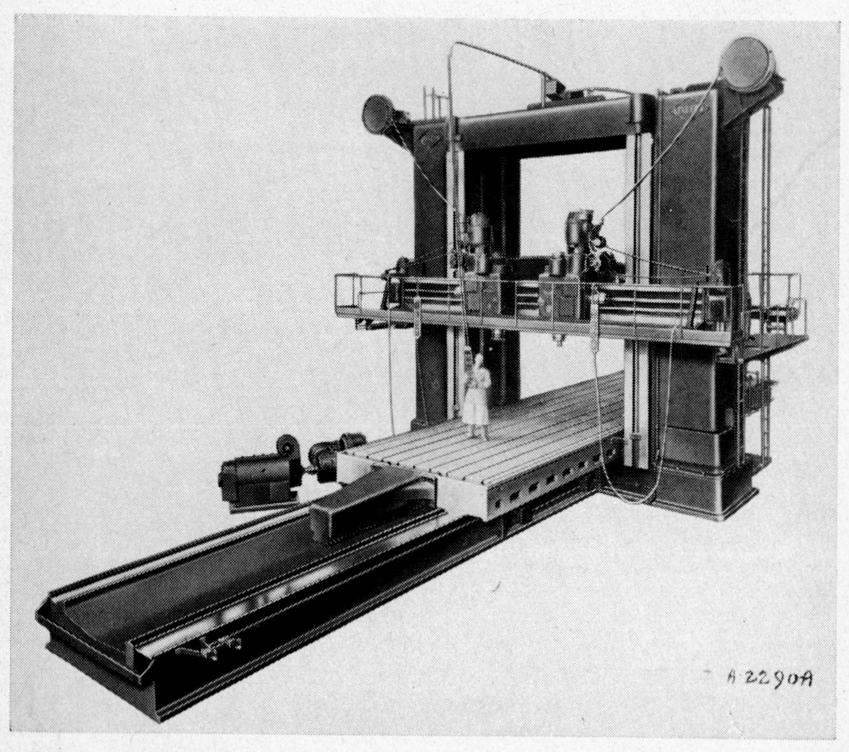

Fig. 2. Machine tools of all sizes are fabricated with welding for lowest cost and best quality.

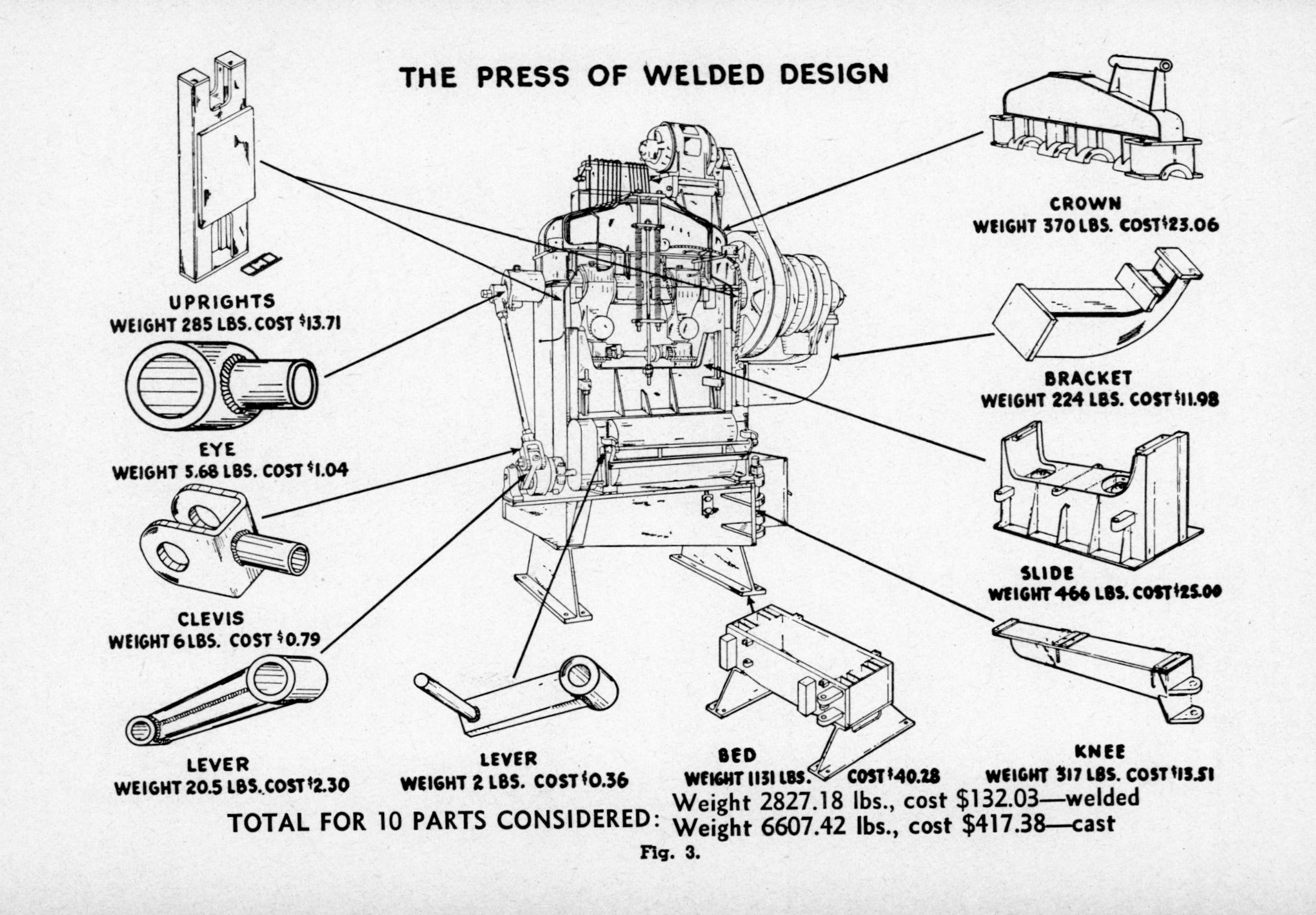
THE PRESS OF WELDED DESIGN
UPRIGHTS
WEIGHT 285 LBS. COST $13.71
EYE
WEIGHT 5.68 LBS. COST $1.04
CLEVIS
WEIGHT 6 LBS. COST $0.79
LEVER
WEIGHT 20.5 LBS. COST $2.30
LEVER
WEIGHT 2 LBS. COST $0.36
BED
WEIGHT 1131 LBS. COST $40.28
KNEE
WEIGHT 317 LBS. COST $13.51
CROWN
WEIGHT 370 LBS. COST $23.06
BRACKET
WEIGHT 224 LBS. COST $11.98
SLIDE
WEIGHT 466 LBS. COST $25.00
TOTAL FOR 10 PARTS CONSIDERED: Weight 2827.18 lbs., cost $132.03—welded
Weight 6607.42 lbs., cost $417.38—cast

Fig. 3.

Bases

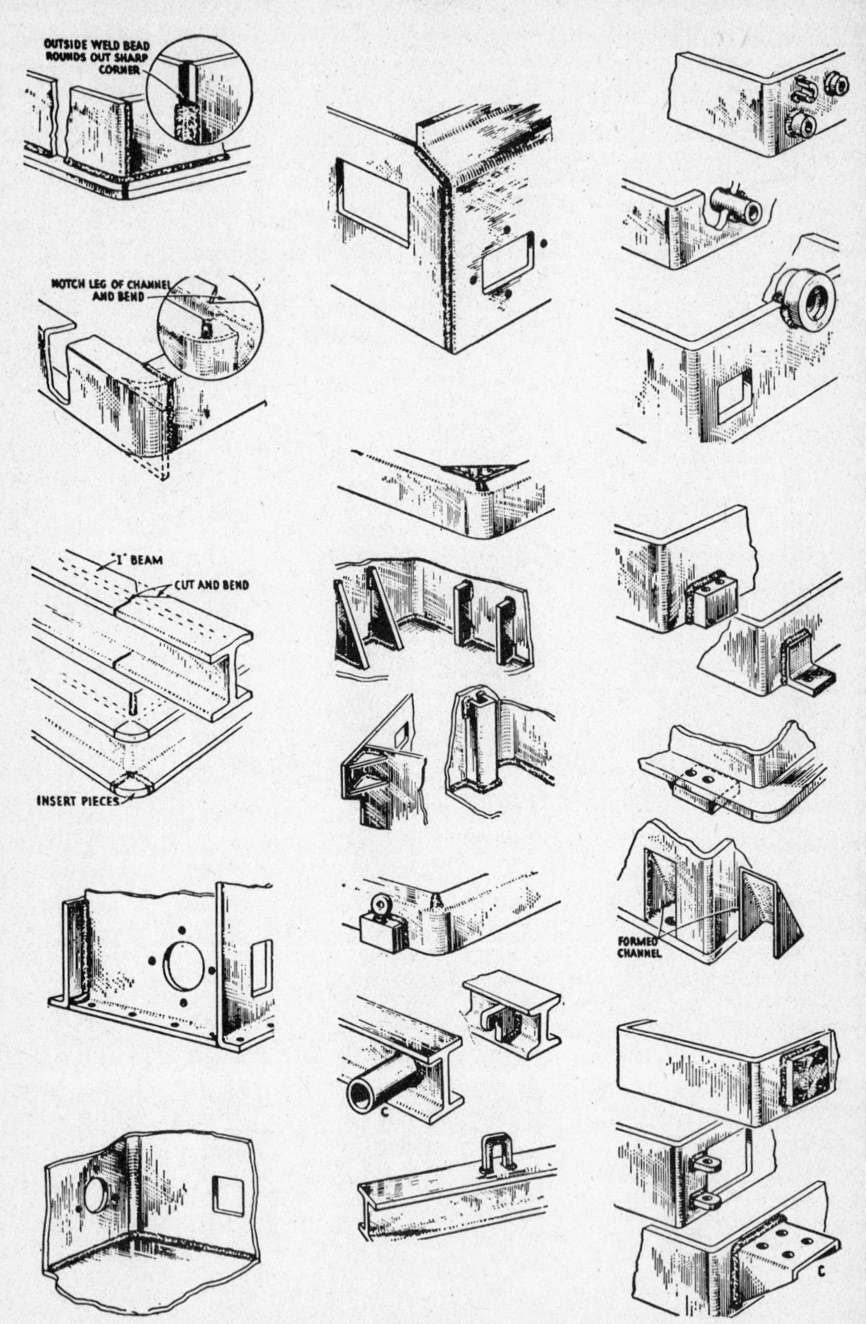

Fig. 5-117

Containers

Fig. 5.

The first thing to remember when considering the possibility of converting a casting to a weldment is that most castings can be simplified into several simpler parts when welding is used. Try to visualize the casting as a group of separate parts and then consider each part individually. Most machines can be broken down into the following main classifications:

1. Bases or frames
2. Covers
3. Containers
4. Wheels
5. Auxiliary parts (brackets, links, levers, etc.)

Figure 3 shows how even a complicated press is actually a group of parts which can be considerably separately.

Bases and frames are the heaviest parts of most machines. Usually this means that they offer the greatest opportunity for conversion, because the cost per pound of the mild steel which goes into a weldment is considerably less than the cost of the cast material. Hence, the greater the weight, the greater the potential savings. Note also that mild steel is stiffer and stronger than cast iron, so that, when a conversion is made from cast iron to steel, less material is required to perform the same job. The conversion of bases and frames is largely one of proper selection of standard steel shapes which can most economically be welded together. The use of ribbing or stiffeners can also be very beneficial. Figure 4 shows some ideas which have been incorporated into base designs.

Covers include gear guards, doors, pulley housings, dust covers, lids, etc. They are usually made of lighter gauge material. Forming operations should be used to eliminate welding wherever possible. The least expensive weldment, particularly on this type job, is the one which has the least amount of welding on it.

Containers is a very broad group and includes all types of closed parts. Though the material may be heavier than that used for covers, the investigation of various methods of forming is extremely beneficial. Figure 5 includes illustrations of containers of welded design.

Wheels are widely used and many are readily converted to welding. Tubing, bar stock and plate are usually used on wheels. Figure 6 shows several wheels of welded design.

Wheels

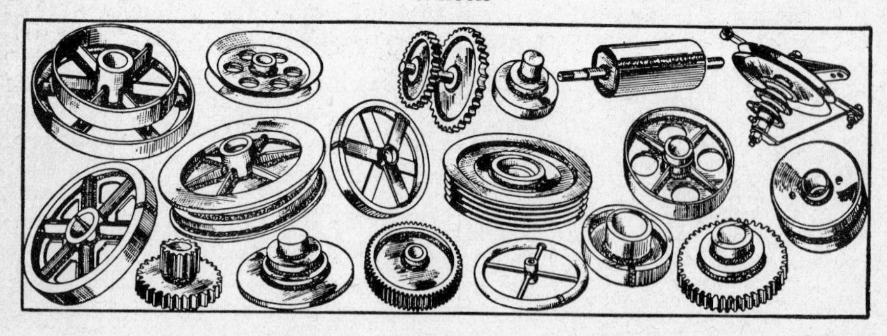

Fig. 6.

Levers

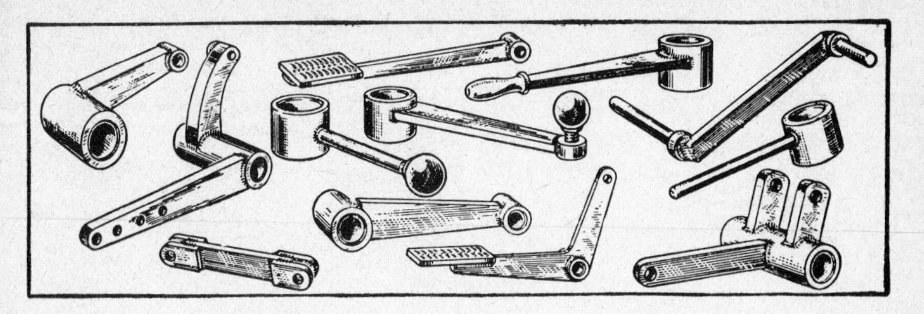

Links and Clevises

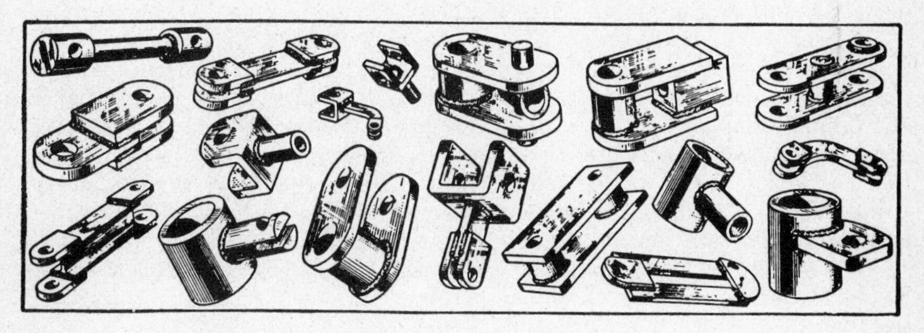

Fig. 7.

Auxiliary parts include brackets, bosses, bearings, cams and eccentrics, levers, cranks and crankshafts, and links and clevises. Originality in the selection of mild steel shapes can produce sizeable cost reductions in conversions of this type. Figure 7 shows examples of some of these parts.

Many concerns have been started on the road to conversion of castings to weldments by the weldor who intelligently pointed out the savings which could be obtained and convinced the skeptics that welding is a reliable and economical process. Not only did this result in savings for the companies, but many of the weldors, who were farsighted enough to make the suggestions, are now supervisory people and continuing to advance their own career in welding. It pays the weldor to promote welding!

PART 7.1 — *Metals:*

Their Identification, Classification, Manufacture, and Weldability.

When planning a career in welding, some provision must be made for spending time and effort to follow a definite program of self-improvement in the technical and scientific aspects of welding. An all-around weldor must know more than just how to deposit weld metal correctly.

It is necessary to know something about the characteristics and identification of metals, their chemical and physical structure, and the effects of heat on them, since welding is a process of joining metals with heat. An acquaintance with some basic metallurgy is helpful in solving problems which may arise. A knowledge of the metal classifications and order specifications as used on working drawings and purchase orders is a necessity in many types of welding jobs.

It is often necessary to first identify the general type of metal used in a part before selecting or working out the proper welding procedure in a maintenance or construction job. This means that the weldor should have one or more dependable, accurate and rapid methods of identifying metals. Most of the metals encountered by the weldor engaged in general construction and repair work will be some type of ferrous metal. A ferrous metal is one which contains a predominant amount of iron, such as carbon steel, low alloy steels and cast iron. Non-ferrous metals are those which contain no iron or very little iron content. Emphasis in metal identification, classification and manufacturing methods, therefore, will be placed on the ferrous metals.

Identifying Metal by Surface Appearance—The shape, texture, and color of a metal will help to classify it. An experienced weldor can often immediately identify a metal by visible means.

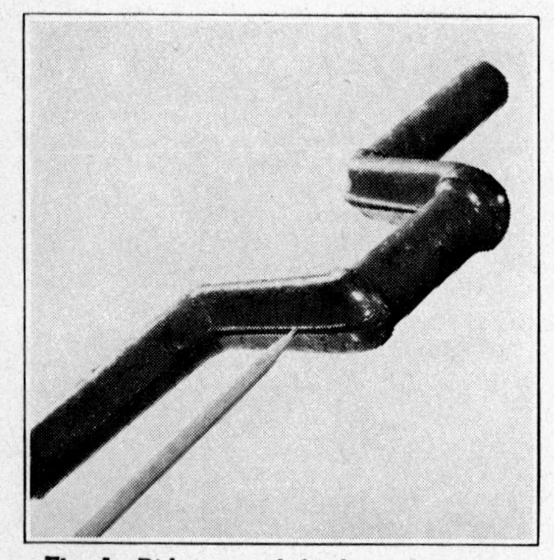

Fig. 1. Ridges are left along the sides of drop-forged parts at points where the two dies overlap. The surface is rough due to iron scales which form on the dies during the forging operation.

Fig. 2. Gates are present on sides of parts which are cast in molds. Cast steel parts have large gates as compared with cast iron.

Most castings are made by pouring molten metal into a mold cavity made by withdrawing a wood or metal pattern from the sand which has been rammed around it. Metal which has been cast in sand has a rough "grainy" appearance, due to the imprint of the sand grains.

Castings have a ridge, called a "flashing," along the edge where the two halves of the mold came together. This may be ground off flush, but its method of removal or a small portion of the flashing will usually be visible. There may also appear one or more enlarged areas along the flashing which are "gates." These are formed by openings cut into the sand mold to allow the metal to flow into the mold cavity during the casting process. On the surface a gate appears as though it were fractured, but may have been smoothed by grinding.

Forgings are made by placing pieces of steel heated to a forging temperature in a die where they are forced into the die under pressure of a heavy hammer or press.

Pieces which have been drop-forged may have a rather rough, scaly surface appearance, while others are tumbled or shot peened to give an even-grained effect. There is usually a flashing left along the edge of the piece where the two dies come together. This is always sheared and sometimes ground smooth. The parts numbers which are stamped on the part during the drop forging operation appear sharp and distinct as compared with those found on castings. Forgings are usually made of medium carbon steel or low alloy steels.

Identifying Metal by Sound—Identification may be made by rapping the metal with a hammer in order to hear the sound given off. It is difficult to explain a sound in words, so it is necessary for each individual to compare the sound from different types of metal a number of times to establish a dependable standard. For example, when one strikes rolled or forged steel with a hammer, it resounds with a tone higher pitched than that given off by cast metal. Gray castings, malleable castings and white castings may be compared by sound. Gray castings have a dull, chalky tone while malleable cast has a higher, clearer-pitched tone, and white cast iron an even higher pitch.

Identifying Metals Using the Spark Test—Most ferrous metals may be roughly classified by observing the sparks given off when the surface or edge of the metal is touched against a grinding wheel. To classify the unknown metal the sparks given off are compared to those of a known metal. Either a portable or stationary grinder may be used. A wheel of medium grit, such as 40 to 60 grain, gives the most satisfactory results. The wheel should operate at a speed of 7,000 to 8,000 surface feet per minute. An 8 inch diameter grinding wheel turning at the rate of 3,600 R.P.M. provides a surface speed of 7,500 feet per minute.

The wheel should be dressed before testing a metal. This removes glazing, sharpens the wheel, and removes traces of metals ground previously.

Sparks can be observed more easily in a diffused light. Avoid bright sunlight or a darkened room. If the sparks are given off against a dark background it is easier to distinguish the spark characteristics.

Different elements in the steel influence spark behavior. The presence of carbon causes a bursting stream of sparks; the higher the carbon content, the more plentiful the spark bursts (Fig. 5). An exception to this is when the metal has alloys such as silicon, chromium, nickel and tungsten which tend to suppress the carbon bursts. As the manganese content is increased the sparks tend to follow the surface of the wheel. Chromium gives off sparks which are orange in color, and make it difficult to get a spark stream. When nickel is added to a metal forked tongues are produced on the end of the spark streams.

Heat treatment also changes spark patterns. The undersurface as well as the surface should be tested since some metals may have a low or high carbon surface due to surface treatment, while the underneath composition is different.

It will take practice to classify metals by the spark test. Spend time observing sparks from common known metals, carefully noting the length, color and end shape of the spark from the time it leaves the wheel until it disappears. It is advisable to keep samples of known metals by the grinder to use in making comparative tests. The color of the spark stream is important. It will vary from white, yellow, orange and straw. Various terms are used to describe the

Fig. 3. A bench grinder is used to grind metal for identifying sparks. A medium grain, 8-inch diameter wheel turning at 3600 R.P.M. is used. Sparks are easier to observe if seen against a slate-colored background.

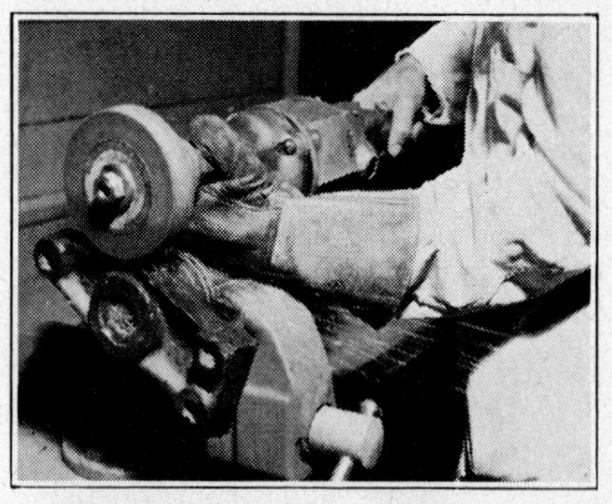

Fig. 4. A portable grinder is used to spark-test pieces before welding operation is begun. It is easier to identify sparks given off in a diffused light.

pattern of the spark stream. The sketches in Fig. 5 illustrate the common patterns and terms.

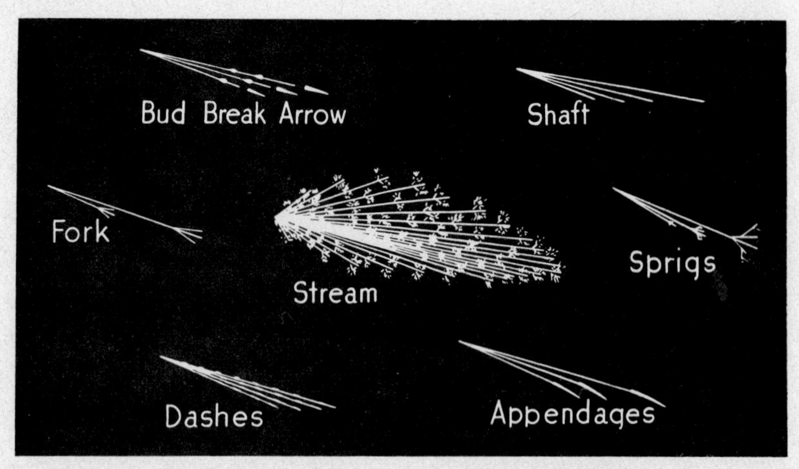

Fig. 5. Illustrations of terms used in spark testing.

Identifying Metals by Fracture—Looking at the broken edge of the part is one of the first steps in metal identification. The fractured surface reveals such items as nature of the break, type of grain and color.

The surface of a fracture on a piece of gray cast iron is dark gray and will usually rub off black (graphite) on the finger tip. White cast iron has a silvery white appearance. Malleable cast iron shows a dark center with a light outer skin, due to its surface treatment.

The way in which the part breaks is of interest. Malleable cast iron parts are ductile and usually bend before breaking. The metal along the edge of the break will indicate this characteristic by showing some distortion. Gray and white cast irons are brittle and make a clean break.

The fractured surface of steel provides a definite grain pattern. Low carbon steel is bright gray, while high carbon is a little darker, but still a light gray.

The fractured edge of a steel axle or rod has a distinctive grain pattern. The metal checks at a point of stress and as the break continues across the piece, the friction of the two surfaces smooths the fractured grain pattern. When the part finally breaks, the portion of metal which broke last has had no rubbing action and shows a crystalline surface. In the presence of oil or grease, this surface will often be free of grease, and the older portion of the break will be darkened by it.

Identifying Metals by Chipping and Filing—It is sometimes impossible to observe a clean fracture in the metal, because it is oxidized, dirty, or the metal is still in one piece. The use of a file and chisel may help the weldor in several ways. The oxidized or dirty surface is removed exposing the true color of the metal, as was discussed when observing a fracture. The relative hardness of the metal can be determined by comparing its resistance to the cutting edge and the type of chip it yields with that of mild steel. A soft metal will cut easily, giving a continuous chip. A tough metal will also give a continuous chip, but will present more resistance to the cutting edge. A brittle metal will give a series of broken chips. A hardened metal will resist the cutting of file or chisel. Mild steel yields a continuous chip, and increases in toughness and hardness as the carbon content is increased. Gray and white cast irons tend to chip off in a series of small pieces, with white cast iron being more resistant to the cutting edge. Malleable cast iron will give a continuous chip, because of the annealed surface.

Identifying Non-Ferrous Metals—Although there are many non-ferrous metals and alloys in use today, they are not encountered as often by the general construction and maintenance weldor as those metals in the ferrous category. A weldor should be able to identify the common non-ferrous metals, as most of them will be confined to a few basic metals or families of metals. The copper family, ranging from pure copper through varying alloys of brass and bronze, aluminum and its alloys, and the nickel alloys are probably the most common. Many other categories such as magnesium, lead, "white metal" and titanium are commercially available and useful, but price, certain physical properties, or their recent development prohibits all but specialized uses.

Most of these metals are formed in the same way as ferrous metals, so that surface appearance will often show the type and form of metal being identified. The same distinguishing marks will identify cast and forged metal, and that which has been formed by rolling. The brasses and bronzes, however, are not usually hot worked. They are known as "hot short," which means that they have no strength above their critical temperatures, and cannot be hammered or rolled, because they break or crumble rather than hold their shape under pressure of a hammer or rolls.

Weight is often a distinctive characteristic of the non-ferrous groups. Aluminum, although it is only slightly whiter in color than low carbon steel and the nickel alloys, it is much lighter in weight. It is approximately one-third the weight of steel, and also approximately one-third the weight of "white metal" or "pot metal" which it is sometimes mistaken for. Alloys high in copper and nickel are slightly heavier than steel, and everyone is familiar with the comparatively heavy weight of lead.

The spark test will not tell anything except to ascertain that the specimen is a non-ferrous metal. Non-ferrous metals and alloys do not yield a spark on the grinding wheel, except for a very small wavy spark from nickel.

Looking at the fracture on non-ferrous metals is a good way to identify them. If the fracture is fresh, the color of the metal is quite true and free of oxides. If the fracture is old, or the metal has not come apart, it may be necessary to take a file or chisel and remove the surface of the metal to observe the undersurface. Many non-ferrous metals quickly lose their true color when exposed to the atmosphere or by handling. Aluminum has a white color. Copper is red, and as it is alloyed with zinc to make brass it becomes yellow, and with the addition of tin or aluminum to make bronze it has a lighter yellow or gold color. Nickel has a nearly white color which turns to a light gray as it is alloyed in Monel and Inconel.

Filing or using a chisel to remove chips is a help in identifying non-ferrous metals and to check their hardness. Aluminum cuts easily with the chisel giving a continuous chip, but as it is alloyed for strength the color becomes more gray and the metal is tougher to cut. A file removes aluminum and its alloys easily, and unless the file is very coarse or has curve cut teeth, the soft aluminum tends to clog and cause "pinheads" in the teeth. Copper and its alloys cut rather easily with the file and chisel, giving a continuous chip, although they are not quite as soft and yielding as aluminum. Nickel is tough and its alloys are tougher. It will cut with the file, and the chisel will give a continuous chip, but it will take more pressure to force the cutting edge through the metal. Lead is the softest of all the common metals and will cut easily with a chisel showing a bright gray-white surface. It will quickly clog up anything but a special file made for it.

Care must be exercised when welding or working non-ferrous metals in the molten state. The fumes given off by these metals or their fluxes are often toxic and should not be breathed into the lungs.

Melting Points for Metals and Alloys—A weldor should know the relative melting points of common metals and alloys. In addition to this it is important to know the temperatures the metal passes through when becoming solid. In repair work the weldor is often required to join two different metals, such as mild steel and cast iron. It is well to know that there is a difference in the melting points of these two metals of about 500 degrees Fahrenheit (F.), cast iron having the lower melting point. A general rule to remember is that as the carbon content is increased, the melting point is decreased.

Bronze is an alloy which melts at a comparatively low temperature. When bronze is applied to steel in the brazing operation, the bronze is molten at 1600 to 1650 degrees F. whereas the steel is heated only to a bright red.

Aluminum and copper melt at definite temperatures. Alloys, being mixtures of two or more metals, melt within a variation of temperatures depending upon the combination. Solder is an example of an alloy of this type. The melting points of the metals and alloys that have a melting-point range narrow

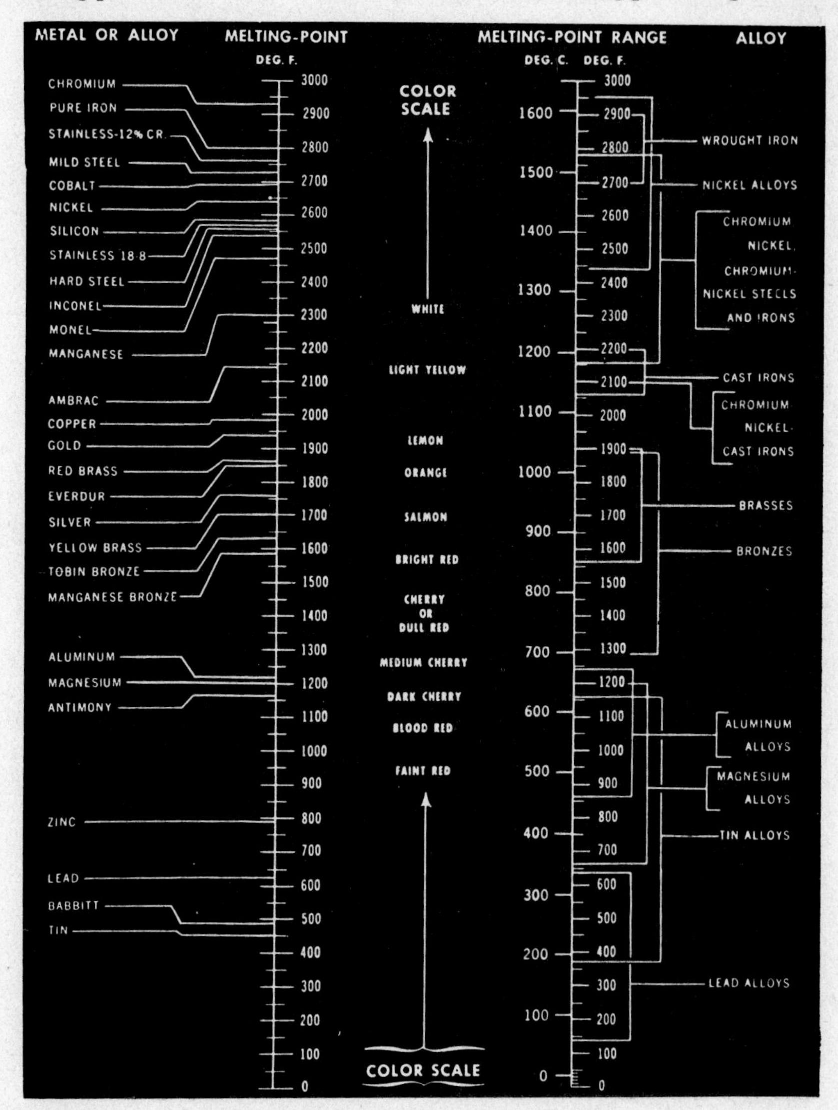

Fig. 6. Melting points of metals and alloys.

enough to be considered a single temperature are given in the left hand column of the chart[1] (Fig. 6). The melting points of various alloys having a wider

[1]Industry and Welding, August, 1946.

range appear in the right-hand column. The color scale in the center shows the approximate temperatures at which color changes appear. The right-hand column with both Centigrade and Fahrenheit scales can be used as a general conversion table between these scales.

The color scale is purposely indefinite as to the border line between colors, because of differences of opinions as to color names, and because the amount of light present effects the visible color. The colors for the temperatures indicated can be seen only in a dark location. In a well-lighted shop, faint red may not appear until the metal has reached a temperature of 1100 or 1150 degrees F.

The Effect of Welding on Grain Structure of Metal—During the arc welding process when metal in the crater is molten such changes in the metal may take place as removal or introduction of gases and formation of oxides. When the metal cools and solidifies, a grain structure is formed which influences the hardness, strength, toughness and impact-resistance of the metal. The rate of cooling influences grain size as well as composition of the metal. Information about the changes which take place in that part of the metal which is heated and cooled during the welding cycle is important. This information is helpful when undertaking new welding applications or in understanding the added precautions necessary when welding certain common metals, such as hardenable steels.

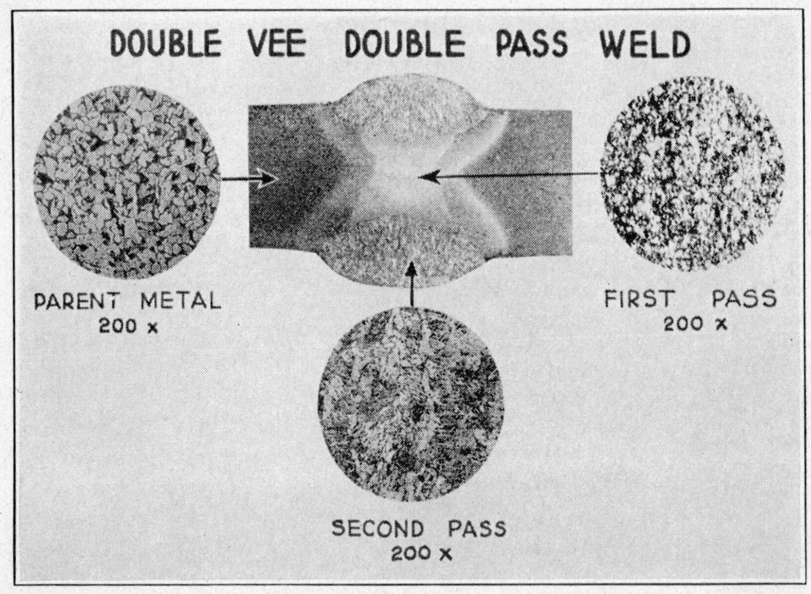

Fig. 7. The annealing effects of a second pass weld are shown by comparing the grain structures of parent metal with those of first and second pass welds in a double-vee butt joint.

Grain structure of metal changes when the second pass of a double-vee butt weld is made. Observe the effects on the grain structure in the magnified sections of the parent metal and the weld metal after the first and second pass is made (Fig. 7). The annealed metal of the first weld pass has a fine grain and characteristics which make it tough. Preheating and postheating will often give a weld added strength and toughness as well as prevent cracking and stress concentration.

Grain Structure—Metals in the solid state are made up of a grain pattern arranged in a precise geometric pattern as they cool and solidify. This is due to the natural arrangement of small sub-microscopic units called atoms. Iron takes a cubic form. This basic cube has one atom at each corner and one at the exact center of the body, and is called body centered cubic. These cubes are the building bricks which fit together and form the grain or crystalline structure of the metal. Many grains, in turn, form the bar or plate of iron with which we are familiar in the shop. The diagram in Fig. 8 is provided as an explanation of the effect of heating on grain growth.

The structure and physical properties of steel are modified or changed entirely when the metal is heated and cooled. The most significant changes take place at temperatures known as critical points (or critical range). The critical point is the point where steel recrystalizes from body centered cubic to face

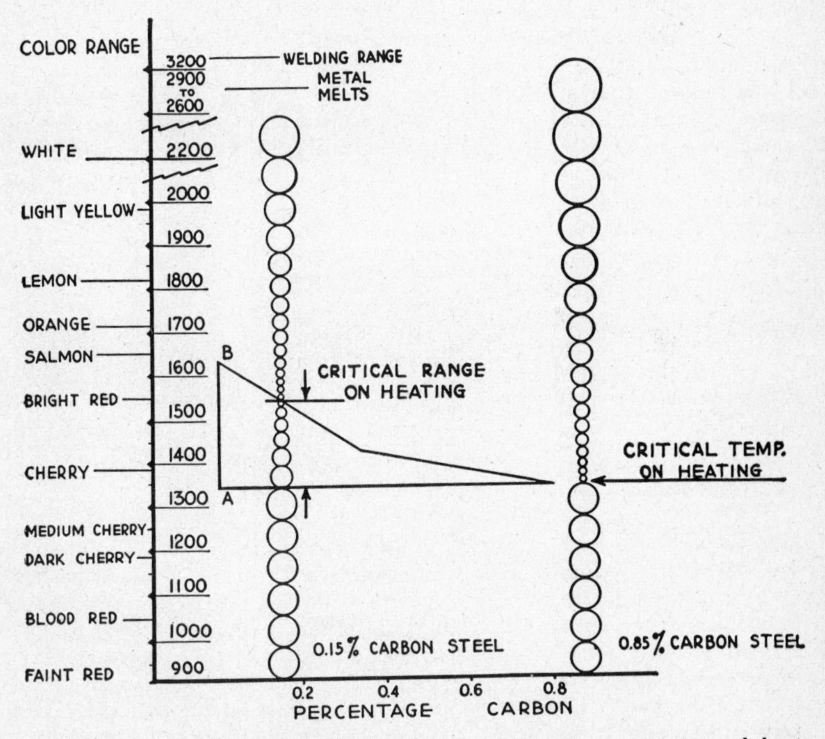

Fig. 8. Steel temperature and grain growth chart. The temperature of heat treated steel is estimated by observing the color. Grain growth increases or decreases depending upon the heat treatment given the steel.

centered cubic on a rising temperature. At this point iron becomes non-magnetic, so that the iron may be roughly checked in the shop with a magnet to see when the point is reached. Grain size increases and decreases depending on heat treatment given the metal. The circles drawn on the diagram in Fig. 8 illustrate the effect of heating on low and high carbon steel grain growth.

When low carbon steel is heated, grain refinement commences at approximately 1325 degrees F. (line A, Fig. 8). Grain size continues to diminish until the metal is heated to a bright red color (line B). When the metal is heated beyond this point the grain begins to grow in size. This continues until the melting temperature is reached.

Grain size in high carbon steel changes immediately when reaching the critical temperature of heating (1325 degrees F., line A). At this point grain size is very small. The size then increases slowly as the temperature of the metal is increased.

The size of grain in steel may be coarse, medium, or fine before heating is begun, depending upon the previous heat treatment which has been given the metal. Regardless of grain size previously obtained, the grain will be fine when the steel is heated to the temperature shown on the chart as line B. It is important that this grain growth be understood and the temperatures be recognized for purposes of welding and heat treating. To obtain maximum strength and impact resistance a refined grain is necessary.

A fine grain structure is reached in low or high carbon steel when the metal is heated to the temperature indicated by line B. Fast or slow cooling will hold this fine grain structure so that it will be present in either a hardened or annealed piece of metal. Rapid cooling of metal from this temperature zone to a temperature of 200 degrees or lower causes the metal to harden. Full hardness in steel, however, is not obtained unless the carbon content exceeds approximately 60 points (0.6%). A piece of steel can not be hardened and used as an edge tool unless there is sufficient carbon content.

Fig. 9. A comparison of grain structure is observed when two pieces of a high carbon, fine grain file are welded together with a butt weld.

Steel is annealed by first heating to a temperature of 50 to 100 degrees F. above that shown by line B. The steel is allowed to cool slowly to room temperature by packing it in sand or some other insulating material, or allowing it to cool off in the furnace after shutting it off. Annealing is used to completely soften steel, relieve internal stresses due to previous heat treatment or mechanical working, and refine the grain. Normalizing is similar to annealing, except that the metal is allowed to cool down from above the critical temperature by leaving it in the air. The resultant metal is softened enough for machining and stress relief.

Manufacture of Ferrous Metals—It may prove helpful to individuals working with metals to have some knowledge of the manufacturing processes used in producing them.

Iron is not found in the free state (except as meteoric iron), but is combined with other elements as an ore. The ore when removed from the mine is mixed with dirt and impurities. Iron ore has the appearance of ordinary reddish soil. Some iron ores are found in surface veins and are mined from open pit mines after the overburden of soil and rock has been stripped off. Deeper veins are mined by underground methods. The iron ore is loaded on railroad cars in the mine and transported to lake ports where it is loaded on boats and shipped through the Great Lakes to blast furnace locations. The blast furnace manufactures pig iron necessary in the later processes to manufacture cast iron and steel.

The raw materials used in the blast furnace to obtain a metallic material are iron ore, limestone and coke. These materials are carefully weighed and charged by layers into the blast furnace. Iron ore as mined is usually high in iron oxide content. The coke derived from soft coal, provides heat and reduces the iron oxides to iron. Limestone forms a flux which unites with the non-metallic content of the ore and forms a slag. Hot air is blown through the bottom of the blast furnace to intensify the heat of the burning coke. When the heating action has been completed, molten slag floats to the top and molten iron settles to the bottom of the furnace (Fig. 12). The iron is tapped off and either transferred to the steel mill for immediate further processing, or poured into molds where it is left to chill and form "pigs" to be used at a later date or at another furnace location. The blast furnace process is continuous. As each layer of coke and limestone is consumed while reducing the layer of iron ore above it into molten metal, other layers follow them into the smelting zone and more are charged into the top of the furnace.

Fig. 10. A view of the Hull-Rust-Mahoning open pit iron ore mine of U. S. Steel's Oliver Iron Mining Company at Hibbing, Minnesota.

Fig. 11. This iron ore carrier is shown being loaded at an upper lake port.

Gray Cast Iron—The properties of gray cast iron such as cheapness, ease of machining and forming, and high compressive strength have long made it an important metal in machine construction, in spite of its low tensile strength of 18,000 to 30,000 p.s.i. (mild steel is about 65,000 p.s.i.). Cast iron is not a pure metal but a mixture of iron, carbon, silicon, manganese, sulphur, and phosphorus. Different foundries will vary these elements to obtain a casting suitable for their particular product. Cast iron has a much higher carbon and silicon content than steel.

Cast iron is produced by melting a charge of pig iron, scrap steel and iron with the necessary alloying ingredients in a cupola (similar to a small blast furnace) with limestone for a fluxing agent and coke for fuel. Cast irons may also be melted in an electric arc furnace. Molten cast iron is poured into sand molds. Cover plates, gears, bases, brackets and machinery housings are a few uses of gray cast iron. Gray iron is allowed to cool slowly in the mold. During this cooling period carbon separates from the iron in the form of tiny graphite flakes. It is this graphite which gives the casting its grayish appearance. The graphite provides sufficient lubrication so that gray cast iron may be drilled and machined dry without cutting oil.

Gray cast iron is very low in ductility and malleability, that is, it will not stretch or distort. Welding heat causes expansion and contraction forces that may result in excessive stress or fractures. The weldor must give special consideration to these factors when welding gray iron castings.

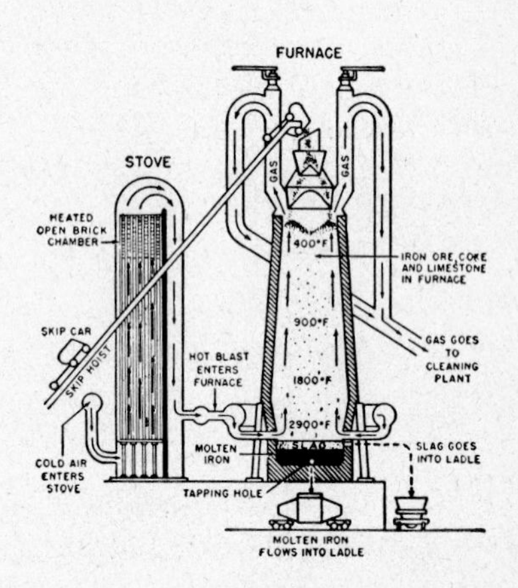

Fig. 12. A view and cross sectional diagram of a blast furnace. The "blast" comes in the form of heated air from stoves, the dome shaped structures to the left of the furnace.

The most satisfactory method of repairing gray iron castings is by fusion welding with the metallic arc. Machinable and non-machinable type electrodes have been developed for use with arc welding equipment. The machinable type electrode is nickel. The weld deposit is soft and easily machined, having characteristics similar to the metal in the casting. A low heat used in making the application reduces the fractures which result from heat stresses. This type electrode is used extensively for repairing broken parts, filling up defects in castings, repairing machining errors, building up worn parts and filling holes which must be drilled and tapped. The non-machinable electrode is a low carbon electrode with a special coating for cast iron. It may be ground to shape. The skin of the casting, formed by the sand mold, should be ground off in the weld area exposing clean metal.

Bronze welding may be used to weld gray cast iron using either the carbon arc or the oxy-acetylene torch. In most cases the casting will have to be preheated to prevent localized stresses. The weld area must be brought to a red heat so that the bronze flux and rod will flow and tin the surface.

Fracture	Spark Test

Dark gray color rubs off black.

Surface Appearance

Evidences of sand mold. Color, dark gray.

Gray Cast Iron

Color-red

Color - straw yellow

Average stream length with power grinder - 25 in.

Volume-small

Many sprigs, small and repeating

Fig. 13. Identification characteristics of gray cast iron.

Fracture	Spark Test

Very fine silvery white crystalline formation.

Surface appearance

Evidences of sand mold. Color, dark gray.

White Cast Iron

Color-red

Color - straw yellow

Average stream length with power grinder-20 in.

Volume-very small

Sprigs-finer than gray iron, small and repeating

Fig. 14. Identification characteristics of white cast iron.

White Cast Iron—This is also known as "chilled" cast iron, due to its method of manufacture. It is produced in the same manner as gray cast iron, except that the molten metal is cooled rapidly in the mold. In some cases the iron and carbon content may be varied to assist in making white cast iron. The sudden cooling or correct balance between iron and carbon causes the carbon to remain in combination with iron instead of separating in a graphitic form. This results in a metal which is harder and more brittle, a fact which must be kept in mind when white cast iron is welded. Due to its brittleness, white cast is not used extensively in machinery, except for parts that require a wearing surface which will withstand abrasion. A white cast iron surface is sometimes put on a gray iron casting by chilling the surface of a heavier casting. This affords a more abrasion resistant surface. Although it can be repaired by arc welding under special conditions, the repair of white cast iron by the welding process is not recommended. Heating and cooling during the welding process tends to change the characteristics of the metal in the weld area. Generally the type of part made of white cast iron is not worth repairing when it is broken or worn out.

Malleable Iron—Many uses have been made of malleable iron castings in the production of machinery during the last century. It is much more impact resistant than gray iron and will withstand some bending without breaking. Its tensile strength ranges from 38,000 to 55,000 p.s.i.

Malleable iron castings are manufactured by further processing hard, brittle white cast iron. These white iron castings are packed in annealing boxes buried in furnace slag, mill scale or sand and annealed in a furnace at a temperature of about 1600 degrees F. for a period of 40 to 60 hours. The temperature is lowered slowly to room temperature. After this operation, the castings are cleaned by blasting or tumbling. The weldor must understand that there is a critical temperature when white cast changes to malleable. Whenever malleable iron is heated above this critical temperature (1325 degrees F.), the metal reverts to one having some of the original characteristics of white cast. Since the desirable physical properties of malleable iron are dependent upon the heat treatment or malleableizing process, heat from fusion welding has a tendency to change the physical properties of the metal unless the critical temperature is observed. The following two methods are commonly used to repair broken malleable parts.

Bronze welding has been one of the standard methods of repairing malleable iron parts. This application is made with either the oxy-acetylene or carbon arc torch. The fracture is veed out if the base metal is $3/16''$ or thicker. The metal is heated to a dull red color, and a brazing flux is applied with the bronze rod to tin the surface. Additional bronze is applied to build the joint up to the desired height.

Arc welding with nickel type electrode is also used on malleable castings. This electrode gives a machinable deposit. The temperature is kept low enough so that the hand can be placed on the casting after each pass. It is important not to overheat the casting to prevent it from reaching the temperature where stresses are developed. This makes the electric arc process ideal for this metal, as the arc heat is concentrated, the puddle is instantaneous and the filler metal may be deposited quickly and the heat removed. Little difficulty is encountered at the hard fusion zone which forms at the line where the alloy electrode deposit and casting join. Post heating is applied to provide an annealing action to the casting if the weldor believes that the critical temperature of the metal has been reached.

Special Cast Irons—In an effort to overcome the disadvantages of low tensile strength and impact resistance, many types of cast irons have been tried. Alloy gray castings are coming in for more frequent use. This casting contains the basic elements of gray iron with the addition of nickel, chromium, molybdenum and vanadium being the most common alloying elements. These alloys raise the tensile strength up in the 40,000 to 60,000 p.s.i. range, and increase the price of the casting two to five times over gray iron.

Fracture	Spark Test
Color, dark gray except for light outer layer. **Surface appearance** Evidences of sand mold. Color, dark gray.	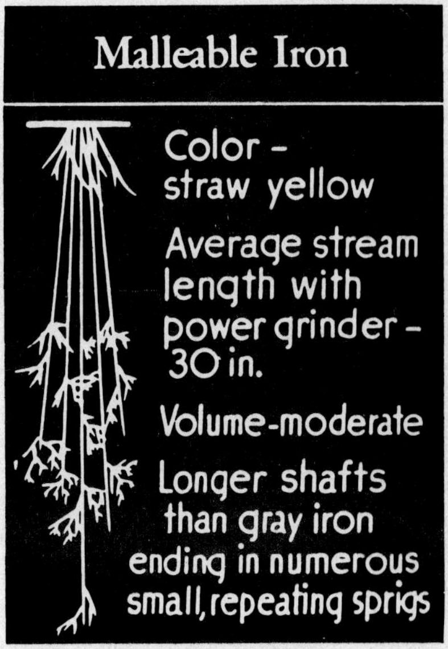 Malleable Iron Color - straw yellow Average stream length with power grinder - 30 in. Volume-moderate Longer shafts than gray iron ending in numerous small, repeating sprigs

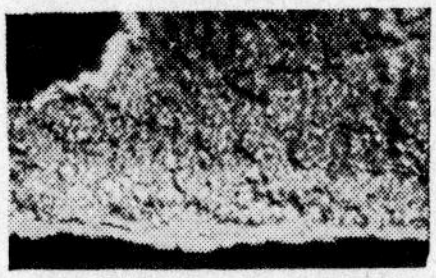

Fig. 15. Identification characteristics of malleable cast iron.

With the addition of special alloys and mixtures a change may be effected in the form of the carbon within the casting. Instead of being in flake form, as in gray cast iron, the carbon is in small spheres with the iron. This is called "ductile iron" or "nodular cast iron". The resultant castings are nearly comparable in tensile strength with alloy steels, and possess even higher impact resistance. The term "semi-steel" is also used for these castings.

These castings will not be encountered too often. Special alloy electrodes are available for welding them. Repairs may usually be made using the nickel electrode, although the strength of the weld may not be as high as the base metal. The same care should be exercised as when working on gray iron.

Wrought Iron—Wrought iron is the oldest form of structural ferrous metal, and is made from pig iron, iron oxide and silica. Modern manufacturing processes mix and work these ingredients together. After the pig iron has been melted and purified, it is carried to the processing machine where the molten pig iron is poured into a ladle which contains a predetermined amount of silicate slag. A reaction takes place between the slag and molten metal, and the resulting product is a sponge-like ball or mass of metal. This sponge-like mass is then worked in a heavy press. The action of the press removes the surplus slag and welds the various particles of slag-coated plastic iron into a solid bar or "bloom". This bloom is rolled into plates, sheets or bars, depending upon the use to be made of the metal.

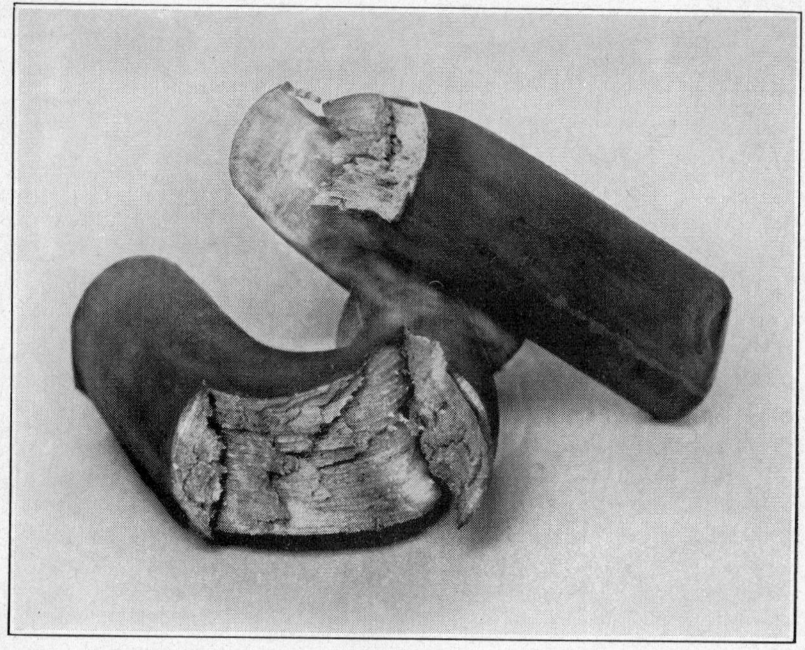

Fig. 16. A broken surface of wrought iron has a fibrous appearance.

When wrought iron is bent so the surface is fractured, it appears to have many fibers running the length of the stock.

The chief advantages of wrought iron are its corrosion resistance and its ability to resist breakage when exposed to sudden or excessive shock. Its corrosion resistance is attributed to the pure iron which is used as a base metal and the presence of glass-like fibers embedded in the metal.

The arc welding procedures used when welding wrought iron are similar to those followed when welding mild steel. A mild steel electrode, such as the E6013, will give satisfactory results. The speed of travel is reduced slightly in order to keep a molten pool of metal in the crater for a longer period of time. This permits the excess slag present in the base metal to flow out on top of the bead. A slightly lower current is used to provide only sufficient penetration to bond the weld and base metal.

Excessive penetration results in freeing large amounts of slag present in the base metal. Remove a minimum amount of base metal during the joint preparation, so that additional filler metal is also kept to a minimum.

Steel—Carbon steel is a mixture of iron and carbon with additional small amounts of manganese, sulphur and phosphorus. Over 99 per cent of carbon steel may be iron. The carbon content varies from 0.05 per cent to 1.5 percent. The amount of carbon, while relatively small, has a very important part in determining the property of the metal. Steel differs from gray cast iron in that the carbon in the steel is all in a chemical combination with the iron and none exists in the free state. Other alloying elements which are sometimes added, depending upon the use to be made of the steel are nickel, chromium and molybdenum. These alloys provide

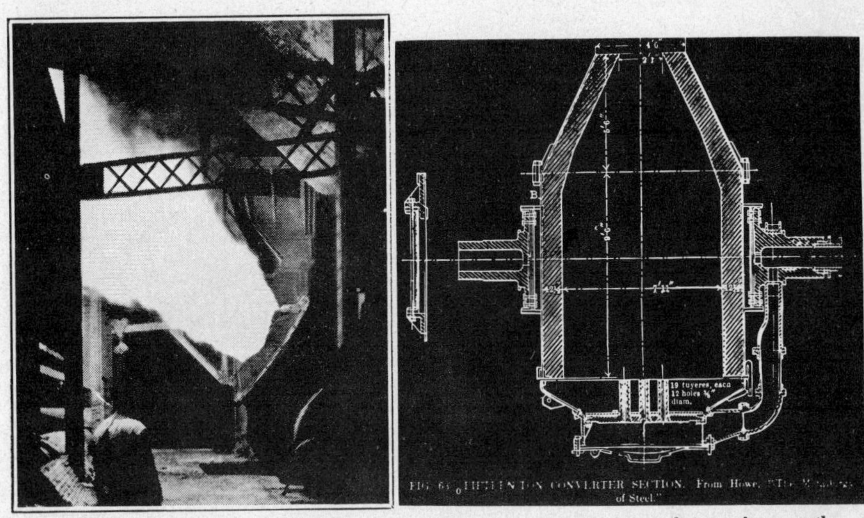

Fig. 17. Bessemer converter furnace. Liquid pig iron is poured into the mouth of the converter while it is turned on its side. An air blast is blown through holes in the bottom when the converter is turned in upright position. The oxygen burns out impurities. Flames and sparks belch forth from the mouth of the converter, producing the spectacular sights of a steel mill.

additional strength and impact resistance. The steels are usually termed "low alloy" or "high tensile", and are used where the extra strength is needed. They may be welded with special high tensile electrodes for maximum weld strength. Mild steel electrodes will join them, but the weld tends to be a lower tensile strength than the base metal.

Pig iron, the raw product turned out of the blast furnace, is used as one of the ingredients in the manufacture of steel. Large quantities of scrap steel may also be used in the manufacturing process. The first step in steel making is the removal of impurities in the pig iron by the process of oxidation. The three most common steel making processes are the Bessemer converter, open hearth furnace and electric arc furnace.

Steel Making Processes—In the Bessemer process, molten pig iron is refined by blowing unheated air through it. The Bessemer converter, named after an Englishman who developed it, is an egg-shaped vessel having a capacity of 15 to 25 tons of metal (Fig. 17). Heat is generated from oxidation of the impurities in the charge. The blowing of the charge

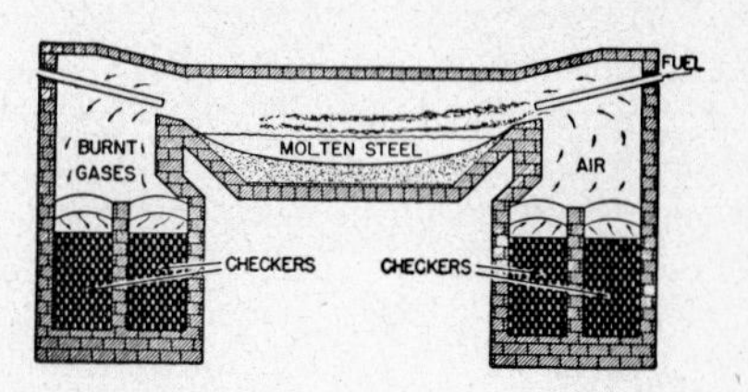

Fig. 18. The open hearth furnace. A number of furnaces are arranged in a row. Doors on front permit use of mechanical ladle to charge furnaces with pig iron, scrap steel, and limestone. Additional alloying elements are added during the melting operation. Fuel is mixed with hot air in burners at each end of the furnace. These operate alternately and the flames sweep down across the open hearth and out at opposite end. The furnaces range in capacity from 50 to 250 tons of steel at one heat. More than 90 per cent of all steel produced is made in the open hearth process. Over 50 per cent of the charge is composed of scrap steel. About half of the scrap is junked steel equipment such as automobile and farm machinery.

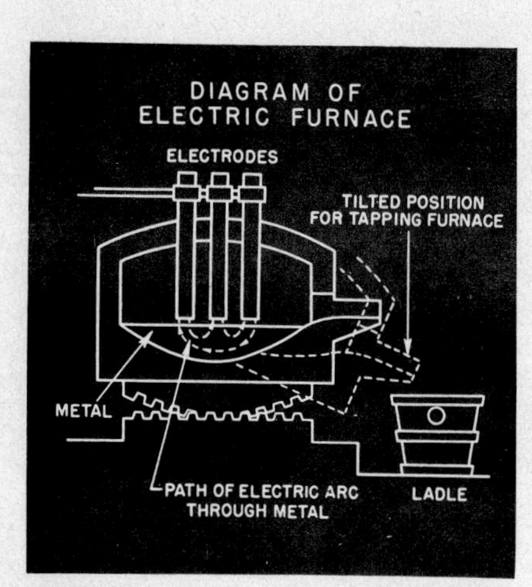

Fig. 19. Electric furnace. High alloy steel from an electric furnace at the South Works, Chicago, Illinois, of U. S. Steel's Carnegie-Illinois Steel Corporation. The entire furnace tilts forward to spill its molten load into the ladle. Note at the top of the furnace the three carbon electrodes that furnish the intense heat.

Fig. 20. Tapping a 30-ton heat of alloy steels and tool steels frequently are made with the electric furnace. Heat to melt the charge is supplied by the arc which leaps between the electrodes. Impurities are removed and alloying elements added at different stages during the heating process. When the metal is ready to be tapped the furnace is tipped forward and the steel flows into a ladle.

causes an intense heat in the mouth of the converter reducing the silicon, manganese and carbon content. Blowing continues until the correct chemical content is reached. The furnace is then tipped forward and the charge is poured off. This process cannot produce a steel as closely controlled in content as will other methods of refining.

The open hearth process uses a larger saucer-like hearth to hold the charge of pig iron (solid or molten), scrap steel and limestone (Fig. 18). It is this furnace that makes use of the tons of scrap iron collected through salvage yards. Heat is supplied by preheated gas and air. The charge is heated from 6 to 14 hours. Carbon, manganese and other elements are added during the heat. The slow heating procedure increases the possibility of controlling the refining action, and samples are taken and tested on the spot during the heat. Many different grades and qualities of steels are made by this process. Approximately 90 per cent of the total steel output is attributed to the open hearth process.

Fig. 21. Steel ingots are reheated in the soaking pit in preparation for rolling. The hot ingots are first rolled in the blooming mill to reduce them to desired size.

The electric arc furnace is used primarily in the manufacture of special steels or alloying elements which go into these steels. The heat for the electric arc process is supplied through the use of three large carbon electrodes placed in the chamber which has been charged with graded scrap steel, pig iron, alloys and fluxing materials (Fig. 19). Heat is generated by the flow of electricity, and purification of the metal is brought about through the action of iron oxide and flux. Welding electrodes which must be very closely controlled in content may be made in the electric furnace.

Cast Steel—Cast steel machinery parts are formed by pouring molten steel into molds similar to those used in making gray iron castings. When the metal solidifies it is the desired shape and further forming is not necessary. The exterior surface of a steel casting resembles gray cast iron;

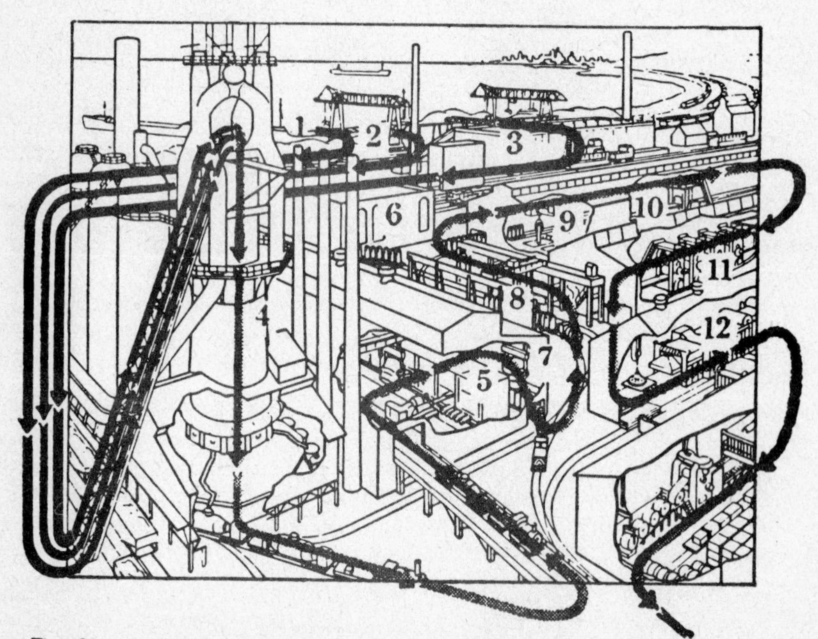

Fig. 22. A cut-away and cross sectional diagram showing the steps in the manufacture of steel in a steel mill.

however, this is the extent of the similarity. Cast steel as compared to cast iron is more ductile, and has fine grain structure after proper heat treat-

Spark Test Characteristics

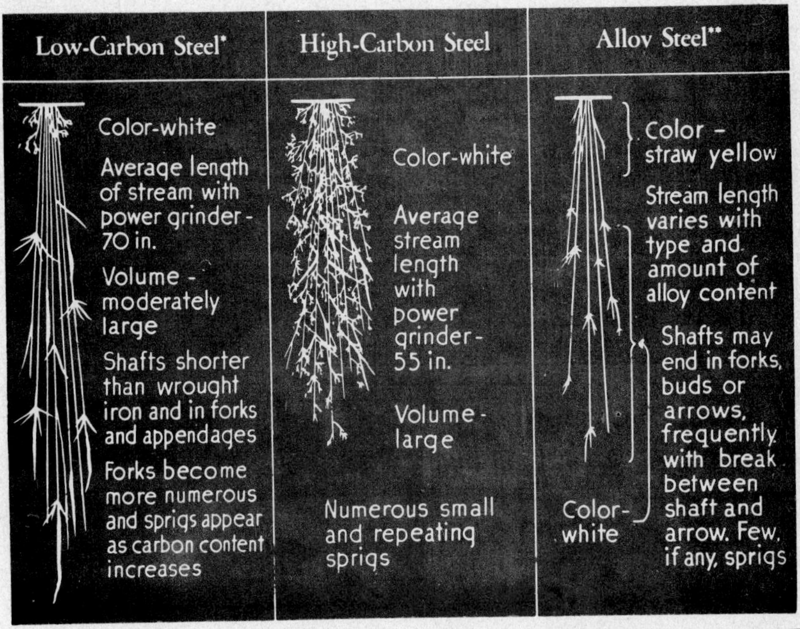

Low-Carbon Steel*	High-Carbon Steel	Alloy Steel**
Color-white Average length of stream with power grinder-70 in. Volume-moderately large Shafts shorter than wrought iron and in forks and appendages Forks become more numerous and sprigs appear as carbon content increases	Color-white Average stream length with power grinder-55 in. Volume-large Numerous small and repeating sprigs	Color-straw yellow Stream length varies with type and amount of alloy content Shafts may end in forks, buds or arrows, frequently with break between shaft and arrow. Few, if any, sprigs Color-white
Fracture Bright gray	**Fracture** Light gray	**Fracture** Medium gray
Surface Appearance Dark gray—rolling or forging marks are present. Evidence of sand mold on cast steel.	**Surface Appearance** Dark gray—rolling or forging lines may be noticeable.	**Surface Appearance** Dark gray—smooth.

*These sparks also apply to cast steel.
**Spark shown is used to identify stainless steel.

Fig. 23. Identification characteristics used to determine low carbon, high carbon, and alloy steel. The spark test is used to distinguish types of steel. An increase in carbon content results in a short spark with luminous bursts. The addition of alloying elements tends to suppress the spark bursts given off from alloy steel.

ment. A fracture shows its color is bright gray rather than dark gray as in the cast iron.

Techniques followed when welding cast steel are similar to those used when welding mild steel in structural shapes. Carbon in steel castings generally runs around 0.3 per cent, although steel castings may be made from various carbon contents as well as alloy steels. No special precautions are needed when welding cast steel unless the carbon content is higher or it is an alloy steel. The spark test should be used to determine what type of steel has been used in the casting. If the part has a high carbon content, preheat to control localized heating and rate of cooling. The sand mold edge or casting skin is removed as recommended for cast iron. Parts are not to be overheated when welded. It is a good plan to cool parts slowly after welding and protect them from drafts. If underbead cracking results after a pass has been run, this is an indication that the carbon content is high. More satisfactory results are obtained if electrodes suitable for welding high carbon steels are used.

Fig. 24. Hot plastic steel is passed between two horizontal rolls to reduce size and to shape the steel to the desired type such as angle, channels, "I" beams.

Steel Forgings—Steel which is worked hot is forged or rolled. Many common machinery parts are made by the "drop forging" process. A large hammer operated by steam, air, hydraulic, mechanical force or gravity is forced down on a piece of hot steel which has been placed on a die. The force of the hammer forms the part. Examples of forged parts are ball-peen hammers, pliers and wrenches, rock guards for a mower and mower knife heads. Most of these parts are made by the "closed die" method for maximum accuracy. The characteristics which help the weldor distinguish forged parts from castings were discussed in a previous paragraph. Forgings are used for pieces which are expected to have a rather high strength and withstand shock and vibration. This calls for medium carbon steels, in some instances high carbon steels, and alloy steels.

Hot and Cold Rolled Steel—Large chunks of steel, called "billets", are heated and rolled into various shapes such as squares, angles, channels, flats, etc. Many different types of rollers and hydraulic presses are used for these processes. Hot rolled steel is usually identified by the black oxide on the surface. This is due to the formation of oxides as the metal travels through the air at a red heat. Hot rolled stock is usually slightly lower in price than cold rolled shapes, although the carbon content may be the same.

Cold rolled steel is produced by descaling hot-rolled steel in an acid pickling solution. The steel is then passed through several sets of rolls or drawn through dies while cold. Cold rolled metal is used whenever close tolerances or metals of smooth finish are needed. Cold rolled metals have a surface which is harder and smoother than hot rolled stock, due to the cold working.

Fig. 25. A hot-rolled strip runs swiftly down the runout table. The strip is immersed in a pickling tank containing hot sulphuric acid, to remove the oxide scale, after which it is cold rolled to convert the hot-rolled strip into a smooth surface cold-rolled sheet of steel.

Classification of Steel for Construction Purposes—Weldors and operators of welding and metalworking shops use new stock in steel construction and repair operations. It is necessary to order or keep a supply of commonly used types of steel on hand for use as the need arises. Classifications and size specifications must be designated correctly for the various shapes when placing orders to insure correct stock. Manufacturers of steel and distributors can supply catalogs of available stock. Stock sizes, together with weight per foot, AISI-SAE classifications, alloy contents, working qualities and heat treatment may also be listed in this catalog. It makes a helpful reference for anyone using steels.

The Society of Automotive Engineers (SAE) have set up a 4 digit (sometimes 5) number classification for steels. The first digit of the number refers to the type or alloy category of the steel: 1, carbon steels; 2, nickel steels; 3, nickel-chromium; 4, molybdenum; 5, chromium; 6, chrome-vanadium, etc. The second digit refers to the series within that category, and the last two digits tell the average carbon content in points (.01%). The classification numbers may or may not have prefix letters designating the melting process used for the steel, such as "C" for Basic open-hearth carbon, or "A" for Basic open-hearth alloy. These were set up by the American Iron and Steel Institute (AISI).

A common type of steel used for general repair and construction work is the C1020. One could specify this steel when ordering angles or flat stock, and know that he would get: C, a basic open-hearth carbon steel; 1, within the carbon steel category; 0, plain carbon steel containing no important alloys; 20, having 18 to 23 points (.18-.23%) of carbon. For forging a chisel or punch a steel classified as C1085 might be used.

Steel is sold on a per hundred-weight basis. It may also be ordered by the running foot, however, the cost is usually billed for weight. The stock lengths of angles, rounds and bars are standardized at 16, 18 and 20 foot lengths. Extra charge is made when stock is cut to special lengths.

Further detailed information on basic metallurgy and weldability of metals may be obtained from *Welding Metallurgy*, *Metals and How To Weld Them* and the *Procedure Handbook* listed in Part 7.6.

Method of Estimating Weight Per Foot of Steel

Welding operators need to know the weight per foot in estimating costs of construction jobs or selling prices of finished projects. The following procedure uses a formula as a method of estimating weight per foot, which can be used when charts are not available. All that is necessary is to multiply thickness times width times 10 and divide by 3. The operator will have to substitute size of stock into formula. Sample problems follow:

Sample Problem: Flat plate (48″ wide, 96″ long, ¼″ thick)

Thickness (¼″) x Length (48″) = 12 square inches.
12 square inches x 10* = 120
120 ÷ 3* = 40 (pounds per linear foot of plate)
40 pounds x 8 feet (length of plate) = 320 pounds (approximate weight of plate)

Sample Problem: Flat Bar Iron (¼″ thick, 2″ wide)

Thickness (¼″) x width (2″) = .5 square inch
.5 square inch x 10* = 5.0
5 ÷ 3 = 1.6 pounds (approximate weight of bar per linear foot)

Sample Problem: Angle (width 1″, width 1″, thickness 3/16″)

Add width of 2 flanges (1″ + 1″) and solve problem the same as if stock were a flat bar.
Flange width (1″) + flange width (1″) = 2 inches
(2″) x thickness (3/16″) = .37 square inch
.37 square inch x 10* = 3.7
3.7 ÷ 3* = 1.2 pounds (approximate weight per linear foot)

Sample Problem: Rounds (¾″ diameter)

If stock is round problem is worked out this way.

¾ x ¾ = .55
.55 x 8* = 4.4
4.4 ÷ 3 = 1.46 pounds (Approximate weight per linear foot)

Sketches illustrating commonly used steel stock are shown. Methods of designating stock sizes accompany each type of steel stock. The method of manufacture and characteristics of hot and cold-rolled steel are discussed on page 7.31.

Bars (Square)	Size Specifications (thickness x width)
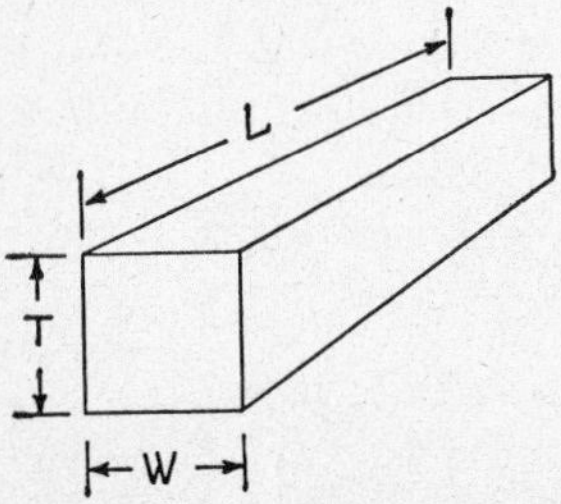 L, T, W	W = Width T = Thickness ¼″ to 2¾″, in graduations of 1⁄16″ and ⅛″. L = Length—16′, 20′, 36′

Flats (Hot and Cold-Rolled)	Size Specifications (thickness x width)
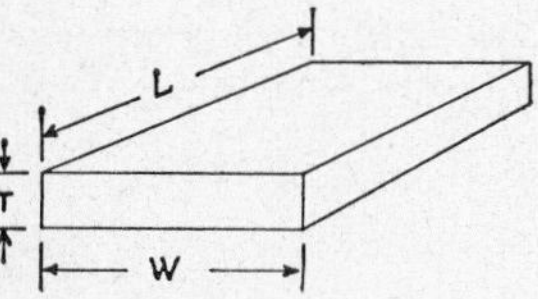 L, T, W	T = Thickness—¼″ to ½″ in graduations of 1⁄16″ ½″ to 1¼″ in graduations of ⅛″ 1¼″ to 2′ in graduations of ¼″ (3⁄16″ and lighter, see Strips) W = Width—⅜″ to 6″ in graduations of ⅛″ to ¼″ L = Length—16′, 36′

H Beams (Structural)	Size Specifications (height and width of beam)
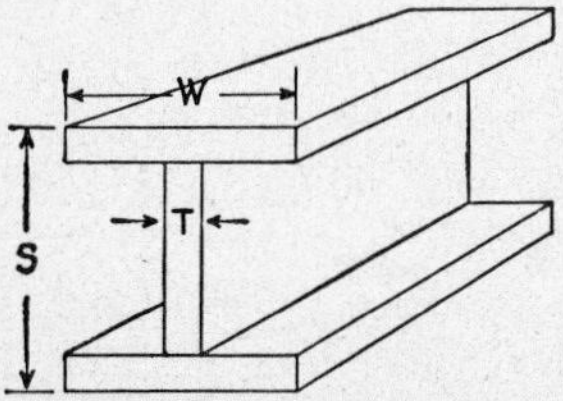 W, T, S	S = Size—3″, 4″, 6″ and larger. W = Width of flange. T = Thickness of web. L = Length—5′ to 60′.

Strips (Hot-Rolled) (3/16″ or less in thickness)	Size Specifications (thickness x width)

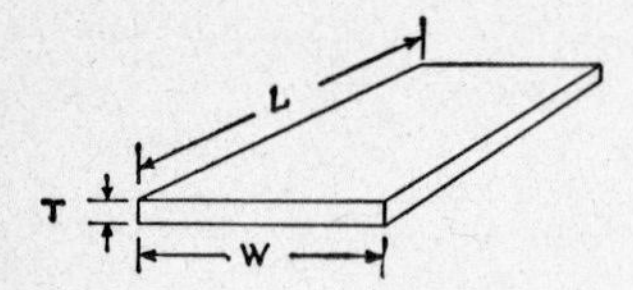

T = Thickness—(thickness measurement expressed in fractions of an inch and gauge number)
3/16″ (.1875″)
No. 10 (.134″); 1/8″ (.125″);
No. 12 (.109″); No. 14 (.083″);
No. 16 (.065″); No. 18 (.049″);
No. 20 (.035″); No. 22 (.028″).

W = Width—3/8″ to 2″ in graduations of 1/8″
2″ to 3½″ in graduations of ¼″
3½″ to 6″ in graduations of ½″
6″ to 12″ in graduations of 1″

L = Length —14′ to 16′

Sheets (Hot and Cold-Rolled)	Size Specifications Rolled steel 3/16″ or less is classified as sheet; over 3/16″, plate.

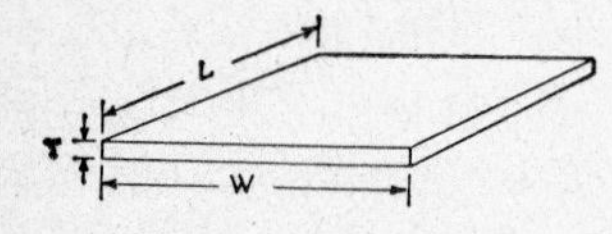

F = Thickness—3/16″ and less expressed according to gauge number.
3/16″
No. 8 (.1644″)
No. 10 (.1345″)
No. 11 (.1196″)
No. 12 (.1046″)
No. 14 (.0747″)
No. 16 (.0598″)
No. 18 (.0478″)
No. 20 (.0359″)
No. 22 (.0299″)
No. 24 (.0239″)
No. 26 (.0179″)
No. 28 (.0149″)

W = Width—measurements given in inches —30″, 36″, 42″, 48″, 54″, 60″, 72″, 84″.

L = Length—measurement given in inches —96″, 120″, 144″, 156″, 240″.

Angles (Bar and Structural) — Size Specifications (width x width x thickness)

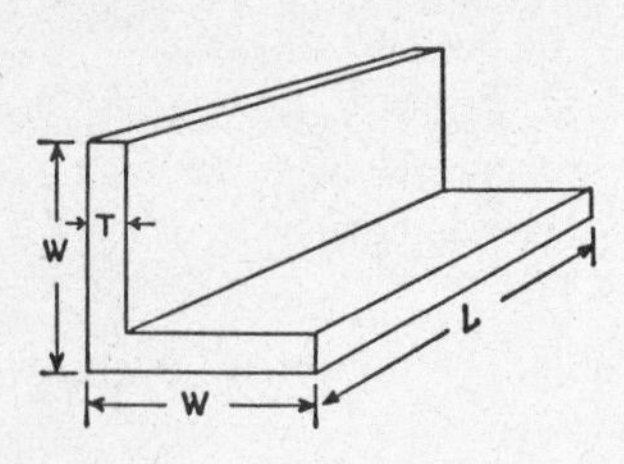

W = Width $\frac{1}{2}''$ to $1\frac{1}{2}''$ in graduations of $\frac{1}{8}''$ of flange.
$1\frac{1}{2}''$ to $2\frac{1}{2}''$ in graduations of $\frac{1}{4}''$
$3''$ and over structural steel

T = Thickness of flange $\frac{1}{8}''$ to $\frac{1}{2}''$ in graduations of $\frac{1}{16}''$.

Angle stock is manufactured in unequal width of flanges, example: $2\frac{1}{2}$ x $1\frac{1}{2}$.

L = Length—16′-18′-20′

Tees (Bar and Structural) — Size Specifications (flange width x stem x thickness)

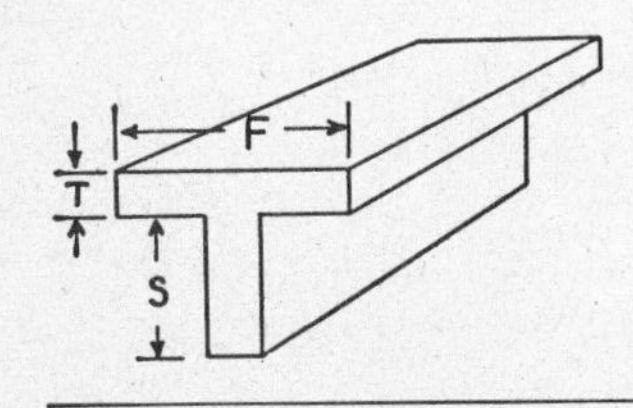

F = Flange width—$\frac{3}{4}''$ to $2\frac{1}{2}''$ in graduations of $\frac{1}{4}''$

S = Stem width—$\frac{3}{4}''$ to $2\frac{1}{2}''$ in graduations of $\frac{1}{4}''$

T = Thickness—$\frac{1}{8}''$ to $\frac{3}{8}''$ in graduations of $\frac{1}{16}''$

I Beams (Structural) — Size Specifications (height and width of beam)

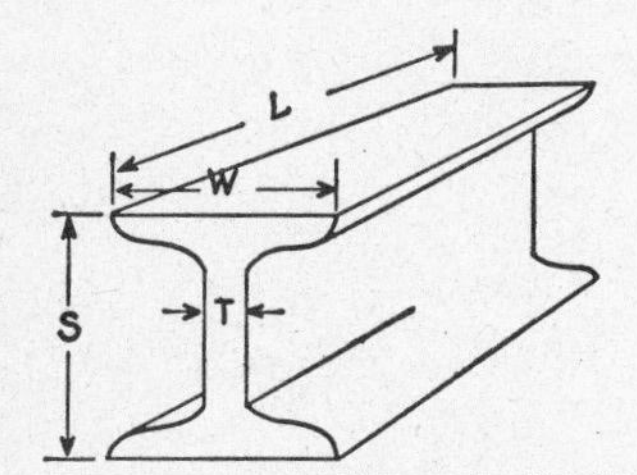

S = Size—4 x 4
5 x 5
6 x 6

W = Width of flange—same as S

T = Thickness of web

L = 5′ to 60′.

Channels—Bar Size — Size Specifications (depth x width of flange x thickness of web)

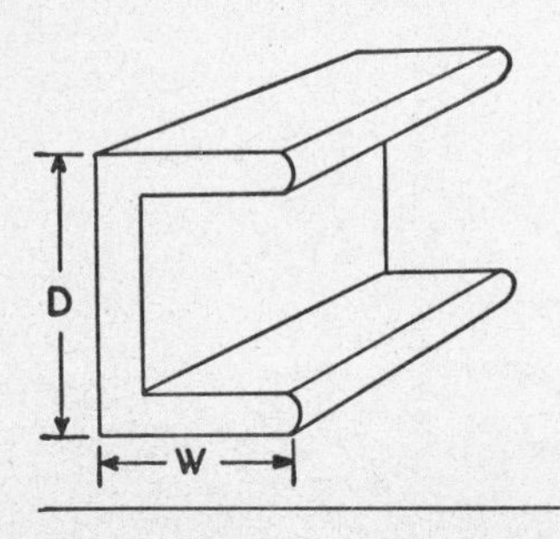

D = Depth—$\frac{1}{2}''$ to $\frac{3}{4}''$ in graduations of $\frac{1}{16}''$
$\frac{3}{8}''$ to $1\frac{1}{4}''$ in graduations of $\frac{1}{8}''$
$1\frac{1}{4}''$ to $2''$ in graduations of $\frac{1}{4}''$

W = Width (of flange)—$\frac{1}{4}$, $\frac{5}{16}$, $\frac{3}{8}$, $\frac{7}{16}$, $\frac{1}{2}$, $\frac{9}{16}$, $\frac{3}{4}$, $1''$.

T = Thickness of web—$\frac{1}{8}$, $\frac{3}{16}$, $\frac{1}{4}''$

Channels—Structural	Size Specification (depth of channel)
	Common sizes are 3″, 4″, 5″, 6″, 7″, and 8″.

Half Ovals (Hot Rolled)	Size Specifications (thickness x width)

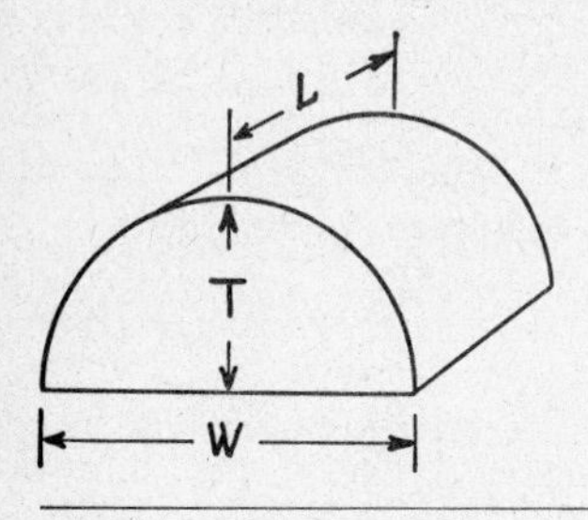

T = Thickness—3/8″ to 1/2″ in graduations of 1/16″
W = Width—3/8″ to 1 1/4″ in graduations of 1/8″
1 1/4″ to 2″ in graduations of 1/4″
2″ to 3″ in graduations of 1/2″
L = Length—16′

Rounds (Hot and Cold-Rolled)	Size Specifications

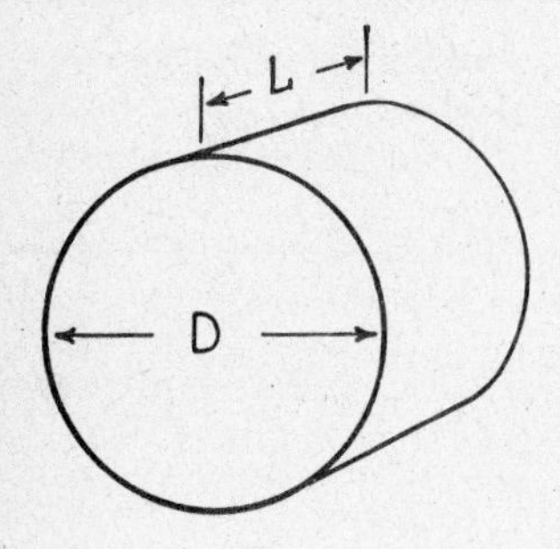

D = Diameter 3/8″ to 2″ or larger in graduation increases of 1/16″ and 1/8″

L = Length—16′, 18′, and 20′.

PART 7.2 — *Manual Welding Procedures*

The tables on Pages 7-48 to 7-74 give general welding procedures for making fillet, lap, butt and plug welds in plate and sheet metal at the lowest cost. They should be used as a general guide to establishing precise procedures for the particular conditions surrounding the welding job. To give procedures to cover all of the variables that could be met in welding would require several thousand pages. For the purposes of this handbook, procedures have been generalized for typical conditions from which variations can be worked out to meet individual needs. In using these tables, the following general information will be helpful.

How to Determine the Size of Weld Needed

When the information in the Procedures regarding plate preparation, fit-up, etc., is followed, the use of the figures for a specified *Plate Thickness* will result in a weld equal to the plate in section and strength, unless otherwise stated in the Procedures. When 50% penetration only is necessary, the feet of joint welded per hour will be doubled.

How to Determine the Amount of Electrode Needed

The figure *Pounds of Electrode Per Foot of Weld* is the amount of electrode to be purchased to weld a joint with the recommended plate preparation, fit-up and a build-up of 1/16" or less. The amount includes normal spatter loss and 2" stub ends. An increase in gap over the recommended fit-up will greatly increase the amount of electrode needed. On plates up to 1/2" thick, a gap 3/16" larger than the recommended gap approximately doubles the amount of electrode needed. Excess build-up will also greatly increase the amount of electrode to be purchased.

Obviously, for best performance, conditions should be right. If results indicated in the procedure tables are not obtained, then the conditions should be checked to ascertain that they met those for which the tables are computed.

The tables on Pages 7-38 & 7-39 will permit the easy calculation of the weight of weld metal (pounds per foot) for all types and variations of joints. Joints are broken down into various portions. To get the total weight of weld metal in any joint, it is only necessary to add the values for these portions.

Determining Amount of Current Carried by Electrode

Often the location of the welding is at considerable distance from the welding machine, and, in such cases, it is impossible or inconvenient to check the arc amperage by reading an ammeter during inspection of the welding in progress. Most modern welding machines are not equipped with meters in any event. A quick and easy method of determining the amount of current carried by the electrode is desirable. A watch with second hand and a chart, see Fig. 1, which may be prepared by the following method, are the only tools needed.

Considering a popular type of shielded-arc electrode, it has been found that, for a given size electrode, the melt-off rate is approximately proportional to the current. The effect of voltage may be neglected, because, under usual conditions, it is negligible. The melting or "melt off" is the same for a given size of electrode and current regardless of position.

TABLE 1

WEIGHT OF WELD METAL—POUNDS PER FOOT

d	14°	20°	60°	45°	flat (Leg Size Increased by 10%)	convex (Leg Size Increased by 10%)	concave (Leg Size Increased by 10%)
1/8"	.0065	.0094	.031	.027	.032	.039	.037
3/16"	.0147	.021	.069	.060	.072	.087	.083
1/4"	.026	.037	.123	.106	.129	.155	.147
5/16"	.041	.059	.192	.166	.201	.242	.230
3/8"	.059	.084	.276	.239	.289	.349	.331
7/16"	.080	.115	.376	.326	.394	.475	.451
1/2"	.104	.150	.491	.425	.514	.620	.589
9/16"	.132	.190	.621	.538	.651	.785	.745
5/8"	.163	.234	.766	.664	.804	.970	.920
11/16"	.197	.283	.927	.804			
3/4"	.234	.337	1.11	.956	1.16	1.40	1.32
13/16"	.275	.396	1.30	1.12			
7/8"	.319	.459	1.50	1.30	1.58	1.90	1.80
15/16"	.367	.527	1.73	1.50			
1"	.417	.599	1.96	1.70	2.06	2.48	2.36
1 1/16"	.471	.676	2.22	1.92			
1 1/8"	.528	.758	2.48	2.15	2.60	3.14	2.98
1 3/16"	.588	.845	2.77	2.40			
1 1/4"	.651	.936	3.07	2.66	3.21	3.88	3.68
1 5/16"	.718	1.03	3.38	2.93			
1 3/8"	.789	1.13	3.71	3.21	3.89	4.69	4.45
1 7/16"	.836	1.24	4.05	3.51			
1 1/2"	.938	1.35	4.42	3.82	4.62	5.58	5.30
1 9/16"	1.02	1.46	4.79	4.15			
1 5/8"	1.10	1.58	5.18	4.49	5.43	6.55	6.22
1 11/16"	1.19	1.71	5.59	4.84			
1 3/4"	1.28	1.84	6.01	5.20	6.29	7.59	7.21
1 13/16"	1.37	1.97	6.45	5.58			
1 7/8"	1.47	2.10	6.90	5.97	7.23	8.72	8.28
1 15/16"	1.56	2.25	7.36	6.38			
2"	1.67	2.40	7.85	6.80	8.23	9.93	9.43

weight	r
.021	1/16"
.083	1/8"
.188	3/16"
.334	1/4"
.531	5/16"
.750	3/8"
1.02	7/16"
1.33	1/2"

EXAMPLE

Find weight (#/ft) of weld metal required in the following joint.

1 1/8" 1/8" 1 1/2" 7° 1" 1/4" rad. 1/4"

ⓐ 1 1/8" × 1/8" • .318#/ft.

ⓑ 1" × 14° • .417#/ft.

ⓒ 1" × 1/2" • 1.70#/ft.

ⓓ 1/4" rad. • .334#/ft.

TOTAL WT.

2.77#/ft. of WELD METAL

or 2.77 × 1 1/2 = 4.16 lbs. rod/ft.

TABLE 1 (Contd.) WEIGHT OF WELD METAL—POUNDS PER FOOT

d	t 1/16"	1/8"	3/16"	1/4"	3/8"	1/2"	1/16"	1/8"	3/16"	1/4"
1/8"	.027	.053	.080	.106	.159	.212				
3/16"	.040	.080	.119	.159	.239	.318	.027			
1/4"	.053	.106	.159	.212	.318	.425	.035			
5/16"	.066	.133	.199	.265	.390	.531	.044			
3/8"	.080	.159	.239	.318	.478	.637	.053	.106		
7/16"	.091	.186	.279	.371	.557	.743	.062	.124	.186	
1/2"	.106	.212	.318	.425	.637	.849	.071	.142	.212	
9/16"	.119	.239	.358	.478	.716	.955	.080	.159	.239	
5/8"	.133	.265	.398	.531	.796	1.06	.089	.177	.266	
11/16"	.146	.292	.438	.584	.876	1.17	.097	.195	.292	.389
3/4"	.159	.318	.478	.637	.955	1.27	.111	.212	.318	.424
13/16"	.172	.345	.517	.690	1.04	1.38	.114	.230	.345	.460
7/8"	.186	.371	.557	.743	1.11	1.49	.124	.248	.372	.490
15/16"	.199	.398	.597	.796	1.19	1.59	.133	.266	.398	.530
1"	.212	.425	.627	.849	1.25	1.70	.142	.282	.418	.566
1 1/16"	.226	.451	.677	.902	1.35	1.80	.150	.301	.451	.602
1 1/8"	.239	.478	.716	.955	1.43	1.91	.159	.318	.477	.637
1 3/16"	.252	.504	.756	1.01	1.51	2.02	.168	.336	.505	.672
1 1/4"	.265	.531	.796	1.06	1.59	2.12	.177	.354	.531	.706
1 5/16"	.279	.557	.836	1.11	1.67	2.23	.186	.372	.557	.743
1 3/8"	.292	.584	.876	1.17	1.75	2.34	.195	.389	.584	.777
1 7/16"	.305	.610	.915	1.22	1.83	2.44	.203	.407	.610	.814
1 1/2"	.318	.637	.955	1.27	1.91	2.55	.212	.425	.636	.849
1 9/16"	.332	.664	.995	1.33	1.99	2.65	.221	.442	.664	.884
1 5/8"	.345	.690	1.04	1.38	2.07	2.76	.230	.460	.690	.920
1 11/16"	.358	.716	1.07	1.43	2.15	2.87	.239	.477	.716	.956
1 3/4"	.371	.743	1.11	1.49	2.23	2.97	.249	.495	.743	.990
1 13/16"	.385	.769	1.15	1.54	2.31	3.08	.257	.513	.770	1.03
1 7/8"	.390	.796	1.19	1.59	2.39	3.18	.266	.531	.796	1.06
1 15/16"	.411	.822	1.23	1.65	2.47	3.29	.274	.549	.823	1.10
2"	.425	.849	1.27	1.70	2.55	3.40	.283	.566	.849	1.13

By securing from the manufacturer of the electrode the recommended current value, also the maximum and minimum current values for this electrode, and then, by actual test, determining the number of seconds required to melt off 12 inches of this electrode at minimum current value, recommended current value and maximum current value, the time may be plotted against the amperes on a chart similar to the chart, Fig. 1. The welding should be done close to a machine with an accurate ammeter or dial controls. The complete curve for any given size of electrode may then be secured by determining the time required to melt off 12″ of electrode at several intermediate amperage values. The results then plotted on the chart will furnish sufficient points from which to secure a complete curve as shown. Curves for all the various sizes of electrodes may be plotted by this method. The chart is then ready for use.

1. Clock the time required to melt off 12″ of the electrode in question. 2. From this time point on the time scale of the chart, project a horizontal line to the curve representing the size electrode used. 3. From the point of intersection of the horizontal line with the curve, project a vertical line to the amperage scale. 4. The point of intersection of this vertical line with the amperage scale indicates the amount of current being carried by the electrode in question.

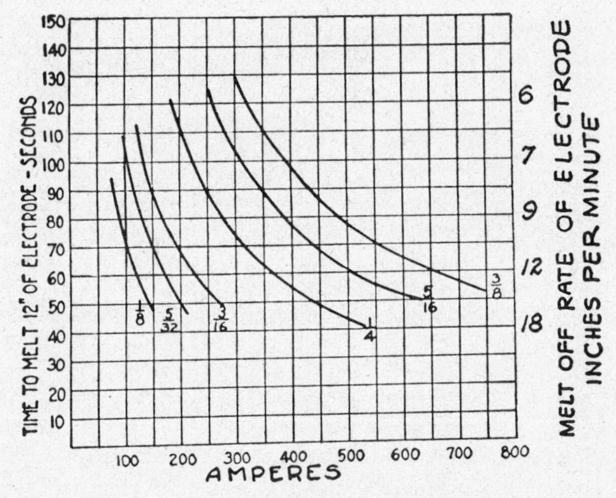

Fig. 1. A typical chart as used in the determination of current carried by an electrode. However, small variation will be found when comparing different makes and different types.

Another method of current measurement is the use of the clip-on, or tong-type ammeter. This instrument clamps over the welding cable and indicates the amperage directly.

If meters or dials of known accuracy are not available, the most satisfactory method of determining the current is by measuring the number of inches of electrode melted off in one minute of welding. In the procedures that follow, this melt-off rate is included under the heading, "Electrode Melt-off Rate."

On many of the procedure sheets, more than one type of electrode or polarity (Negative, Positive, or AC) is specified for each joint. For a given current, the melt-off rate may change when the type of electrode or polarity is changed (Electrode Negative has the highest melt off). The melt-off rates given in the procedure tables are approximately those which will be obtained when the recommended current is used. It may be found that a melt-off rate up to 10% higher or lower than the value given may be more desirable due to arc blow, fit-up, etc.

As an example, when the procedure sheet for the joint to be welded recommends 3/16" E-6010 at 150 amperes, the column on Electrode Melt Off will indicate that the melt-off rate should be approximately 8" per minute. Set the current controls of the welding machine for 150 amperes and weld with a new length of electrode for exactly one minute. If less than 8" of electrode has been melted off, the current setting of the machine should be changed to increase the current until at least 8" is melted off in one minute. If more than 8" has been melted off, it will indicate that the current is higher than specified; however, it is not necessary to reduce the current until the melt-off rate is exactly 8" a minute, unless difficulty is encountered with poor surface appearance or burn-through.

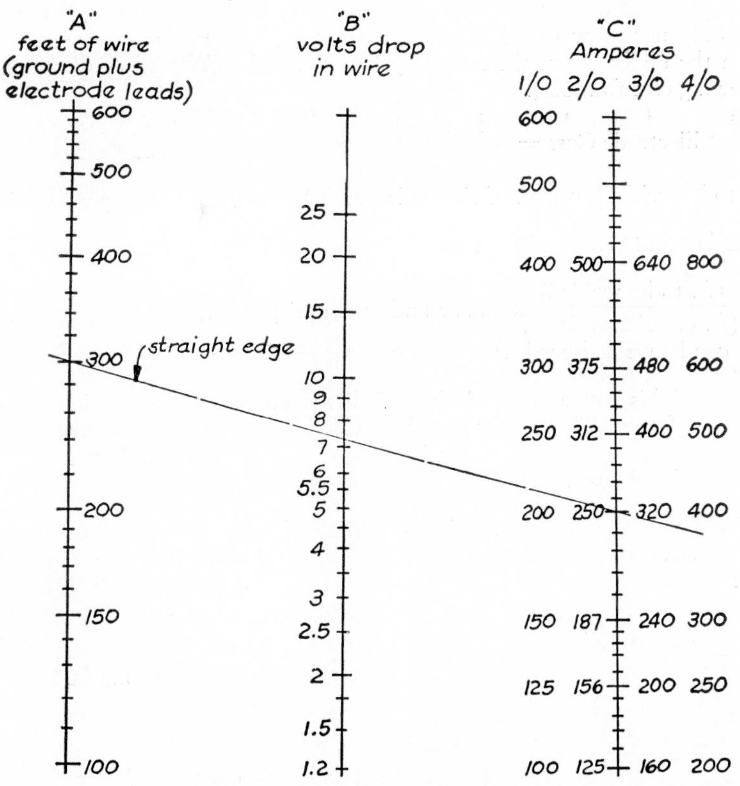

Fig. 2. Nomograph to Determine Voltage Drop in Welding Circuit. Example: 300 feet of 3/0 copper wire carrying current of 320 amperes. To determine voltage drop, lay straight edge, as shown, from 300' point to "A" to 320 amperes point on "C" for 3/0 wire. Drop is read under "B" as 7.2 volts.

Speed and Number of Passes

The welding speed is given in the tables in two ways: *Arc Speed in Inches Per Minute* and *Feet of Joint Welded Per Hour.*

Arc Speed in Inches Per Minute is given for single pass welds and for the first pass only in multiple pass welds, since the speed of this first pass is important in obtaining the proper penetration. The arc speed for succeeding passes and the total number of passes to make the joint will vary with individual operators, but the *Feet of Joint Welded Per Hour* will be the same, regardless of the arc speed per pass or the number of passes, as long as the recommended plate preparation, fit-up, current, and build-up are used.

Feet of Joint Welded Per Hour is based on actual welding time only. No factor has been included for setup, electrode changing, cleaning, or other factors which will vary greatly with the type of work to be done. In order to use the figures in this column in cost calculations, they must be multiplied by an operating factor which can be estimated or determined by trial for the job in question.

How to Calculate Welding Costs

Labor and overhead generally account for something like 80% to 86% of the cost of making any joint. Electrode cost may run from 8% to 15%. Power and equipment costs usually are as low as 2%. For most purposes of control, it is only necessary to calculate electrode cost and labor and overhead costs. From the tables given, the necessary figures can be obtained to work out the following cost formulae.

Electrode Cost =

(ft. to be welded) x (lbs. of electrode per ft.) x (electrode cost per lb.)

Labor and Overhead Cost =

$$\frac{\text{(ft. to be welded)}}{\text{(ft. of joint which can be welded per hr.)}} \times \text{(labor and overhead per hr.)}$$

> This cost as figured above is based on a 100% operating factor. The answer must be divided by the operating factor. If the operating factor is 40%, divide the labor and overhead cost by .4.

To figure the cost of making a horizontal fillet on ½″ plate, using an E-6024 electrode, the amount of welding being 750′, consult Page 7-61 to obtain:

$$750 \times .50 \times \$.20 = \$\ 75.00 \text{ electrode cost}$$

$$\frac{750}{31} \times \$5.00 = \$121.00 \text{ labor and overhead}$$

$196.00 total cost at 100% operating factor

$ 75.00
$242.00
$317.00 total cost 50% operating factor

Calculators and nomographs are available to aid in cost calculation. A nomograph is given in Figure 3.

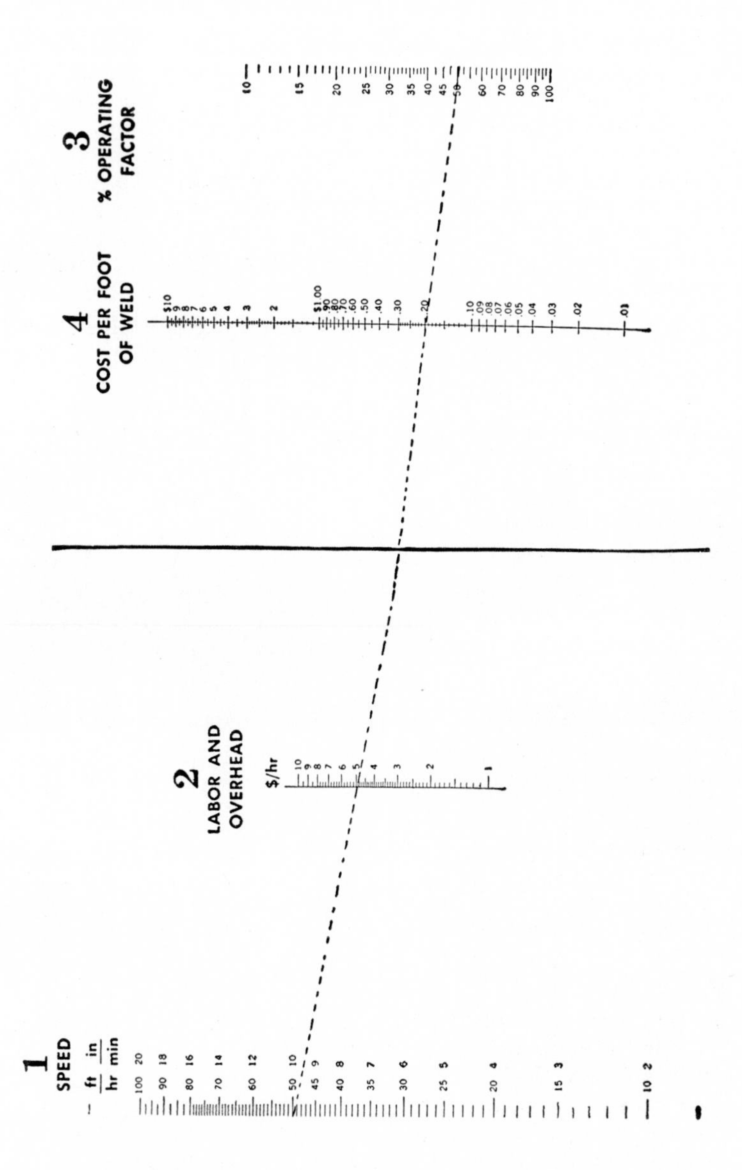

Fig. 3. Nomograph for Calculating Welding Costs. Example: Find cost of making ¼" single pass fillet weld. From tables, determine arc speed is 10"/min. Labor and overhead are $5.00. Operating factor is 50%. Draw straight line from 10" on scale 1 (speed) to $5.00 on scale 2 (labor and overhead) and extend to center reference line. Draw line from this point on center reference line to 50% on scale 3 (operating factor). Read answer on scale 4—$0.20 per foot.

BUTT WELD PROCEDURES*

Square-Edge Butt Welds

For square-edge butt welds, procedure used can vary considerably the amount of penetration that will be obtained. Even a very shallow molten pool under the arc will form an insulating layer and absorb some of the penetrating power of the arc force. The molten pool under the arc also tends to cause the operator to raise the tip of the electrode to keep it out of the molten metal, and the resulting long arc dissipates more of its heat into the air and, in flaring out, tends to widen the bead. The operator should, therefore, hold a short arc with electrode at proper angle which will automatically tend to force the molten pool from under the electrode tip, especially when aided by rapid arc travel speed. With the heavily coated electrodes recommended, the correct arc length will be obtained when the electrode is dragged along the joint with the coating touching the plates. See Fig. 4.

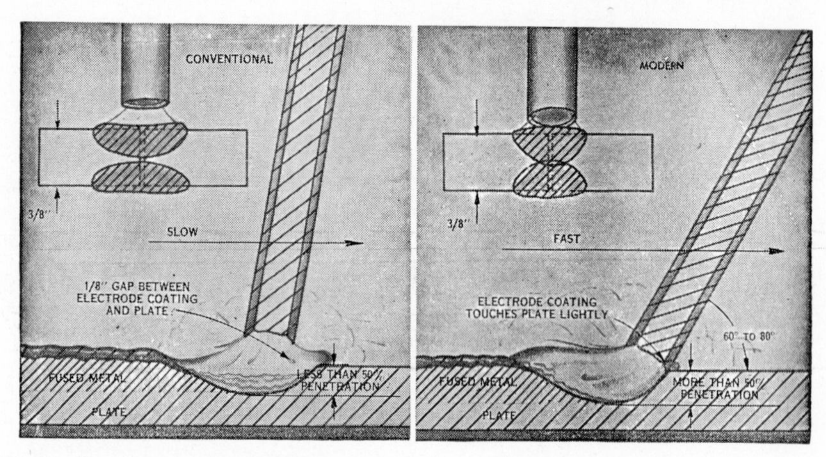

Fig. 4. On square edge butt welds made with an E-6010 type electrode, high speed, with electrode coating dragging gives better penetration than long arc and slow speed.

Vee Butt Welds

For vee butt welds, still greater variation in penetration exists between slow and fast travel speeds. The penetration can be increased as much as 40% by proper speed and arc length as compared with a slow speed.

Fig. 5 shows two welds made at the same current, one run slowly in one pass and the other in two fast passes. By going slow and weaving, the molten metal and slag got under the arc and left an unpenetrated section at the root of the joint. By merely using faster arc travel speed and keeping the molten pool from under the arc, this can be avoided.

The tip of the electrode should be held down in the groove so that the coating touches both sides of the joint lightly. Travel as fast as possible on the first pass with good slag coverage. A slight undercut on any but the last pass will not be objectionable, as the succeeding passes will fill in the undercut.

*For Butt Weld Procedures for Sheet Metal, See Page 7-66.

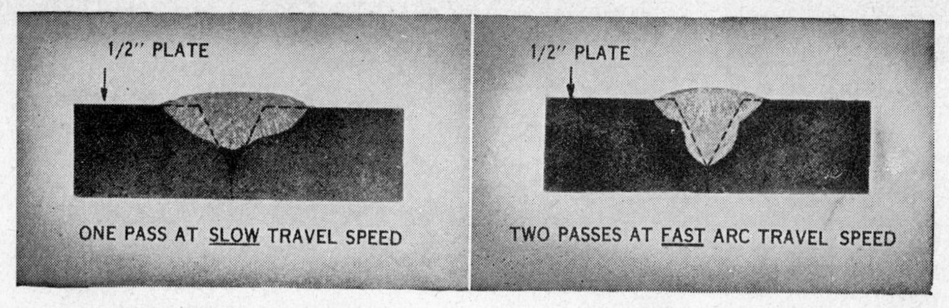

3/16" E6010—150 Amps. **3/16" E6010—150 Amps.**

Fig. 5. Two V-butt welds made at same current, the left-hand one run slowly in one pass, the right-hand one run in two fast passes.

For the last few passes on heavy plate with wide grooves, weaving is permissible, but is necessary only where fusion at both edges cannot be obtained with a straight bead. For multiple-pass, butt welds, the surface will be much smoother if enough room is left on the next-to-last pass so that the last pass will be just flush with the plate. This also assures the ideal cross section for the weld.

In General

(1) Use the electrode, plate preparation and current recommended in procedures. (2) In welding first pass, hold the electrode as shown in Fig. 4. The arc should be so short the coating drags lightly on the surface of the plates. In welding succeeding passes, hold a short arc, if desired, but dragging of coating is preferred. (3) Keep just ahead of the molten pool, traveling as fast as possible, still maintaining uniform slag coverage to assure good appearance and watching for undercut. Weaving is not recommended except on wide beads.

Selection of Joint—Flat Position

Five welding procedures are included for butt welds in the flat position. Factors that determine which procedure should be used include:

1. Plate thickness.
2. Equipment available for preparing and fitting up the edges of the plate.
3. Whether or not the joint can be welded from only one side:
   (a) With back-up strip.
   (b) Without back-up strip.

Wherever possible, it is recommended that the joint be welded from both sides because the welding speeds generally are higher and the costs lower.

If the joint must be welded from one side, the procedure for welding into a back-up strip (Page 7-51 is recommended, because less care is required for fitting up and the welding speeds are higher than for joints without back-up strip. If it is impossible to use a back-up strip, use procedures on Page 7-52 (welded from one side without back-up).

When it is possible to turn the plates so that they can be welded downhand from both sides, the following procedures are recommended:

Plate Thickness	Procedure
1. Up to and including 5/16″	Page 7-48 (Square Edge)
2. 3/8″ up to 5/8″	Page 7-49 (Single V)
3. Over 5/8″	Page 7-50 (Double V)

Fit-Up

In order to obtain maximum welding speeds, proper fit-up is important. Except when the joint must be welded from one side only (see Page 7-52), a gap of 1/16″ or less is recommended. A 1/32″ to 1/16″ gap results in less distortion than no gap and, therefore, is recommended.

If the gap exceeds 1/16″, one or more beads should be deposited with 3/16″ size electrode to seal the gap.

If the gap can be sealed with one pass, the welding time per foot of joint will be increased by 2 or 2½ minutes above that which is required with a gap of 1/16″ or less. Correspondingly more time will be required if more than one bead is necessary to seal gap.

Edge Preparation for Single V Butt Weld

The 1/8″ shoulder is preferred to the feather edge from the standpoint of welding economy. Therefore, the procedures are based on the preparation with the 1/8″ shoulder. If the feather edge preparation is used, it is recommended that the first pass, or passes required to seal a gap, be welded at approximately 150 amperes with 3/16″ E-6011. Where structure is extremely rigid and of heavy plate or off-analysis steel, an iron-powder, low-hydrogen type electrode is recommended for seal beads. See Fig. 6.

If the feather edge preparation is used and one pass is sufficient to seal the gap, the welding time per foot of joint will be increased by approximately

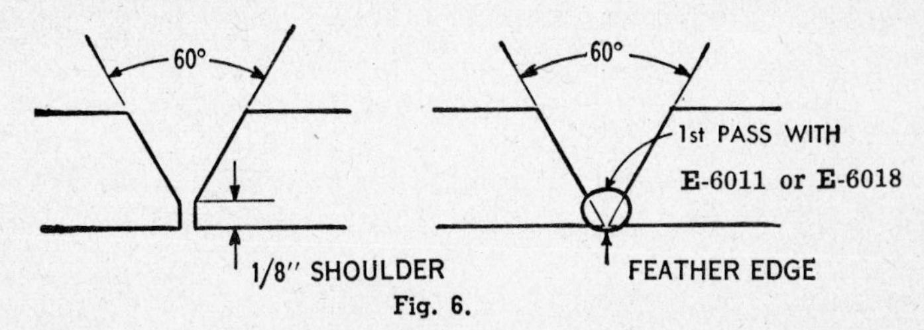

Fig. 6.

2 to 3 minutes above that required with 1/8″ shoulder and 1/16″ gap when welding plate of 3/8″ to 1/2″ thickness and by 4 to 6 minutes when welding plate of 1″ thickness. The welding time is based on 100% operating factor.

Electrode

The electrodes recommended for butt joints welded in the flat position are indicated in tables for specific procedures.

Technique

When the joint preparation and fit-up are as recommended, weld the first pass by dragging the tip of the coating in the groove and pointing the electrode back as shown in Fig. 4. The travel speed or arc speed should be governed to keep the electrode just ahead of the molten pool. A small ball of molten slag or metal rolling along ahead of the electrode may be encountered, but will cause no difficulty.

When a sealing bead or beads must be run with E-6010 or E-6011, it may be necessary to whip the electrode back and forth along the seam to prevent burn-through.

Iron Powder Technique—Iron powder electrodes are contact electrodes, and the best results generally will be obtained by using this technique. However, if the slag follows too closely, as may be true in the case of initial passes on deep groove welds or under some arc blow conditions, holding an arc will usually give better results.

To obtain the full advantage of this type electrode in deep groove welding:

(1) The weldment should be flat—not uphill or downhill.

(2) Use recommended current which is generally somewhat higher than with other types.

(3) Travel should be such to insure complete slag coverage.

(4) Use contact technique—don't hold an arc.

(5) Use stringer beads—don't use a wide weave.

(6) Keep the electrode as nearly perpendicular to the work as possible.

(7) AC welding current is recommended—it will invariably produce the best results.

(8) Because of the greater coating diameter and tendency for slag interference, it is generally necessary to use one size smaller than the conventional electrode.

This procedure will improve the cover pass appearance with any electrode, but, with iron powder electrodes, the results will be excellent—comparable to automatic welds.

Back Chipping

(Chipping into back side of 1st bead before welding any beads from the back side.) For butt welds made in flat position:

Back chipping normally is not required to obtain a joint as strong as the plate when welding in flat position. If it is necessary to use currents lower than those recommended in the procedure pages for the 1st pass made on either side of the joint, back chipping is recommended. Currents lower than those recommended may be required by variation in fit-up or where certain corrective measures (Page 7-75) must be used. When welding butt welds, if complete fusion is required, back chipping is recommended for all positions except flat position.

Back chipping may also be required because it is specified as part of a code requirement.

BUTT WELDS

FLAT POSITION—SQUARE EDGE—WELDED FROM BOTH SIDES

For high welding speed and economy of joint preparation. Use on steels within the preferred analysis range where plates can be turned for downhand welding on both sides and where joint is level.

PREPARATION:
Square Edge

FIT-UP:
Maximum gap recommended, 1/16"†

ELECTRODE:
E6010

POLARITY:
Electrode Positive

CHIPPING NOT NECESSARY

Fig. 7.

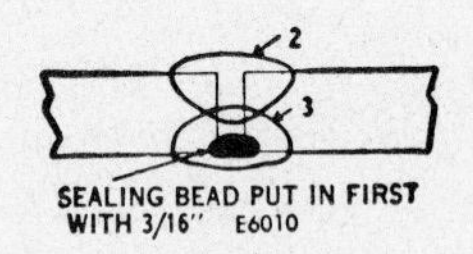

Fig. 8.

Plate Thickness (In.)	Electrode Size (In.)	Current (Amps.)	Electrode Melt Off Rate (In. per Min.) ‡	Arc Speed (In. per Min. per Pass)	Passes or Beads	Ft. of Joint Welded per Hr. (100% Operating Factor)	Lbs. of Electrode per Ft. of Weld
3/16	1/4	190	7	18	2	45	.16
1/4	5/16	325	7	18	2	45	.27
5/16	5/16	375	8	18	2	45	.31

†If gap is over 1/16", put in flush sealing bead as shown in Fig. 8. The sealing bead need not be chipped out before putting in the second pass.

‡This is the minimum melt-off rate which should be used to assure complete penetration.

BUTT WELDS

Flat Position* — Single V-Groove

Welded From Both Sides

For use where plates can be turned for downhand welding on both sides.*

PREPARATION:
60° V-groove with 1/8" shoulder

FIT-UP:**
Recommended gap 1/32" to 1/16"

ELECTRODE & POLARITY
E6011—AC
E6027—AC

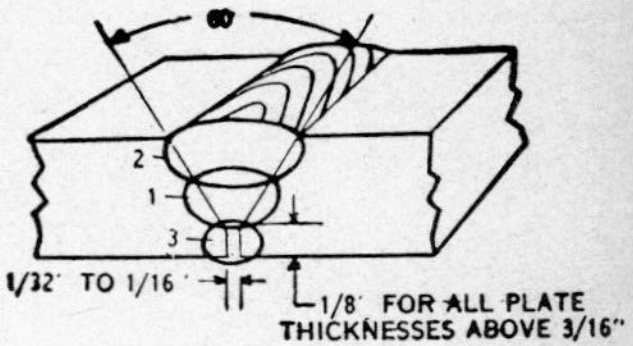

Fig. 9.

Plate Thickness (In.)	Electrode Size (In.)	AC Current (Amps.)	Electrode Melt-Off Rate (In. per Min.)	Arc Speed (In. per Min. for First Pass)	Passes or Beads	Ft. of Joint Welded per Hr. (100% Operating Factor)	Lbs. of Electrode per Ft. of Weld
3/8	3/16 E6011	175	10.0	9.0	1st	20.1	.17
	3/16 E6027	280	12.3		(2nd & Back)		.34
							Total .51
1/2	1/4 E6011	275	9.0	8.2	1st	17.4	.23
	7/32 E6027	315	10.8		(2nd & Back)		.52
							Total .75
5/8	1/4 E6011	275	9.0	7.7	1st	13.2	.24
	1/4 E6027	375	10.0		(2nd & 3rd)		.80
	7/32 E6027	315	10.8		Back		.23
							Total 1.27

*Where the plates cannot be turned over, chip into bead 1 (Figure 9) and put the last pass in overhead with 3/16" E6010 or E6011 with the current adjusted to produce an electrode melt-off of **approximately 8½" per minute.**

**See Page 7-46.

BUTT WELDS

Flat Position — Double V-Groove

Welded From Both Sides

For use where single V-groove cannot be used because of plate thickness or where it is essential to keep distortion to a minimum.

PREPARATION:
Double 60° V-Bevel

FIT-UP:
See Sketch

ELECTRODE & POLARITY:
E6011—AC
E6027—AC

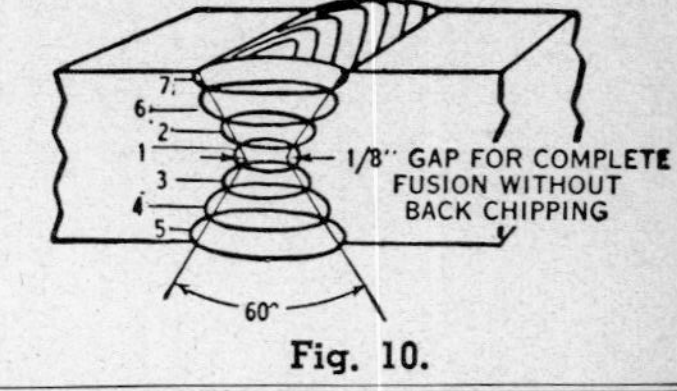

Fig. 10.

Plate Thickness (In.)	Electrode Size (In.)	AC Current (Amps.)	Electrode Melt-Off Rate (In. per Min.)	Arc Speed (In. per Min. for First Pass)	Passes or Beads*	Ft. of Jt. Welded per Hr. (100% Operating Factor)	Lbs. of Electrode per Ft. of Weld
3/4†	3/16 E6011	135	8.5	6.0	1st	8.2	.19
	1/4 E6011	275	9.0		2nd & 3rd,		.40
	1/4 E6027	400	11.0		4th & 5th		.71
							Total 1.30
1†	3/16 E6011	135	8.5	6.0	1st	6.1	.19
	1/4 E6011	275	9.0		2nd & 3rd		.40
	1/4 E6027	400	11.0		4, 5, 6, 7		1.40
							Total 1.99
1 1/4	3/16 E6011	135	8.5	6.0	1st	4.6	.19
	1/4 E6011	275	9.0		2nd & 3rd		.40
	1/4 E6027	400	11.0		4, 5, 6, 7, 8 & 9		2.44
							Total 3.03
1 1/2	3/16 E6011	135	8.5	6.0	1st	4.1	.19
	1/4 E6011	275	9.0		2nd & 3rd		.40
	1/4 E6027	400	11.0		4, 5, 6, 7, 8, 9 & 10		3.04
							Total 3.63

*The number of passes will depend upon the operator, but the feet of joint welded per hour should be the same.

†It is recommended that 3/4" and 1" plate be welded with single V-groove preparation except where it is desirable to use double V-groove to minimize distortion. For most effective control of distortion, the passes should be put in alternately, first from one side and then from the other. However, this necessitates turning the plates over several times (6 times for 3/4" plate). Distortion usually is controlled effectively if the plates are turned twice, as follows: Weld half the passes on the 1st side; turn the plates over and weld all of the passes on the 2nd side; turn the plates over and complete the passes on the first side.

Important Note:
The current values given above are *actual* currents. The actual currents with E6027 invariably will be lower than the dial setting indicated on the machine. This is because of the higher than normal arc voltage of E6027.

Multiple Passes:
For 3 or more passes, the 2nd and subsequent layers should be put in 2 beads side by side.

BUTT WELDS

Flat Position — Single V-Groove

Welded From One Side With Back-up Strip

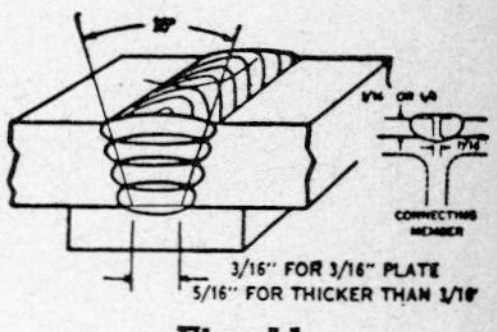

Use where plates cannot be welded from two sides or where joint is welded through to a connecting member.

PREPARATION:
$\frac{3}{16}$" & 1/4" Plate: Square Edge
$\frac{5}{16}$" Plate & Over: 30° V-Bevel

FIT-UP:
See sketches.

ELECTRODE:
E6027—AC

3/16" FOR 3/16" PLATE
5/16" FOR THICKER THAN 3/16"

Fig. 11.

Plate Thickness (In.)	Electrode Size (In.) E6027	Current (Amps.)	Electrode Melt-Off Rate (In. per Min.)	Arc Speed (In. per Min. for First Pass)	Passes or Beads	Ft. of Joint Welded per Hr. (100% Operating Factor)	Lbs. of Electrode per Ft. of Weld
1/2	3/16 1/4	300 400	13.4 11.2	13 10	1 2, 3, & 4	15	1.30
5/8	3/16 1/4	300 400	13.4 11.2	13 10	1 5	11.5	1.71
3/4	3/16 1/4	300 400	13.4 11.2	13 10	1 7	9.2	2.11
7/8	3/16 1/4	300 400	13.4 11.2	13 10	1 8	7.8	2.40
1	3/16 1/4	300 400	13.4 11.2	13 10	1 10	6.9	2.91

Important Note:
The current values given above are *actual* currents. The actual currents with E6027 invariably will be lower than the dial setting indicated on the machine. This is because of the higher than normal arc voltage of E6027.

Multiple Passes:
For 3 or more passes, the 2nd and subsequent layers should be put in 2 beads side by side.

Butt Welds

Flat Position — Single V-Groove

Welded From One Side Without Back-up Strip

Used only where plates cannot be welded from two sides and where back-up strip cannot be used.

PREPARATION:
8/16" Plate: Square Edge
1/4" Plate & Over: 60° V with 1/8" Shoulder†

ELECTRODE & POLARITY:
E6027—AC
E6011—AC

FIT-UP:
See Sketches

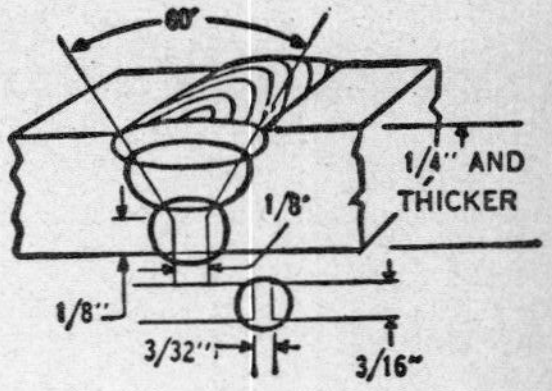

Fig. 12.

Plate Thickness (In.)	Electrode Size (In.)	AC Current (Amps.)	Electrode Melt-Off Rate (In. per Min.)	Arc Speed (In. per Min. for First Pass)‡	Passes or Beads	Ft. of Joint Welded per Hr. (100% Operating Factor)	Lbs. of Electrode per Ft. of Weld
5/16	5/32 E6011	135	11.0	6.0	1st	20.5	.17
	5/32 E6027	240	15.0		2nd		.20
							Total .37
3/8	5/32 E6011	135	11.0	6.0	1st	16.0	.17
	5/32 E6027	240	15.0		2nd & 3rd		.32
							Total .49
1/2	5/32 E6011	135	11.0	6.0	1st	13.2	.17
	1/4 E6011	275	9.0		2nd		.22
	1/4 E6027	400	11.0		3rd		.33
							Total .72

†See Page 7-46 if plain 60° V-bevel must be used.

‡Gaps larger than 1/8" may cut the arc speed for 1st pass to less than half the value given in the table. Gaps smaller than 1/8" make complete penetration extremely difficult.

BUTT WELDS

Horizontal Position—V-Groove
Welded From Both Sides

PREPARATION:
See Fig. 13 and footnote*.

FIT-UP:
See Fig. 13 and footnote*.

May also be welded from just one side with or without a back-up strip. For complete fusion, welding from both sides is recommended.

ELECTRODE & POLARITY:

E6010—Electrode Positive

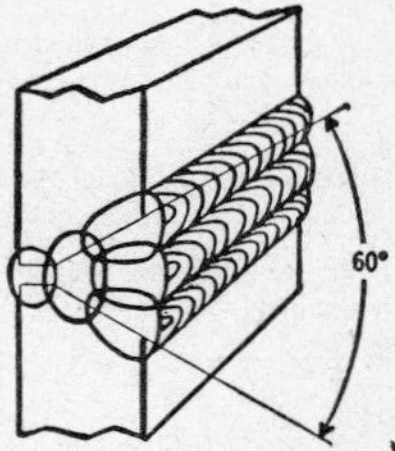

Fig. 13(a).

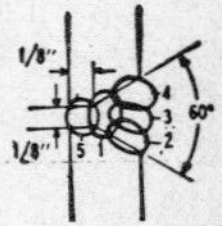

Fig. 13(b).

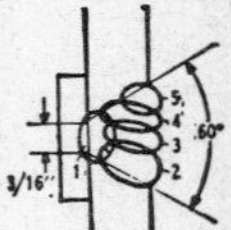

Fig. 13(c)

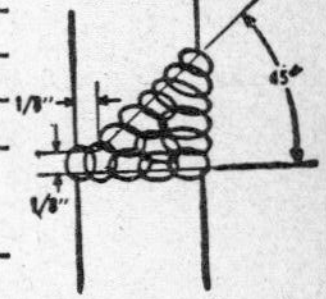

(Fig. 13(d).

Plate Thickness (In.)	Electrode Size (In.)	Current (Amps.)	Electrode Melt-Off Rate (In. per Min.)	Arc Speed (In. per Min. for First Pass)	Passes or Beads	Ft. of Joint Welded per Hr. (100% Operating Factor)	Lbs. of Electrode per Ft. of Weld
3/16	5/32	130	9	10	2	25	.15
1/4	5/32	130	9	7	2	17.5	.24
5/16	5/32	140	9½	9	3	14	.38
3/8	3/16	170	8½	7	4	10	.50
7/16	3/16	170	8½	7	‡	8	.68
1/2	3/16	170	8½	7	‡	6.2	.85
5/8	3/16	170	8½	7	‡	4.0	1.3
3/4	1/4†	250	8	7	‡	3.9	1.8
	3/16	170	8½				
1	1/4†	250	8	7	‡	2.4	3.1
	3/16	170	8½				

*For plates 5/8" and thinner, where the joint is being welded from 2 sides or one side without back-up strip, the 60° V with 1/8" shoulder (Fig. 13(b) is recommended. If for ease of preparation it is desired to use a plain 60° feather edge bevel, the plate preparation shown in Fig. 13(c) with a 1/16" maximum gap may be used when a back-up strip is not used but the feet of joint welded per hour will be approximately 3/4 of the value given in the procedure table.

For plates thicker than 5/8", one plate may be square-edged and the other bevelled 45° as shown in Fig. 13(d). This preparation facilitates the use of 1/4" or 7/32" electrode.

The plate preparation shown in Fig. 13(c) is recommended when the welding is done from one side into a back-up strip or connecting member. Edges of plates should be spaced approximately 3/16" to allow fusion into the back-up strip. The feet of joint welded per hour will be approximately 2/3 of the value given in procedure table.

‡Number of passes will depend upon the operator but the feet of joint welded per hour should be the same.

†When welding all passes from one side use 3/16" electrode for 1st and last pass. When welding from both sides, weld last pass on groove side and back pass with 3/16" electrode.

Note: Under certain circumstances the above times can be exceeded **by the use of low-hydrogen type electrodes.**

Butt Welds

Vertical Position—Welded Up

Without Back-Up Strip Welded From One Side or Both Sides***

PREPARATION:

Square-edge for 3/16″ plate
60° V-groove with 1/8″ shoulder for heavier plate†

FIT-UP:

See Sketch. When welding from one side only, 1/8″ minimum gap is recommended for 1/4″ and thicker plates.

ELECTRODE & POLARITY:

E6010—Electrode Pos.

Fig. 14.

Plate Thickness (In.)	Electrode Size (In.)	Current (Amps.)	Electrode Melt-Off Rate (In. per Min.)	Arc Speed (In. per Min. for First Pass)	Passes or Beads	Ft. of Joint Welded per Hr. (100% Operating Factor)	Lbs. of Electrode per Ft of Weld
3/16	5/32	130	9	10	2	25	.15
1/4	5/32	130	9	7	2	17.5	.24
5/16	5/32	140	9½	5	2	14	.38
3/8	3/16	150	8	3	2	10	.50
7/16	3/16	170	8½	5	3	8	.68
1/2	3/16	170	8½	4	3	6.2	.85
5/8	3/16	170	8½	3	‡	4	1.3
3/4	3/16	170	8½	3	‡	2.9	1.9
1	3/16	170	8½	3	‡	1.7	3.3

*If complete fusion is necessary, welding from both sides is recommended. When welding from both sides, it is recommended that 3/16″ plate be welded vertically down.

**If it is necessary to weld from one side into a back-up strip or connecting member, prepare the edges of the plates to a "feather edge" instead of with a 1/8-inch shoulder and space the edges of the plates 3/16″. Approximately 2/3 as many feet of joint will be welded per hour as when no back-up strip is used.

†The plates may be prepared to a feather edge in which case the feet of joint welded per hour will be about 3/4 as much as when 1/8″ shoulder is used.

‡The number of passes will depend upon the operator but the feet of joint welded per hour should be the same.

Butt Welds

Overhead Position—V-Groove
With Back-Up Strip*

PREPARATION:
60° V-Groove

FIT-UP:
$\frac{3}{16}$" Min. Gap

ELECTRODE & POLARITY:

E6010—Electrode Pos.

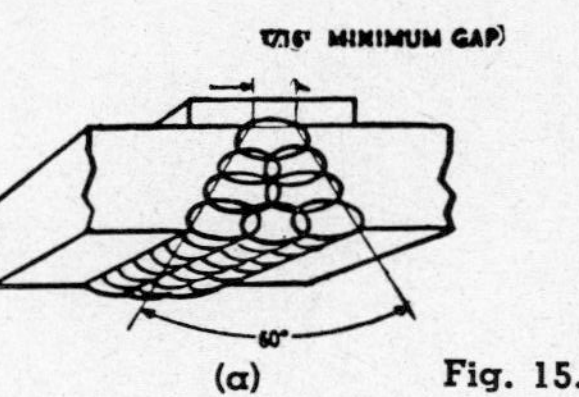

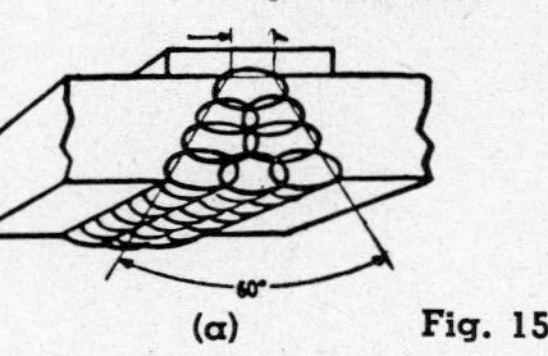

(a) Fig. 15. (b)

Plate Thickness,	Electrode Size (In.)	Current (Amps.)	Electrode Melt-Off Rate (In. per Min.)	Arc Speed (In. per Min. for First Pass)	Passes or Beads	Ft. of Joint Welded per Hr. (100% Operating Factor)	Lbs. of Electrode per Ft. of Weld
$\frac{3}{16}$	$\frac{3}{16}$	150	8	7	2	15	.28
$\frac{1}{4}$	$\frac{3}{16}$	150	8	7	3	12.5	.42
$\frac{5}{16}$	$\frac{3}{16}$	170	$8\frac{1}{2}$	7	4	9	.62
$\frac{3}{8}$	$\frac{3}{16}$	170	$8\frac{1}{2}$	7	5	6.7	.80
$\frac{7}{16}$	$\frac{3}{16}$	170	$8\frac{1}{2}$	6	†	5.2	1.0
$\frac{1}{2}$	$\frac{3}{16}$	170	$8\frac{1}{2}$	6	†	4.1	1.20
$\frac{5}{8}$	$\frac{3}{16}$	170	$8\frac{1}{2}$	6	†	2.9	1.90
$\frac{3}{4}$	$\frac{3}{16}$	170	$8\frac{1}{2}$	6	†	2.1	2.5
1	$\frac{3}{16}$	170	$8\frac{1}{2}$	6	†	1.3	4.0

*When back-up strip cannot be used. Preparation and fit-up as shown in Fig. 15. A uniform gap of approximately $\frac{1}{16}$" should be maintained between plate edges. Weld 1st pass with $\frac{1}{8}$" or $\frac{5}{32}$" E6010. Drag coating in groove and adjust current and travel speed so that two edges just melt together. For succeeding passes, use procedures as given above.

†The number of passes will depend upon the operator but the feet of joint welded per hour should be the same.

FILLET AND LAP WELD PROCEDURES

The American Welding Society defines fillet weld size, for equal leg fillet welds, as the leg length of the largest isosceles right triangle which can be inscribed within the fillet weld cross section. For unequal leg fillet welds, the size is the leg lengths of the largest right triangle which can be inscribed within the fillet weld cross section. The theoretical or effective throat thickness is the distance from the beginning of the root of the joint perpendicular to the hypotenuse of the inscribed right triangle. (The effective throat thickness of an equal leg, 45° fillet weld is 0.707 times the normal leg size of the weld.)

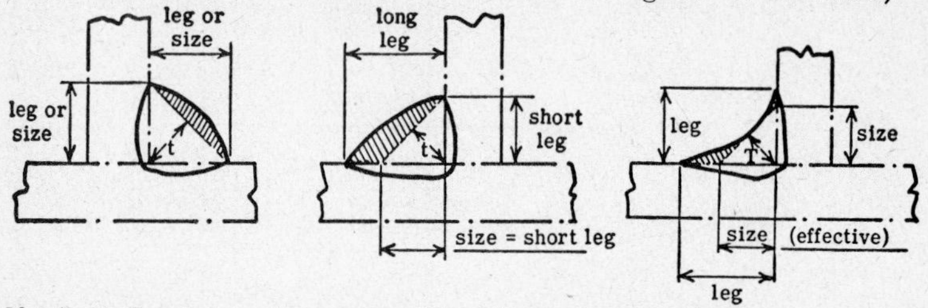

Fig. 16. (Left) Equal leg size convex fillet. (Center) Unequal leg size convex fillet. (Right) Unequal leg size concave fillet.

Although a concave fillet weld produces a smooth change in section at the joint, it is more susceptible to shrinkage cracks, especially in higher carbon steels. Because it is concave, the critical dimension which must be maintained is the throat size. Concave fillets are measured with gauges which measure the theoretical throat section of the weld. See Figure 17.

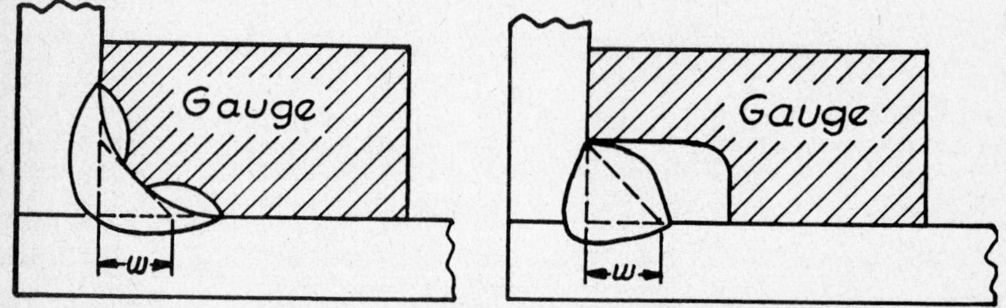

Fig 17. Gauges for measuring fillet size.

A convex fillet weld will have much less tendency to crack as a result of shrinkage upon cooling. It is relatively free from undercut. However, excessive convexity will result in excessive weld metal which will decrease welding speed and add nothing to the strength of the weld. Because it is convex, the critical dimension which must be maintained is the leg size. Convex fillets are measured with gauges which measure the length of the smallest leg.

The ideal fillet weld would be a flat or slightly convex 45° fillet.

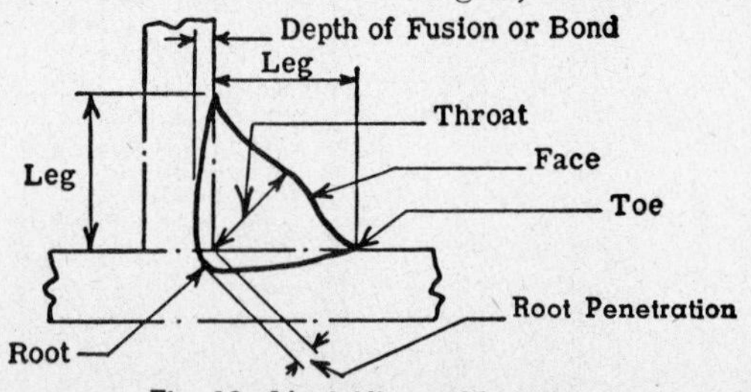

Fig. 18. Ideal fillet weld shape.

The American Welding Society will allow a maximum shear stress on the throat of the fillet weld of 13,600 psi. For an equal leg, 45° fillet weld, the throat is .707 times the leg size.

In accordance with this specification, the following table of allowable strengths of fillet welds has been made:

Fillet Size	Welded One Side. Allowable Design Load Per Inch of Joint	Welded Both Sides. Allowable Design Load Per Inch of Joint	Welded Both Sides. Typical Breaking Load Per Inch of Joint
3/16	1800	3600	15,000
1/4	2400	4800	20,000
5/16	3000	6000	24,000
3/8	3600	7200	28,000
7/16	4200	8400	32,000
1/2	4800	9600	37,000

When the load is applied at right angles to the fillet weld, the throat of the weld is still assumed to be stressed in shear, even though this load causes tensile stresses in the plate. The same table of strength of fillet welds applies, although a fillet weld loaded at right angles is about 30% stronger than if loaded parallel.

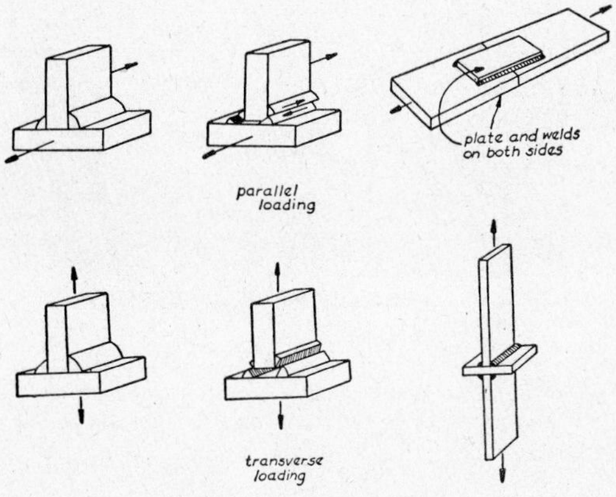

Fig. 19. Fillet weld loading.

As a result of the allowable values for fillet welds set up by the American Welding Society and actual testing of fillet welds to determine their ultimate strength, the following table has been set up which will give the fillet weld size for a given plate thickness, so that the weld will equal or exceed the plate strength. These are called full plate strength welds or 100% plate strength welds. The fillet welds are on both sides of the plate and extend the full length of the plate.

Plate Thickness	Fillet Weld Size
3/16"	5/32"
1/4"	3/16"
5/16"	1/4"
3/8"	5/16"
7/16"	3/8"
1/2"	7/16"
5/8"	1/2"
3/4"	5/8"
1"	3/4"

Techniques—Where possible fillet welds should be made using iron powder electrodes specifically designed for making fillets. The following techniques used with these electrodes will produce the best results:

(1) In general, use one size smaller diameter electrode than is used with conventional techniques for a particular job. However, if the same size can be used, still greater savings can be effected.

(2) These are AC-DC electrodes, but AC is recommended over DC because greater speeds are obtained.

(3) These are contact electrodes. The coating touches the plate and the weld is made using a simple drag technique. The electrode should be held into a fillet with light pressure so that the electrode is pressing against both legs of the fillet. The electrode is held at an angle of approximately 45° to the horizontal and leans in the direction of welding at an angle approximately 30°. (Figure 20.) If the electrode leans back towards the bead, slag will run under the arc causing slag holes.

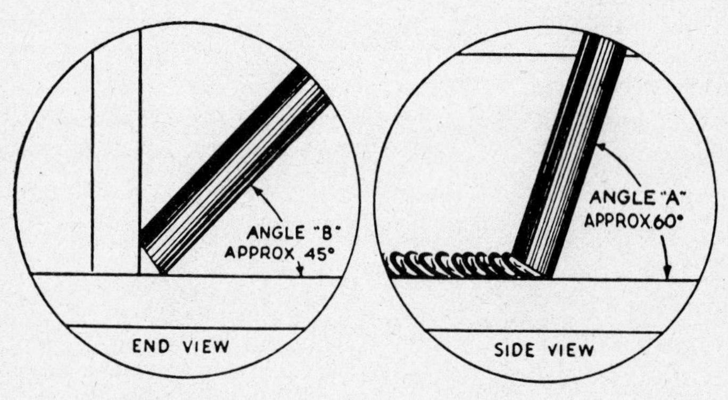

Fig. 20. Angle of electrode for maximum speed.

(4) Too low current will permit slag to follow too closely. Excessive current will adversely affect bead appearance and slag removal.

(5) Maintain a travel speed such that the arc is 1/4" to 3/8" ahead of the slag. The natural leg sizes for fillets resulting from this technique are: 3/16" electrode produces 1/4" leg fillet, and 1/4" electrode produces 5/16" leg fillet. The bead shape will be flat or slightly convex. If the bead appears to have an overhang, increase the travel speed.

(6) Slag will be almost self-removing, if most favorable conditions are created: smooth, uniform bead; reasonable quench effect; use of moderate current.

(7) When welding on short sections, traveling toward the ground will help keep the slag from under the arc. This is especially true of the 1/4" electrode size.

(8) Correct welding currents are very important to obtaining maximum results with iron powder electrodes. Since machine efficiency varies and conditions, such as cable length, vary, current should be deter-

mined by a tong meter, if possible. The following currents will produce excellent results at maximum speeds:

5/32"	225 amps. AC	30-31 arc volts
3/16"	275 amps. AC	31-32 arc volts
1/4"	350 amps. AC	34-36 arc volts

If a meter is not available, current can be determined by melt off. If the melt-off rate of a given electrode is known, the current and amount of electrode deposited in a minute can be correlated. A typical electrode produces the following:

5/32"	225 amps. AC	13.4"/min.
3/16"	275 amps. AC	12.1"/min.
1/4"	350 amps. AC	9.5"/min.

Burn-off rates will give close approximation of correct welding current, and the amount of current indicated by the machine dial should be disregarded.

Another way of approximating current is by observing slag formation. At proper welding current with an electrode angle as indicated in Figure 20, slag formation should begin from 1/8" to 3/8" behind the electrode. Burn-off determination is more accurate, however, since the variable of travel speed influences slag formation.

The reason for the current variation with iron powder electrodes is because the electrode core burns high inside the coating, thus producing a relatively high arc voltage (10 to 20% higher than conventional electrodes used for fillet welding). Many machines in use are calibrated for conventional E-6012 and E-6020 electrodes which operate at substantially lower voltages. Electrodes with lower voltages for a given machine setting will receive more current than those operating at higher voltages.

Multiple Pass Horizontal Fillets

Multiple pass fillets are made with either E-6012, E-6014 or E-6024 electrodes.

The first bead is laid in the corner with a fairly high current and speed and with little attention paid to undercutting. Subsequent beads should be made with the electrode held at an angle of 70° to 80° with the horizontal plate and line of weld, except the beads against the vertical plate, in which case, the electrode should be at about a 45° angle. See Figs. 21 and 22. The beads are laid from the bottom upward, as shown in Fig. 23. The idea here is to provide a flat horizontal surface upon which to place succeeding beads, permitting higher currents, resulting in faster welding.

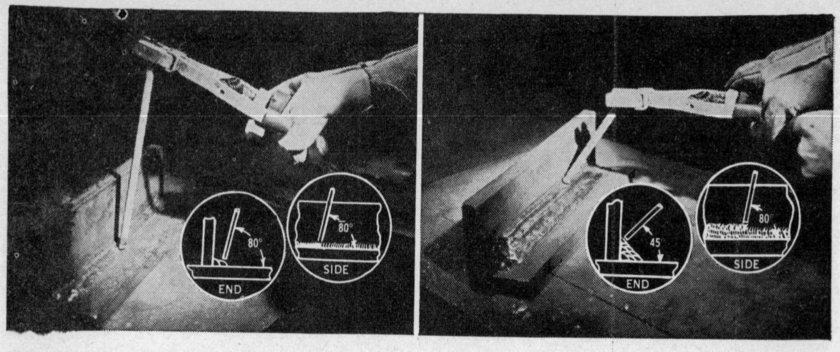

Fig. 21. Angle of electrode for beads not against vertical plate.

Fig. 22. Angle of electrode for beads against vertical plate.

The slag is to be left on the bead in order to provide a dam to keep the metal from running off the edge of the previous bead. This is illustrated in Fig. 23. The slag is not removed until after each layer of beads is completed. For example, slag of the weld shown in Fig. 23 is removed only after completion of bead No. 1 and bead No. 4. This procedure not only saves man-hours in the cleaning of the weld, but it facilitates and speeds up the welding operation and makes possible a smoother weld. Any number of layers of beads may be built up in this manner. See Fig. 24.

When a weld of two passes is required, the first bead can be put in as shown in Fig. 25. Here, the first bead is deposited mostly on the bottom plate, then the second bead is applied without removing slag from the first bead—holding the electrode at about 45° and fusing into vertical plate and the first bead. The slag may be removed after first bead for inspection, if desired.

Fig. 23.

Fig. 24.
Multiple Pass Fillets

Fig. 25.

Fig. 23. Dotted line above bead 2 shows where slag was before bead 3 was put on; dotted line above bead 3 is where slag was before bead 4 was put on. Cross-hatched area shows slag after completion of weld. . . . **Fig. 24.** A 16-pass fillet weld made with E6012 electrode. Plate is 1¼″ thick. . . . **Fig. 25.** Location of beads for two-pass fillet.

FILLET AND LAP WELDS

Horizontal and Flat Position

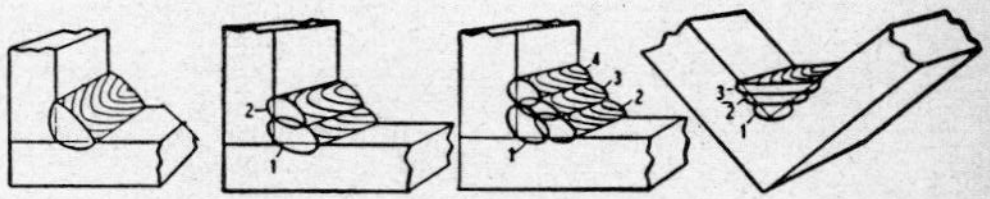

HORIZONTAL POSITION FLAT POSITION

Fig. 26.

PREPARATION:
Square Edge

ELECTRODE & POLARITY*
Iron-Powder E6024—AC

FIT-UP:
Recommended: 1/32" Gap

Plate Thickness (In.)	Gauge Size of Fillet (In.)	Electrode Size (In.)	Current (Amps.)	Electrode Melt-Off Rate (In. per Min.)	Arc Speed (In. per Min. for First Pass)	Passes or Beads	Ft. of Joint Welded per Hr. (100% Operating Factor)	Lbs. of Electrode per Ft. of Weld
3/16	5/32	1/8	170	15.0	15-16	1	75-80	.10
1/4	3/16	5/32	225	14.0	15-16	1	75-80	.15
5/16	1/4	3/16	275	12.0	14-15	1	70-75	.19
3/8	5/16	1/4	350-375	9.6-10.2	12-13	1	60-65	.30
1/2	7/16	1/4	350-375	9.6-10.2	10-11 (1st pass)	2	28-31	.57
3/4	5/8	1/4	350-375	9.6-10.2	10-11 (1st pass)	3-4	14–16	1.34
1	3/4	1/4	350-375	9.6-10.2	10-11 (1st pass)	5	9-10	1.66

Important Note:
These currents are measured by an ammeter. The indicated current on the machine will be lower, (due to the higher arc voltage of the iron powder type electrodes) so the dial setting should be increased slightly.

Multiple Passes:
For 3 or more passes, the 2nd and subsequent layers should be put in 2 beads side by side.

***Note: Use E6014 electrodes when the weldment cannot be positioned close enough to the flat position for E6024.**

Fillet and Lap Welds

Vertical Position

Welded Up*

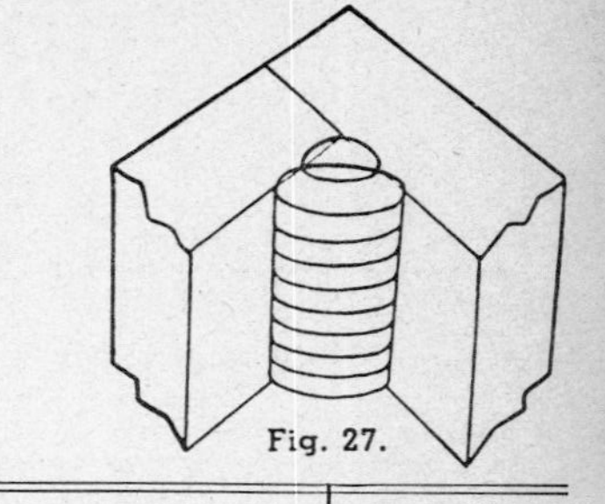

Fig. 27.

PREPARATION:
Square Edge

FIT-UP:
Recommend Gap: $\frac{1}{32}$"
Maximum Gap: $\frac{1}{16}$". If gaps greater than this must be welded, use same procedure but add width of the gap to the fillet size required

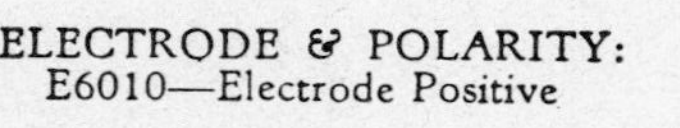

ELECTRODE & POLARITY:
E6010—Electrode Positive

Plate Thickness (In.)	Gauge Size of Fillet (In.) ‡	Electrode Size (In.)	Current (Amp.)	Electrode Melt-off Rate (In. per Min.)	Arc Speed (In. per Min. for First Pass)	Passes or Beads	Ft. of Joint Welded per Hr. (100% Operating Factor)	Lbs. of Electrode per Ft. of Weld
$\frac{3}{16}$	$\frac{5}{32}$*	$\frac{5}{32}$	140	9½	10	1	50	.08
¼	$\frac{3}{16}$	$\frac{3}{16}$	150	8	7	1	35	.14
$\frac{5}{16}$	¼	$\frac{3}{16}$	170	8½	6	1	30	.17
⅜	$\frac{5}{16}$	$\frac{3}{16}$	170	8½	3.9	1	19.5	.26
$\frac{7}{16}$	⅜	$\frac{3}{16}$	170	8½	2.7	2	13.5	.38
½	$\frac{7}{16}$	$\frac{3}{16}$	170	8½	6	2	10	.52
⅝	½	$\frac{3}{16}$	170	8½	4	2 or more	7.7	.67
¾	⅝	$\frac{3}{16}$	170	8½	4	§	5	1.1
1	⅞	$\frac{3}{16}$	170	8½	4	§	2.5	2.1

*It is recommended that $\frac{5}{32}$" size of fillet be welded vertically down.

‡A tee joint welded on both sides with this size fillet will have strength equal to plate strength.

§The total number of passes required will depend on the operator but the feet of joint welded per hour will be the same regardless of number of passes.

Fillet and Lap Welds
Overhead Position

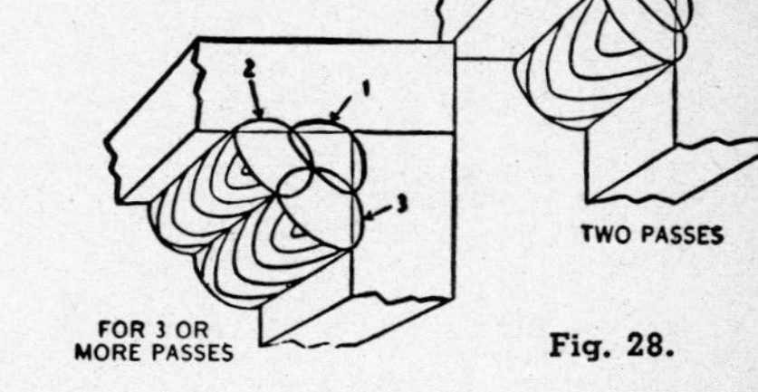

Fig. 28.

PREPARATION:
Square Edge

FIT-UP:
Recommended Gap: 1/32"
Maximum Gap: 1/16". If gaps greater than this must be welded, use same procedure but add width of the gap to the fillet size required

ELECTRODE & POLARITY:
E6010—Electrode Positive

Plate Thickness (In.)	Gauge Size of Fillet (In.)†	Electrode Size (In.)	Current (Amp.)	Electrode Melt-off Rate (In. per Min.)	Arc Speed (In. per Min. for First Pass)	Passes or Beads	Ft. of Joint Welded Per Hr. (100% Operating Factor)	Lbs. of Electrode per Ft. of Weld
3/16	5/32	5/32	140	9½	9	1	45	.09
¼	3/16	3/16	160	8¼	7	1	35	.14
5/16	¼	3/16	160	8¼	5.7	1	28.5	.18
⅜	5/16	3/16	160	8¼	3.7	2	18.5	.27
7/16	⅜	3/16	160	8¼	7¾	3	12.5	.40
½	7/16	3/16	160	8¼	7	§	9.2	.55
⅝	½	3/16	160	8¼	6	§	7.0	.72
¾	⅝	3/16	160	8¼	6	§	4.5	1.2
1	⅞	3/16	160	8¼	6	§	2.2	2.3

†T-joint welded on both sides with this size fillet will have strength equal to plate strength.

§The total number of passes required will depend on the operator but the feet of joint welded per hour will be the same regardless of number.

Corner Welds

Flat Position Half Open Joint

Throat equal to ½ the plate thickness

Use where a throat section equal to ½ the plate thickness produces sufficient strength. This type of joint is welded faster and is less affected by variation in fit-up than the full open corner weld.

When greater throat section than ½ the plate thickness is required and the joint can be welded from one side only, use the procedure for the full open joint. When throat section equal to plate thickness is required, add fillet weld to inside corner (see last column in procedures).

PREPARATION:
Square Edge: Plates overlapped half of plate thickness.

FIT-UP:
Recommended Gap: None
Maximum Gap: $\frac{1}{16}$" for $\frac{3}{16}$" and ¼" plate; ⅛" for $\frac{5}{16}$" plate and thicker.

ELECTRODE: E6024.

POLARITY: AC

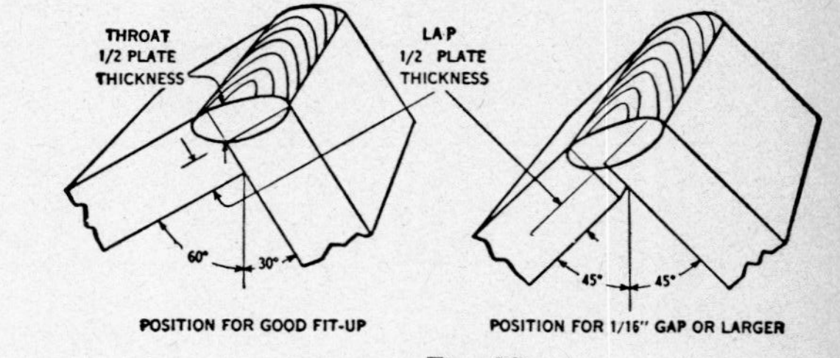

Fig. 29.

Plate Thickness (In.)	Electrode Size (In.)	Current (Amps.)*	Electrode Melt-Off Rate (In. per Min)	Arc Speed (In. per Min. per Pass)	Passes or Beads	Ft. of Joint Welded per Hr. (100% Operating Factor)	Lbs. of Electrode per Ft. of Weld	Size of Fillet to Be Added to Inside Corner to Make Weld Throat Equal to Plate Thickness
3/16	5/32	215	12.0	24.5	1	121	.08	3/16
¼	3/16	275	12.0	21	1	105	.11	3/16
5/16	7/32	300	10.5	20.5	1	102.5	.16	¼
⅜	7/32	325	11.5	18.0	1	90	.18	¼
½	¼	375	10.2	14.5	1	72.5	.26	5/16

Note: If burn-through is encountered, weave electrode to decrease penetration.

*When the gap with the best fit-up obtainable exceeds the maximum indicated above, a sealing bead should be put in.

Corner Welds

Flat Position*—Full Open Joint
Throat equal to ¾ plate thickness

Use where weld can be made from one side only and a greater throat than that obtained with the Half Open Joint is required.

PREPARATION:
Square Edge

FIT-UP:†
Recommended Gap: None
Maximum Gap: 1/32″ on 3/16″ plate; 1/16″ for ¼″ plate and thicker

ELECTRODE:
E6024

POLARITY:
AC

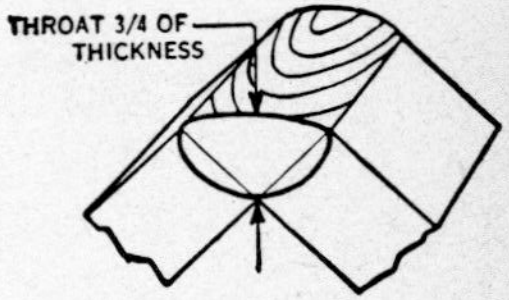

THROAT 3/4 OF THICKNESS

Fig. 30.

Plate Thickness (In.)	Electrode Size (In.)	Current (Amps.)	Electrode Melt-Off Rate (In. per Min.)	Arc Speed (In. per Min.)	Passes or Beads	Ft. of Joint Welded per Hr. (100% Operating Factor)	Lbs. of Electrode per Ft. of Weld
3/16	3/16	250	10.8	23.0	1	115	.10
¼	7/32	280	9.8	20.0	1	100	.15
5/16	7/32	300	10.5	16.5	1	82.5	.20
⅜	¼	340	9.2	14.2	1	72.5	.25
½	¼	350	9.5	13.0	2	65	.53

Note: Since the full open corner weld joint comes to a feather edge at the seam, difficulty with burn-through may be encountered if the usual deep penetrating technique is used. The following special technique is recommended if burn-through is encountered: Weave the electrode from one side of the joint to the other, touching the top corner of one plate and hesitating slightly until the edge is just about to melt, then moving across to the other plate.

*If it is necessary to weld corner welds in the vertical or overhead position, fit-up as recommended above and use the welding procedure for vertical or overhead fillet welds.

†When the gap with the best fit-up obtainable exceeds the maximum indicated above, a sealing bead should be put in.

SHEET METAL PROCEDURES

For 18 Gauge to 10 Gauge Sheets

The following are the main factors governing the speed of sheet metal welding:

1. Type and size of electrode.
2. Current.
3. Fit-up of joint.
4. Position in which the joint is welded.

General

Hold a short arc with the electrode coating almost touching the plate. In the flat position, it may be desirable to lightly touch the coating against the plate for greater ease of welding. Angle of electrode should be about the same as in welding heavier plate for the same type of joint.

Current, Speed, and Fit-Up

Within the limits of good weld appearance, the speed of welding will increase as the current is increased. As the gap in the joint becomes greater, the current must be decreased to prevent burn-through and the welding speed will be reduced. Hence, it is important to fit up the joints as closely as possible in order to obtain maximum welding speeds. A clamping fixture is a practical aid in maintaining fit-up of sheet metal joints, and, on production runs, such equipment soon pays for itself. Such clamping fixtures, if equipped with a copper backing strip, will result in easier welding by decreasing the tendency to burn through. Such a fixture also aids materially in reducing time for aligning joints and tacking, and in minimizing warpage.

The current values on the sheet metal procedure sheets are given merely as a guide in choosing the proper current. The actual current used should be determined by trial, using the highest current possible without burning through or melting away the edges of the joint.

The speeds given in the sheet metal procedures are for high-speed production welding where uniform tight fit-up can be maintained. A few hours training on any certain joint in a production setup will enable an experienced operator to obtain these speeds.

The decrease in speed resulting from poor fit-up is shown on the graph of Fig. 31. The current will also have to be reduced to prevent burn-through.

Feet of Joint Welded per Hour is based on actual welding time only. No factor has been included for setup, electrode changing, cleaning, or other factors which will vary greatly with the type of work to be done. In order to use the figures in this column in cost calculations, they must be multiplied by an operating factor which can be estimated or determined by trial for the job in question.

Control of Welding Current

It is desirable to have some means by which the operator can regulate current while actually welding, preferably a foot-controlled device which leaves both hands free. The operator then will be able to lower his current when he comes to a section of the joint with poor fit-up to prevent burning through and then raise the current again when he comes to a section with good fit-up. Thus, instead of setting his current at a low value that will not burn through at the

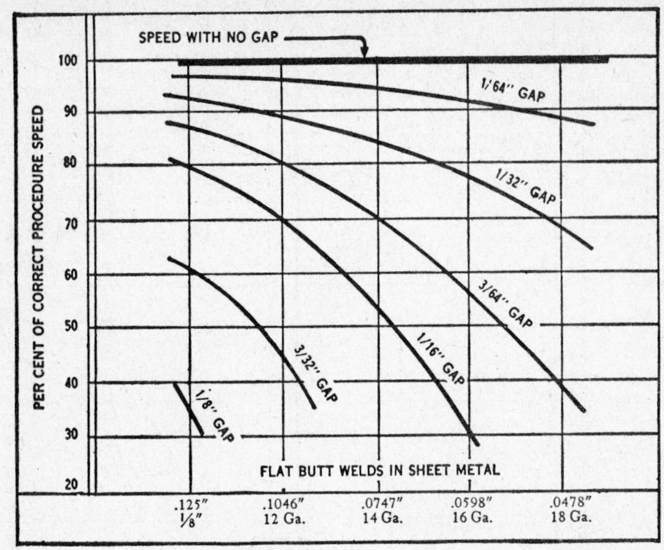

Effect of joint fit-up on welding speed.

Fig. 31.

point with the poorest fit-up and handicapping his travel speed on the entire joint, he can use the highest practical current at all times.

Position in Which the Joint Is Welded

The sheet metal procedures have been set up in the assumption that, at times, sheet metal must be welded in all positions. However, for maximum welding speeds, to minimize distortion and low electrode consumption, sheet metal joints should be welded downhill about 45° with the same or slightly higher currents that are used in the flat procedures. By welding downhill, it is generally possible to increase the speed of welding over the speed obtainable when welding in the flat position. Welding downhill also decreases the tendency to burn through a joint that has poor fit-up and generally produces a flatter and better appearing weld.

Experience of Operator

Because of small electrodes and light currents, the molten pool behind the arc is small and the beads are narrow. Therefore, the speed at which a joint can be made depends a great deal on the operator's ability to stay on the seam and travel at a uniform rate. Welds made with the procedures outlined will have good physical qualities and good appearance when made by a reasonably good operator after a few days' experience in the specialized field of light gauge welding.

Pin Holes in Sheet Metal Welds

On watertight joints, difficulty with small pinholes in the weld is sometimes encountered—these are often due to foreign matter in the joint and will usually be eliminated by removing the rust, scale or other foreign material from the edges of the joint.

Butt Welds—Sheet Metal

Sheet Thickness	Electrode Type and Size (In.)	Electrode Polarity	Current (Amps.)*	Arc Speed In. per Min.*†	Arc Speed Ft. per Hr. (100% Operating Factor)	Lbs. of Electrode per Ft. of Weld
Flat Position						
18 Ga. (.0478")	3/32 E6011	Neg.	45	25	125	.016
16 Ga. (.0598")	1/8 E6011	Neg.	80	30	150	.022
14 Ga. (.0747")	1/8 E6011	Neg.	100	30	150	.026
12 Ga. (.1046")	5/32 E6011	Neg.	140	30	150	.041
10 Ga. (.1345")	3/16 E6011	Neg.	180	24	120	.055
Vertical Position						
18 Ga. (.0478")	3/32 E6011	Neg.	45	25	125	.016
16 Ga. (.0598")	1/8 E6011	Pos.	80	30	150	.021
14 Ga. (.0747")	1/8 E6011	Pos.	80	20	100	.025
12 Ga. (.1046")	1/8 E6011	Pos.	100	20	100	.035
10 Ga. (.1345")	5/32 E6011	Pos.	140	15	75	.053

RECOMMENDED FIT-UP*
No gap for 12 gauge and thinner. 1/16 in. gap for 10 gauge if complete penetration is required.

Fig. 32.

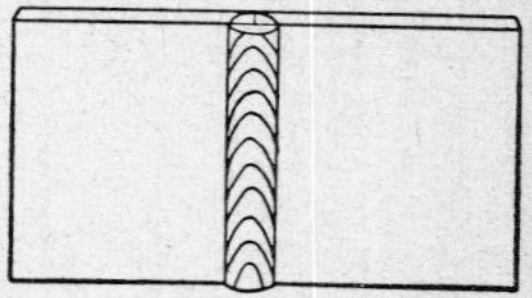

Fig. 33.

In vertical (down) welding with tight fit-up, over 50% penetration can be obtained when using E6011. When using E6013 greater speed will be obtained but the penetration will be less than 50%. Up to 85% penetration can be obtained with E6011 when a gap of 1/2 the sheet thickness is used. (Speeds given with E6011 can be obtained with the gap of 1/2 plate thickness.) Where 100% penetration is necessary, the joint should be welded from both sides with E6011.

*See discussion on Current, Speed and Fit-Up, Page 7-66.

†If complete penetration is required govern travel speed so that the joint is fused through to the back side, or weld the joint from both sides.

Sheet Thickness	Electrode Type and Size (In.)	Electrode Polarity	Current (Amps.)*	Arc Speed In. per Min.*	Arc Speed Ft. per Hr. (100% Operating Factor)	Lbs. of Electrode per Ft. of Weld	Recommended Fit-Up* No Gap
Flat Position or Horizontal Position							
18 Ga. (.0478")	3/32 E6013	Neg.	65	14	70	.027	
16 Ga. (.0598")	3/32 E6013	Neg.	75	14	70	.036	
14 Ga. (.0747")	1/8 E6012	Neg.	115	14	70	.048	
12 Ga. (.1046")	1/8 E6012	Neg.	120	14	70	.054	
10 Ga. (.1345")	5/32 E6012	Neg.	175	14	70	.068	
Vertical Position							
18 Ga. (.0478")	3/32 E6013	Neg.	75	18	90	.024	
16 Ga. (.0598")	1/8 E6012	Neg.	100	18	90	.035	
14 Ga. (.0747")	1/8 E6012	Neg.	120	18	90	.041	
12 Ga. (.1046")	5/32 E6012	Neg.	160	18	90	.052	
10 Ga. (.1345")	3/16 E6012	Neg.	190	18	90	.067	

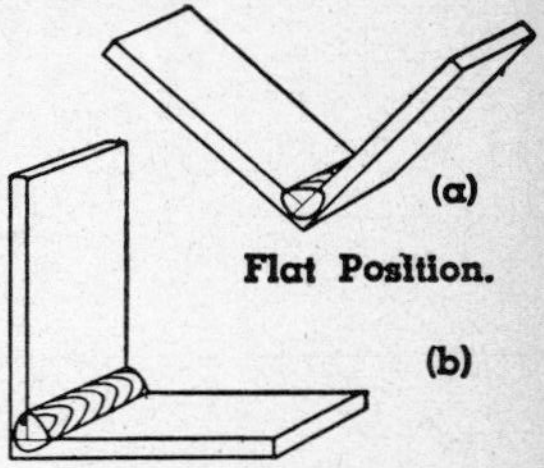

(a) Flat Position.

(b) Horizontal Position.

Fig. 34.

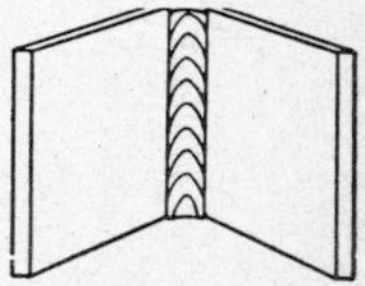

Vertical Position.
Welded Down
Fig. 35.

*See discussion on Current, Speed and Fit-Up, Page 7-66.

Lap Welds—Sheet Metal

Sheet Thickness	Electrode Type and Size (In.)	Electrode Polarity	Current (Amps.)*	Arc Speed In. per Min.*	Arc Speed Ft. per Hr. (100% Operating Factor)	Lbs. of Electrode per Ft. of Weld	Recommended Fit-Up* No Gap
Flat Downhand							
18 Ga. (.0478")	3/32 E6013	Neg.	75	20	100	.022	Fig. 36.
16 Ga. (.0598")	1/8 E6012	Neg.	115	20	100	.030	
14 Ga. (.0747")	5/32 E6012	Neg.	175	20	100	.045	
12 Ga. (.1046")	3/16 E6012	Neg.	220	20	100	.053	
10 Ga. (.1345")	3/16 E6012	Neg.	230	18	90	.070	
Vertical Down							
18 Ga. (.0478")	1/8 E6012	Neg.	90	25	125	.022	Fig. 37.
16 Ga. (.0598")	1/8 E6012	Neg.	110	25	125	.027	
14 Ga. (.0747")	5/32 E6012	Neg.	150	22	110	.043	
12 Ga. (.1046")	3/16 E6012	Neg.	170	22	110	.049	
10 Ga. (.1345")	3/16 E6012	Neg.	220	22	110	.063	

*See discussion on Current, Speed and Fit-Up, Page 7-66.

Edge Welds—Sheet Metal

For work where ease of fit-up and speed of welding are the major factors, the edge weld will be the most economical joint. Where strength is the determining factor, a butt, fillet or corner weld will provide the greatest strength.

Sheet Thickness	Electrode Type and Size (In.)	Electrode Polarity	Current (Amps)*	Arc Speed In. per Min.*†	Arc Speed Ft. per Hr. (100% Operating Factor)	Lbs. of Electrode per Ft. of Weld
Flat Position						
18 Ga. (.0478″)	3/32 E6011	Neg.	60	40	250	.010
16 Ga. (.0598″)	1/8 E6011	Neg.	100	50	250	.016
14 Ga. (.0747″)	1/8 E6011	Neg.	110	45	225	.020
12 Ga. (.1046″)	5/32 E6011	Neg.	120	45	225	.024
10 Ga. (.1345″)	1/8 E6011	Neg.	160	40	200	.037
Vertical Position						
18 Ga. (.0478″)	3/32 E6011	Neg.	60	55	275	.010
16 Ga. (.0598″)	1/8 E6011	Neg.	80	55	275	.012
14 Ga. (.0747″)	1/8 E6011	Neg.	110	55	275	.018
12 Ga. (.1046″)	1/8 E6011	Neg.	120	55	275	.020
10 Ga. (.1345″)	5/32 E6011	Neg.	160	55	275	.030

Recommended Fit-Up† No Gap

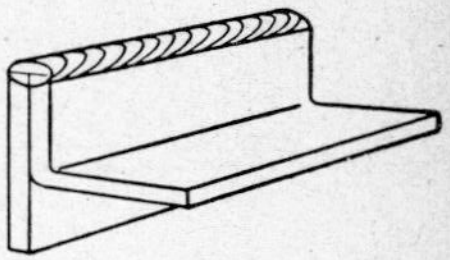

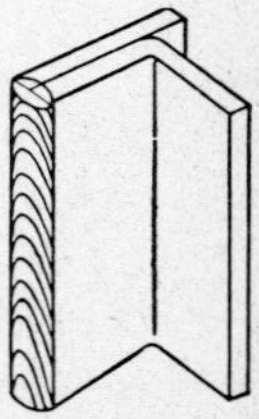

Welded Down.

Fig. 38.

*For edge welds the current should be high enough so that the arc has a definite "hissing" sound.
†See Page 7-66 paragraphs on Current, Speed and Fit-Up.

Corner Welds—Sheet Metal

Sheet Thickness	Electrode Type and Size (In.)	Electrode Polarity	Current (Amps.) *	Arc Speed In. per Min. *	Arc Speed Ft. per Hr. (100% Operating Factor)	Lbs. of Electrode per Ft. of Weld
Flat Position						
18 Ga. (.0478")	3/32 E6011	Neg.	70	40	200	.013
16 Ga. (.0598")	1/8 E6011	Neg.	90	40	200	.018
14 Ga. (.0747")	5/32 E6011	Pos.	120	30	150	.024
12 Ga. (.1046")	3/16 E6011	Pos.	170	30	150	.035
10 Ga. (.1345")	3/16 E6011	Pos. or AC	180	26	130	.040
Vertical Position						
18 Ga. (.0478")	3/32 E6011	Neg.	70	40	200	.013
16 Ga. (.0598")	1/8 E6011	Neg.	90	40	200	.018
14 Ga. (.0747")	1/8 E6011	Neg. or AC	100	40	200	.022
12 Ga. (.1046")	5/32 E6011	Neg.	140	30	150	.032
10 Ga. (.1345")	5/32 E6011	Neg.	150	30	150	.038

Recommended Fit-Up* No Gap (See Sketches)

18, 16, 14, 12 GAUGE
10 GAUGE
HALF SHEET THICKNESS

Recommended Fit-up Fig. 39.

18, 16, 14, GAUGE
12, 10 GAUGE
HALF SHEET THICKNESS

Recommended Fit-up Fig. 40.

*See discussion on Current, Speed and Fit-Up, Page 7-66.

Plug Welds

These are a special type of fillet welds made by fusing the metal of one plate to the side of a hole (generally round) in another plate, the plates being held closely together. Or there may be holes in both plates and the sides of these holes fused together.

Specific types of plug welds are as follows:

(1) Round hole in one plate only. Here the diameter of the hole is from $1\frac{1}{2}$ to 3 times the plate thickness, the larger value being used on the thinner plates. See Figure 41.

Fig. 41.

(2) Scarfed hole in one plate only. The scarfing corresponds to a backed-up butt joint with more than usual root opening, as shown in Fig. 42.

Fig. 42.

(3) Round holes in both plates, as shown in Fig. 43.

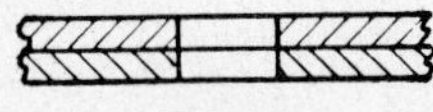

Fig. 43.

(4) Scarfed holes in both plates, as shown in Fig. 44.

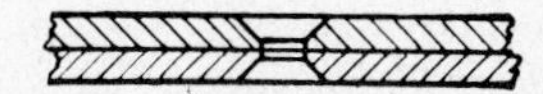

Fig. 44.

Plug welds are used to advantage primarily in cases where access to the work is from one side only, such as flooring, in cover plates for girders, additions to existing structures, or to provide additional strength or stiffness in cases where there is not sufficient space or accessibility available to use the usual fillet welds. An example of the latter case is a lap joint welded from one side only, as shown in Fig. 45.

Fig. 45.

It is to be noted that plug welds cause practically no distortion and are, therefore, particularly useful in cases of plate fabrication, where distortion is encountered.

The procedure for plug welds is unique, inasmuch as the direction of welding changes constantly. The corner all around must be completely fused. For any given instant, the electrode position is approximately standard, or as near 30°—60° as the hole will permit.

The inside corner all around is generally welded first. Weld in a circle, either clockwise or counter-clockwise, whichever is the easier. This welding in a circle should continue without breaking the arc until metal at least one half the depth of the plug is filled in. If the arc is broken, then the weld must be cleaned by chipping to remove the slag. If the arc is not broken, the slag will float to the surface and may be cleaned after half the plug is completed. Low hydrogen electrodes with AC are best.

Use shielded-arc electrodes at high currents. Bare electrodes are not satisfactory.

Use of Fillers

When both plates have holes, metal fillers are often used as shown in Figs. 46 and 47. Complete fusion must be obtained, thus the effect is the same as if all weld metal were deposited. The filler is an aid to speed and ease of welding.

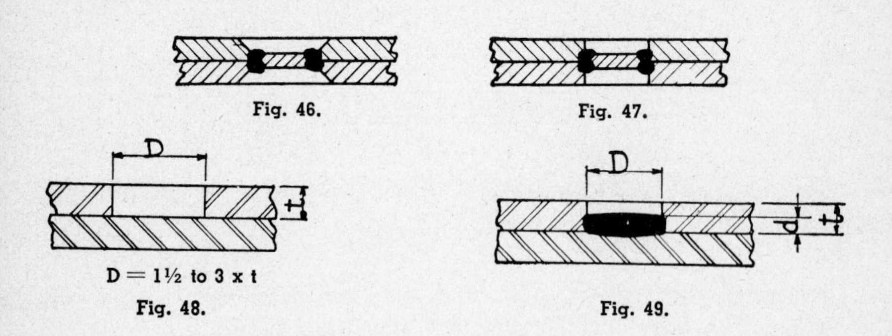

Fig. 46. Fig. 47. Fig. 48. Fig. 49.

SIZE OF PLUG WELDS (SEE FIGS. 48 AND 49)

Plate Thickness (t)	Dia. (D) Hole Inches	Depth (d) of Plug	Lbs. of Electrode per Plug (*)	Time per Plug — Seconds (†)	Design Load (lbs. sq. in.) (††)
¼	¾	¼	.06	33	6,000
⅜	1	⅜	.14	76	10,700
½	1⅛	7/16	.23	125	13,500
⅝	1¼	½	.29	160	16,700
¾	1⅜	9/16	.46	250	20,200
1	1½	9/16	.53	290	24,000

*Based on 3/16″ electrode—includes stub ends.

†Based on 3/16″ electrode at 225 amps. (approx.). No set-up fatigue, etc.

††Based on shear strength of 13,600 lbs./sq. in. on area of plug hole.

Note that the Design Load values given above are for one plug weld. Care must be taken that, in combination with other plug welds or with other types of joints, they are fairly well spaced—and that the load distribution of the combination is taken into account.

CORRECTIVE SUGGESTIONS

Every weld should be a good weld. This means that it should be made at the lowest possible cost to adequately perform the function for which it is designed. A weld that is expensively overwelded is equally as poor as a weld that is underwelded.

To make a good weld requires not only a knowledge of the proper procedures, but also a knowledge of how to recognize a good weld and how to recognize faults in a weld and how to correct them. Failure at this point can be harmful to the quality of welding and substantially increase its cost.

Inspection

If standard procedures are used as specified, the end result, the weld, can be guaranteed. To get the best results, therefore, the best time to inspect a weld is while it is being made. A check as to edge preparation, electrode type and size, current and travel speed used, and surface appearance of the completed weld can tell a qualified inspector all he need to know about the strength of the weld.

A good surface appearance is determined by the following factors: no cracks, no serious undercut, overlap, surface holes or slag inclusions. The ripples and width of bead should be uniform with butt welds flush or slightly above the plate surface without excessive build-up. Fillet welds should have equal legs on each plate. If there is more than a slight variation from these standards, a check should be made on plate preparation, gap limits, polarity, current, speed, electrode angle and other techniques of welding.

Training the Inspector

To employ visual inspection effectively, some training is necessary, both to be able to observe during welding, as well as after welding. The method: Maintain all conditions except one fixed and note the effect of variation of that one condition. The conditions are: (1) arc current; (2) arc length; and (3) arc speed, for a certain plate thickness, type of joint, and size of electrode. The results to be observed are:

(1) Consumption of electrode. How it melts off—smoothly or evenly.
(2) Crater. Its size, shape and appearance of surface.
(3) Bead. Its size, shape and fusion.
(4) Sound of the arc.

Observing these four items and noting the effect of variation of one of the three conditions, the initial observation is made under normal conditions, i.e.

Observation A is made with:

(1) arc current = 100% normal
(2) arc length = 100% normal
(3) arc speed = 100% normal

Observation B is made with:

(1) arc current = 50% normal
(2) arc length = 100% normal
(3) arc speed = 100% normal

Observation C is with:

(1) arc current = 150-250% normal
(2) arc length = 100% normal
(3) arc speed = 100% normal

Observation D is with:

(1) arc current = 100% normal
(2) arc length = 50% normal
(3) arc speed = 100% normal

Observation E is with:

(1) arc current = 100% normal
(2) arc length = 150-200% normal
(3) arc speed = 100% normal

Observation F is with:

(1) arc current = 100% normal
(2) arc length = 100% normal
(3) arc speed = 50% normal

Observation G is with:

(1) arc current = 100% normal
(2) arc length = 100% normal
(3) arc speed = 150-250% normal

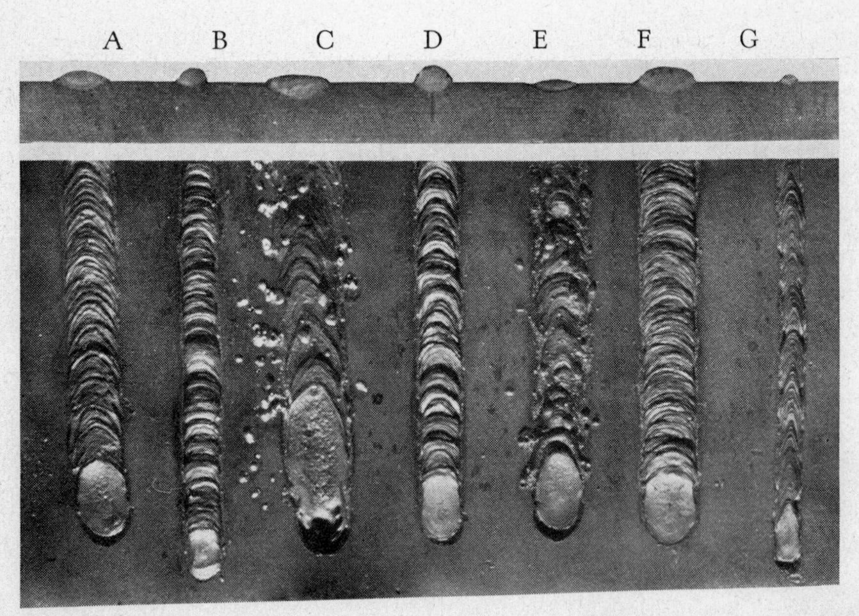

Fig. 50. Plan and elevation views of welds made with an E6010 type electrode under various conditions. Conditions are accentuated to illustrate differences. Iron powder type electrodes when used minimize variations shown here. (A) Current, voltage and speed normal. (B) Current too low. (C) Current too high. (D) Arc length too short. (E) Arc length too long. (F) Speed too low. (G) Speed too high.

Tabulation of Resultant Weld Characteristics obtained on fillet welds when proper welding procedure is used, except as indicated. This tabulation applies to welding of mild rolled steel with coated, E6012 electrodes in position, as shown in Figs. 51, 52, 53.

RESULTING WELD CHARACTERISTICS

	"A"	"B"	"C"	"D"	"E"	"F"	"G"	"H"	"I"
Arc current	Normal	Low	High	Normal	Normal	Normal	Normal	Normal	Normal
Arc length	Normal	Normal	Normal	Short	Long	Normal	Normal	Normal	Normal
Arc speed	Normal	Normal	Normal	Normal	Normal	Low	High	Normal	Normal
Electrode Angle	Normal	Normal	Normal	Normal	Normal	Normal	Normal	Angle "A" too small	Angle "A" too large
Appearance of Bead	Good surface No undercut. Good slag coverage. Throat proper size.	Surface apt to be rough due to poor slag coverage. Not well fused to both plates. Throat much too small	Rough surface. Bad undercut. Slag uneven. Too much metal on bottom plate.	On some steels, surface rough and slag may fall away from vertical plate or "island out." On most mild steels, it is satisfactory. Surface holes on some plates.	Bad undercut. Too much metal on bottom plate. Surface rough because of poor slag. coverage.	Too much metal on bottom p ate. Slag may fall away from vertical plate.	Surface rough due to poor slag coverage. Bad undercut. Small surface holes. Throat too small.	Too much metal on bottom plate. Legs uneven. Throat too small.	Ridge down center of bead. Slag falls away from vertical plate. Undercut. Legs uneven.

The effects of variation in conditions were more apparent when bare electrodes were used. Modern shielded-arc electrodes are more automatic and less sensitive to adverse conditions, but the results above can be observed. The latest iron-powder electrodes still further minimize the possibility and effect of variations. Arc length control is automatic when the contact technique is used. Control of variables is designed into the electrode itself.

Careful observation of the operations by *actual performance* in the shop will enable a good observer to become a trained welding inspector. This training and experience will be of great assistance to the inspector in judging welds both during and after welding.

Inspection During Welding

The method outlined above applies to the inspection of welds during their making.

Inspection After Welding

As in the inspection during welding, certain telltale signs will reveal considerable information to a qualified inspector after the welding is done. Items to consider in inspecting after welding include size and shape of bead, appearance of bead, appearance of slag, undercut, overlap, location of craters (indicating where operator started and stopped welding). A study of the weld and proper interpretation of these telltale signs will disclose other conditions of welding. These conditions are illustrated in Figures 50 and 51, 52, 53 (A to I) for an E-6012 type electrode. The illustrations show the appearance of beads deposited under different procedures; some good, some poor. A study of these will indicate clearly what normal conditions are and the comparison to abnormal conditions. Welds were *not* made with iron powder type electrodes. With an E-6025 electrode, appearance would be improved and be less susceptible to the effects of variations.

Weld Spatter

Weld spatter is an appearance defect of no consequence to the structural function of the weld. Excessive spatter is not necessary, however, and its appearance on a weldment is not pleasing.

Cause. (1) Using too high a welding current. (2) Wrong electrode. (3) Wrong polarity. (4) Too large an electrode. (5) Wrong electrode angle. (6) Arc blow.

Cure. (1) Select the proper current setting for the diameter electrode and plate thickness. (2) Be sure that the electrode does not have an inherent spatter-producing characteristic. (3) Check the polarity switch to determine that the polarity is correct for the electrode. (4) Use an electrode of the proper diameter for the plate thickness. (5) Correct electrode angle for procedure used. (6) See Page 1-40 for corrections for arc blow.

Undercut

Undercut, unless it is serious, is more of an appearance defect than a structural detriment. Unfortunately, however, some inspection agencies will not accept undercut of any type and demand that it be chipped out and the joint rewelded. For this reason, undercut should be avoided.

Cause. (1) Welding current too high. (2) Improper electrode manipulation.

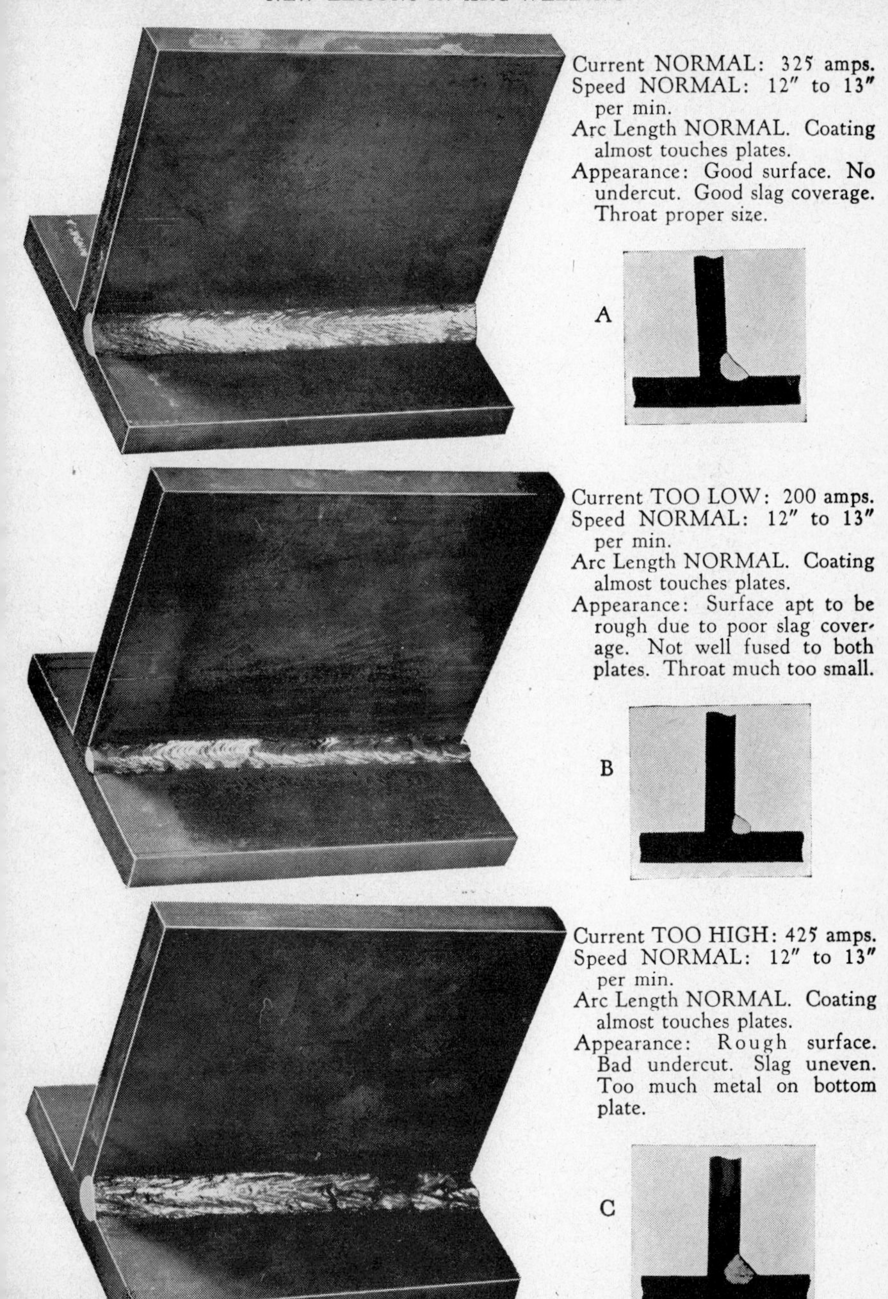

Current NORMAL: 325 amps.
Speed NORMAL: 12″ to 13″ per min.
Arc Length NORMAL. Coating almost touches plates.
Appearance: Good surface. No undercut. Good slag coverage. Throat proper size.

Current TOO LOW: 200 amps.
Speed NORMAL: 12″ to 13″ per min.
Arc Length NORMAL. Coating almost touches plates.
Appearance: Surface apt to be rough due to poor slag coverage. Not well fused to both plates. Throat much too small.

Current TOO HIGH: 425 amps.
Speed NORMAL: 12″ to 13″ per min.
Arc Length NORMAL. Coating almost touches plates.
Appearance: Rough surface. Bad undercut. Slag uneven. Too much metal on bottom plate.

Fig. 51. (A, B, C). Fillet weld specimens made with E6012 electrode. See table on Page 7-77 for explanation. Current and speed values will be substantially different, if iron powder electrodes are used.

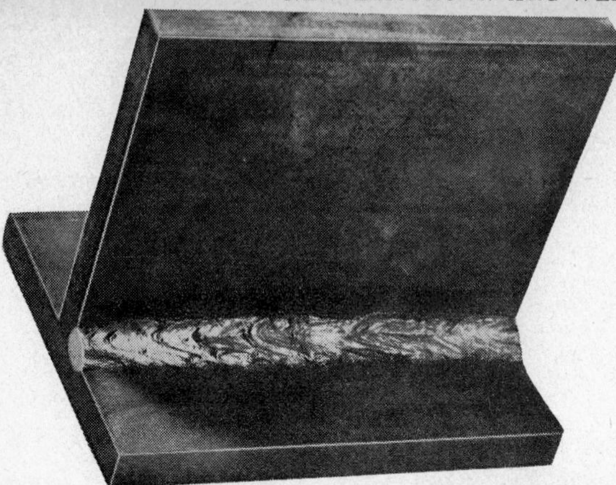

Current NORMAL: 325 amps.
Speed NORMAL: 12″ to 13″ per min.
Arc Length TOO SHORT: Rod jammed into corner.
Appearance: On some steels, surface rough and slag may fall away from vertical plate or "island out." On most mild steels, it is satisfactory. Surface holes on some plates.

D

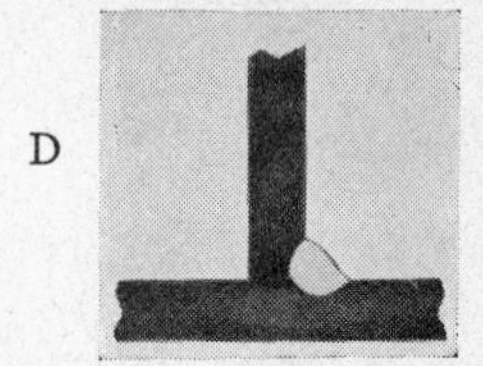

Current NORMAL: 325 amps.
Speed NORMAL: 12″ to 13″ per min.
Arc Length TOO LONG: 1/8″ to 3/16″ away from vertical plate.
Appearance: Bad undercut. Too much metal on bottom plate. Surface rough because of poor slag coverage.

E

Current NORMAL: 325 amps.
Speed TOO SLOW: 6″ per min.
Arc Length NORMAL. Coating almost touches molten pool.
Appearance: Too much metal on bottom plate. Slag may fall away from vertical plate.

F

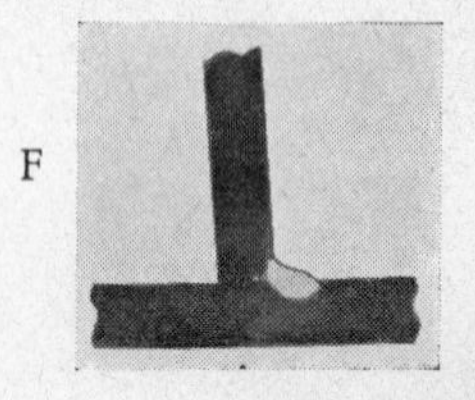

Fig. 52. (D, E, F). Fillet weld specimens made with E6012 erlectrode. See table on Page 7-77.

(Concluded on next page)

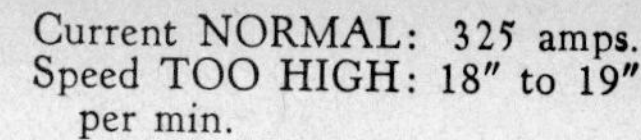

Current NORMAL: 325 amps.
Speed TOO HIGH: 18″ to 19″ per min.
Arc Length NORMAL. Coating almost touches plates.
Appearance: Surface rough due to poor slag coverage. Bad undercut. Small surface holes. Throat too small.

G

Current NORMAL: 325 amps.
Speed NORMAL: 12″ to 13″ per min.
Arc Length NORMAL. Coating almost touches plates.
Electrode held TOO HIGH: Held at 80° to flat plate (instead of normal 45°-60°).
Appearance: Too much metal on bottom plate. Legs uneven. Throat too small.

H

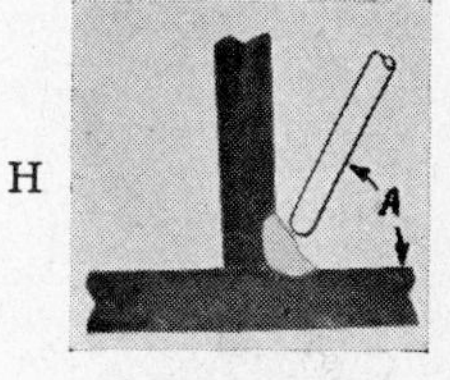

Current NORMAL: 325 amps.
Speed NORMAL: 12″ to 13″ per min.
Arc Length NORMAL. Coating almost touches plates.
Electrode held TOO LOW: Held at 30° to flat plate (instead of normal 45°-60°).
Appearance: Ridge down center of bead. Slag falls away from vertical plate. Undercut. Legs uneven.

I

Fig. 53. (G, H, I). Fillet weld specimens made with E6012 electrode. See table on Page 7-77.

Cure. (1) Use correct welding current and correct travel speed. (2) Undercut may result from using too large an electrode. It may also result if the molten weld puddle is too large. A uniform weave of the electrode will tend to prevent undercutting when making butt welds. Excessive weaving will cause undercut and should be avoided.

Poor Fusion

Poor fusion is sometimes associated with incomplete penetration and likely it is a structural fault. Proper fusion is essential to making full strength welds. It should be the concern of both weldor and inspector to assure that correct procedures are used to obtain the required fusion.

Cause. (1) Current setting improper. (2) Welding technique improper. (3) Failure to prepare joint properly. (4) Wrong size welding electrode used.

Cure. (1) Remember, heavier plates require more current for a given electrode than small plates. Be sure to use sufficiently high welding current to insure correct deposition of weld metal with a good penetration of the base metal. (2) In connection with welding technique, be sure to thoroughly melt the sides of the joint. (3) In preparing the joint, be sure the face of the groove is clean and free of foreign material. Deposit the weld metal in such a manner as to insure good fusion between the plates. (4) Use an electrode sufficiently small to reach the bottom of the groove in making the weld.

Cracks

There are different kinds of cracks in welds, some of which are more serious than others. All types of cracks should be examined to determine what corrective measures, if any, are needed.

The most common cracks in and about a weld joint are crater cracks, underbead cracks and longitudinal cracks. Cracks in the piece of metal along the edge of the weld are sometimes referred to as toe cracks, then there are also hairline cracks across the weld, and micro cracks.

While these various cracks appear in different parts of the weld and result from different causes, in general, the basic fault which leads to such structural defects, if eliminated, will result in crack-free welds.

Cause. (1) The base metal is not of a weldable grade material. (2) Improper preparation of the weld joints. (3) Wrong welding procedure used. (4) The weld joint too rigid. (5) Welds are too small or wrong shape.

Cure. (1) Avoid a high sulfur, high phosphorous steel. If it is necessary to weld this type of base metal, use a low hydrogen electrode. High alloy or high carbon steels should be preheated prior to welding. (2) In preparing joints for welding, space the members uniformly so the gap is even. In some instances, this may mean there is a 1/32″ gap in the welding groove, while, in other instances, the parts may be welded closely together. The size of the weldment and the welding problem at hand will determine the gap spacing. (3) Be sure that the welding procedure is such to provide sound welds of good fusion. The welding sequence should be such as to allow the open ends of the weldment to move as long as possible. Avoid stringer bead welding if cracking is a problem, using a weaving technique to make a full-sized weld, doing the job by sections 8″ or 10″ long. Crater cracks may be eliminated by exercising care to fill the weld crater at the end of each weld or using back-step method to end weld on top of a finished bead instead of plate metal. Change to less penetrating electrode. Weld uphill 4° on first pass to increase weld section. Decrease welding current and speed. Use low hydrogen electrode. (4) Be sure that the structure to be welded has been designed properly and a welding procedure developed to eliminate rigid joints. (5) Always be sure that the weld bead is of sufficient strength to withstand the stress which might develop during the heat of welding. Do not use too small a weld bead between heavy plates. Be sure to use welds of sufficient size on all joints. Make bead shape slightly more convex. Concave beads crack more readily than convex beads. A short arc length helps make beads more convex.

Porosity

Porosity in welds does not present too serious a problem from a strength standpoint, unless the weld is extremely porous. Surface holes in the weld bead are undesirable from an appearance standpoint. The other common forms of porosity, aside from surface holes, commonly referred to as blow holes, are gas pockets and slag inclusions.

Cause. (1) One of the major causes of porosity is poor base metal. (2) Improper welding procedure also results in porosity of weld metal. (3) Porosity may be an inherent defect of the welding electrode being used.

Cure. (1) Be sure the base metal is one that will produce a porosity-free weld. High sulfur, phosphorus and silicon steel sometimes produce gaseous combinations which tend to make blow holes and gas pockets. Non-ferrous material, high in oxygen, also tends to result in porous welds. Base metal, containing segregations and impurities, contribute to porosity. (2) Change welding procedure. Do not use excessive welding currents, but be sure that each layer of weld metal is completely free of slag and flux before depositing another layer. Puddle the weld, keeping the metal molten sufficiently long to allow entrapped gases to escape. Decrease current and use a short arc. (3) Most low hydrogen electrodes will be found helpful in eliminating porosity.

Moisture Pick-Up

Electrodes exposed to damp atmosphere may pick up moisture which, when excessive, may cause undercut, rough welds, porosity, or cracking. This condition is usually corrected by storing the electrodes in a cabinet or room heated to about 10°F. above the surrounding atmosphere. If the electrodes have become wet, they may be dried by removing from box and spreading out to dry at a temperature of 200°F. for one hour.

PART 7.3 | ***Testing Procedures***

Fig. 1. Many welding applications require operator qualification and weld testing.

TESTING WELD METAL AND WELDED JOINTS

In the early days of welding, there was great fear that the welded joints were not going to be strong and that they would not meet the service requirements for which they were designed. Thus, elaborate tests were set up to analyze both the joint and the weld metal itself. In the past fifteen years, the high physical properties of weld metal deposited in the normal way and the amazingly successful performance of welded joints and welded structures of all types has resulted in arc welding being used in most cases without testing of the weld metal and joints in any way. However, some work is so critical that it must be tested for safety reasons, and many different tests are still being made on work of this type. These tests are described in the "Procedure Handbook of Arc Welding Design and Practice." (See Part 7.6)

QUALIFICATION OF WELDING OPERATORS

There is some confusion about the terms "certification" or "qualification" of welding operators. Many people seem to think that there is some all inclusive setup by which an operator can take tests and become certified from there on, for all kinds of work, anywhere. This is not true.

Tests to determine the abilities of welding operators to do a certain kind of work are conducted by many independent laboratories. The ability of the operator to do work of these specific grades is then certified. However, this is not a general certification, but a certification for a definite type of work or employer, and for a specified time.

The following suggestions pertain to quick, low-cost methods of selecting operators for most general classes of work and operators who would later be able to take detailed tests, such as A.S.M.E. requirements, if required.

No standard set of tests for the qualification of welding operators can be devised which is applicable for all types of work. For example, the quality of welding required in a small underground storage tank is not of as high a value as the quality of welding required in the construction of pressure vessels. Therefore, it is impractical to require operators to qualify for A.S.M.E. Boiler Code welding when such quality is not required.

It should be borne in mind that qualification tests of operators should not be confused with qualification tests of electrodes or of weld metal.

Qualification tests should be devised simply on the basis of quality required in actual construction operations. Operators should be tested in making all types of welds usually encountered in the regular line of the particular shop's work. These tests should be so set up that they will as closely as possible approximate the same conditions as are actually encountered in production welding.

Qualifying tests should be such as would qualify the operator for the particular work which he will have to do, and should not impose upon the welding operator a quality greater than is necessary to meet the design requirements.

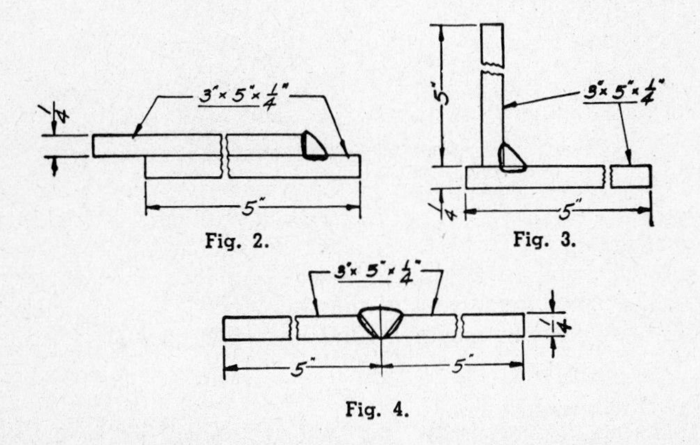

Fig. 2.

Fig. 3.

Fig. 4.

A series of test pieces, such as those illustrated herewith, serve to qualify operators for most of the ordinary applications of welding. A lap weld made in a flat position (see Fig. 2) will reveal a great deal about an operator.

There should be good penetration to the root of the weld and good fusion at both sides of the bead. Examination for these requirements can be made by breaking the weld through the throat.

If the operator fails to pass this test, he is disqualified. If he does pass it, the qualification tests can be continued by making a fillet weld, Fig. 3. This would be judged on the basis of absence of undercutting, shape of the bead, lack of overlap, and proper fusion, as shown by fracture or bending of the test piece. The next test should be a butt weld made in a flat position (see Fig. 4). The weld should show good fusion without overlapping or undercutting.

After this test specimen has been examined for external condition of bead, it should be given a nick-break test. Two saw cuts, in line, and approximately at the center of the weld, will cause the joint to break through this weakened section when specimen is subjected to a sharp blow of sufficient intensity. Examination of the fracture will disclose the quality of the deposited metal.

The deposited metal should show uniform structure, be free of slag inclusions, and be completely fused. There may be some variation in color due to the variety of stresses set up by application of the blow.

Another excellent and inexpensive test is a double butt strap joint. Welds are placed along the sides of the straps, but not extending to the ends. The specimen is prepared in this way so as to place the welds in longitudinal shear only, without any parts in transverse shear. This results in a single load condition. The tongue or pull bars and straps should be of such size in reference to the welds as to cause failure to occur in the welds. The usual precedure followed in the shop can be used. The joint is subjected to a load and broken, failure generally starting at one end of a weld. The weld is accurately measured

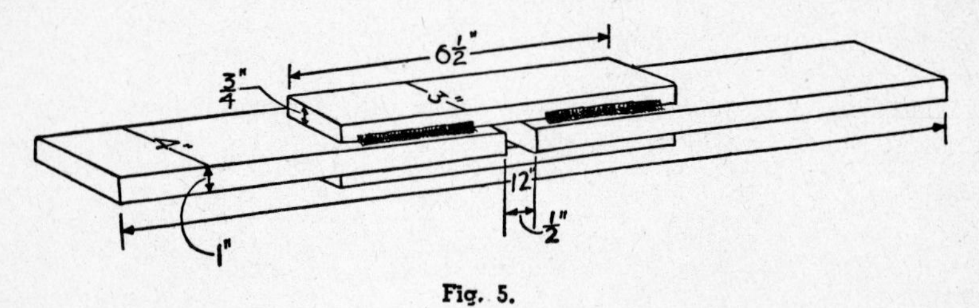

Fig. 5.

and the ultimate value of the joint determined. This figure is then compared with the figure for a perfect joint and the operators rated accordingly.

As an illustration of the method used by a prominent structural steel fabricator to test operators for heavy structural work, a butt joint is made up as shown in the sketch Fig. 5. The joint is made by joining two pull bars by butt straps. The pull bars are 1″ thick, 4″ wide and 12″ long. A space of ½″ is left between the connected ends of the pull bars. The beads, each having approximately 2½″ effective length, are in longitudinal shear, and, since the

Fig. 6.

beads are approximately ⅜″, the ultimate strength of the joint is practically 200,000 lbs. As evidence of improvement in operator's skill, joints of the type mentioned failed at 120,000 lbs. a decade ago. Due to correct instruction, adequate and continued inspection, the value increased to 170,000-185,000 lbs. The test specimen shown in Fig. 6 was pulled to 185,000 lbs.

If the work for which the operator is to be qualified requires vertical welding, then the tests should be made welding the aforementioned joints in vertical position. For example, a typical test specimen of a vertical lap weld is shown in Fig. 7. If the work requires welding a joint in a "hard to get at" position, then the test should simulate such positions. A typical example of such a condition is illustrated in Fig. 8 which shows a test weld of a vertical lap joint which requires greater skill on the part of an operator to make a satis-

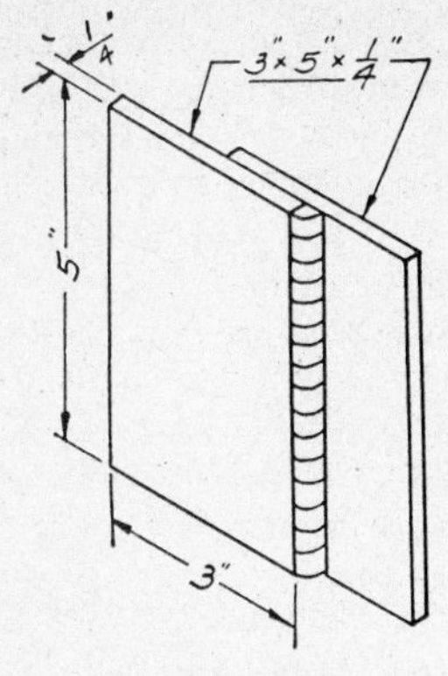

Fig. 7.

factory weld than the test, Fig. 7. Obviously, it is unnecessary to test an operator under conditions presented in Fig. 8, unless such conditions are actually encountered in the work for which the operator is required to qualify. If overhead welding is also required, then the tests should be continued to include welding the same types of joints in the overhead position, as mentioned previously.

In all cases it must be remembered that the purpose of the qualification test is to ascertain the ability of the operator to make a good joint in the field, not just a good test specimen. An operator who makes good test specimens will

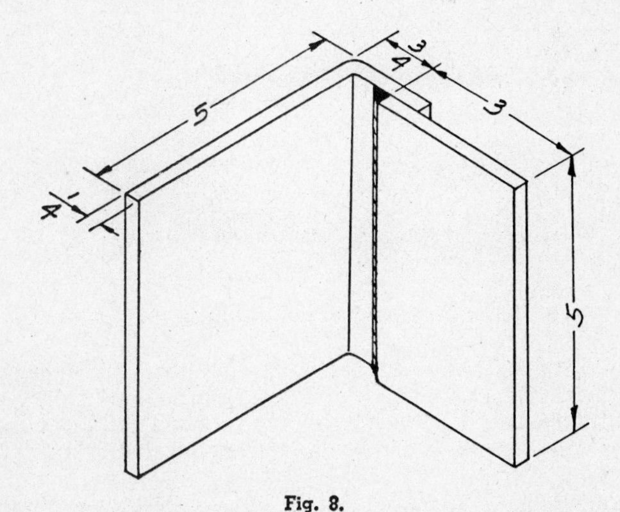

Fig. 8.

usually make good field specimens. This has been brought out in a number of cases where comparative checks on test and field specimens were made. The ability and skill of the operator as indicated by test specimens generally produces these same results in the field work.

Most Code or governing authorities have rules for testing or qualifying of welding operators which vary in minor details. It is beyond the scope of this book to give information and anyone interested should write any one of the government bodies in any particular field.

For most cities of any size, there are various engineers and laboratories which have facilities for testing welds and certifying as to the ability of welding operators to meet requirements of various Codes, tests, etc.

Analysis of Operator Qualification Tests

Is the Guided Bend Test a True Measure of Operator Ability? Although the guided bend test is a standard method included in the codes for testing the ability of the operator to make sound welds, there is considerable doubt whether it should not be modified. Root, Face and Side Bend guided-bend tests are required in the Operator Qualification Tests of the A.W.S., A.S.M.E., A.B.S., and other qualifying agencies. Some authorities stoutly criticize these tests as a measure of operator ability and with good reason.

The guided-bend test consists of bending the joint 180° around a pin which is 1½″ in diameter. The test specimen is ⅜″ thick by 1½″ wide (the length is not important but is normally between 4″ and 8″), and is being bent in such a direction that the weld tends to pull away from the two edges of the joint. This bending operation puts the outside surface in tension and the inside surface in compression with the resulting lengthening or elongation of the outside surface, and shortening of the inside surface. Fig. 9 is the photograph of a specimen which successfully passed the guided-bend test. If a piece of steel ⅜″ thick is bent in the guided-bend fixture and if the assumption is made that the inside surface compresses by the same amount that the outside surface lengthens, and that the piece does not get thinner, and that there is no stretching of the piece other than around the bend, the outside surface must elongate 20% in order to have the piece of steel bend 180° without failure.

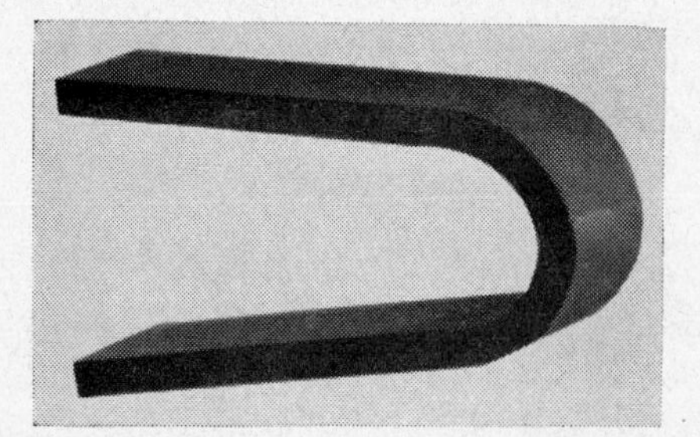

Fig. 9. Guided-bend specimen which met test requirements.

In calculating this, it is assumed that the piece of steel has a uniform tensile strength throughout its length.

By actual test, it has been found that a piece of uniform tensile strength steel elongates 20% in bending the 180° in the guided-bend fixture. This figure of 20% is obtained regardless of the tensile strength of the plate. It is obtained with a 60,000-psi. tensile-strength plate or a 100,000-psi. tensile-strength plate, provided the plate bends without failure.

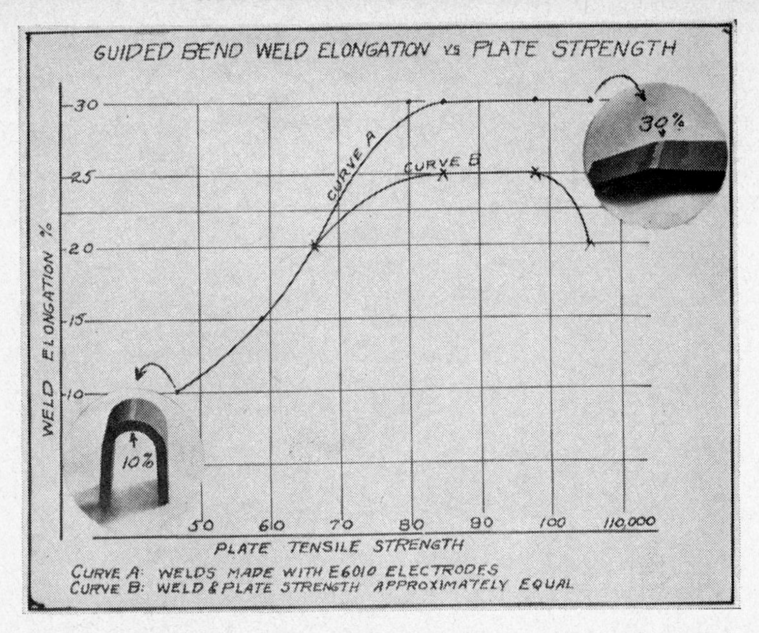

Fig. 10. Curve showing that weld-metal elongation in guided-bend test must increase as the plate tensile strength increases if specimen is to pass the guided-bend test.

Photo at lower left of curve shows specimen which bent 180° with only 10% weld-metal elongation. Photo at upper right of curve shows specimen which bent only a few degrees and the weld elongated 30%.

In making a guided-bend test of a welded joint, the tensile strength of the piece being bent, however, is *not* normally uniform throughout its length. This would be true only when the tensile strength and yield strength of the weld metal were exactly equal to the tensile and yield strength of the plate, and this seldom happens.

If the plate has a lower tensile strength than the weld, the weld will elongate less than 20% in order for the test specimen to bend 180°. If the plate has a higher tensile strength than the weld, the weld will elongate more than 20% in order for the test specimen to bend 180°. In other words, the section having the lower tensile strength, whether it is the plate or the weld, will have the greater elongation.

In order to determine exactly how much the elongation of the weld would be affected by varying the tensile strength of the plate, a series of tests were made. A 60,000-lb. tensile-strength (E6010) electrode of 5/32″ size was used to weld 3/8″ thick plates having the following tensile strengths: 47,000-, 59,000-, 67,000-, 80,000-, 85,000-, 98,000- and 106,000-psi. tensile strength. Root-bend, guided-bend specimens were machined from these plates. Curve A of Fig. 10 is a plot of the test results.

As can be seen from the curve, the amount of weld-metal elongation required for the specimen to bend 180° increased as the plate tensile strength increased.

When testing the specimens made from the 80,000- and 85,000-psi. tensile-strength steel, some of the specimens bent 180° and some of them failed. However, the welds all elongated 30%. All specimens made on plates having a tensile strength higher than 85,000 psi. failed at 30% elongation. All failures started at the fusion zone and tore into the weld metal, if testing was continued, until the specimen broke into two pieces.

Fig. 11. Guided-bend specimen of plate having 59,000 psi. (no weld), which was bent without failure with a notch in the surface.

Fig. 12. Broken cross section of guided-bend specimen. Cross section showed no porosity, cracks, lack of fusion or other defects.

A series of tests were then run using the 67-, 80-, 85-, 98- and 106,000-psi. tensile-strength plates, but an attempt was made to pick an electrode which had approximately the same tensile strength as the plate. The lower tensile-strength plates were not used because it was impractical to make a sound weld which had less than 65,000-psi. tensile strength in the as-welded condition.

If the tensile strength of the weld metal had exactly matched the tensile strength of the plate, all of the specimens should have elongated 20% in bending 180°. Actually, the elongation of all specimens bending 180° varied from 20 to 25%, as is shown by Curve B of Fig. 10.

All welds tested were sound welds. Fig. 12 is a representative sample. The welds were free of cracks, porosity, undercut and spots of lack of fusion, and yet many failed. Not because of the operator, electrode or steel being used, but it failed because the guided-bend test, *as it is now used, is an impractical soundness test* when welding on plates having a tensile strength greater than 70,000 psi. In fact, it is an impractical soundness test.

Test specimens were carefully prepared on the 59,000-psi. plate so that there was a continuous void, approximately 1/16" in diameter, and which was located in the center of the weld and ran the entire length of the weld. This specimen bent 180° without failure. This specimen, although far from sound, did not fail because there is no stress on the metal at the center of the weld during the bend test.

Some of the higher-strength plates which were welded with electrodes of approximately the same tensile strength failed at as low as 10% elongation even though there were no flaws. This is not surprising, since there is an abrupt change in tensile strength at the fusion zone if the plate and weld tensile strength are different. This produces the same effect on a bend specimen as a notch.

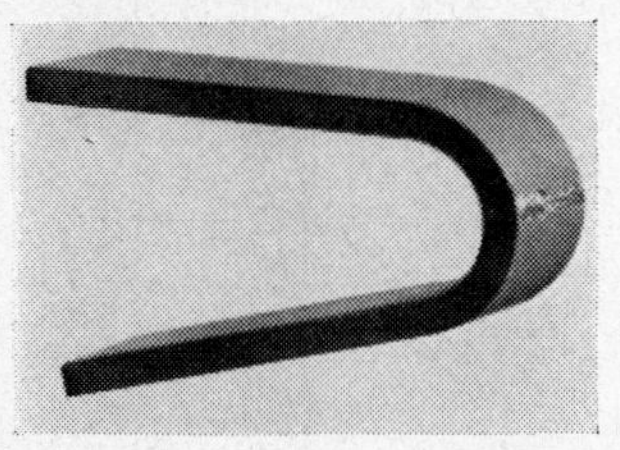

Fig. 13. Guided-bend specimen of weld made in plate having 59,000-psi. tensile strength. Specimen bent without failure even though fusion was incomplete.

Fig. 14. Guided-bend specimen of plate having 85,000-psi. tensile strength. Failure started through a surface notch as soon as the plate started to bend.

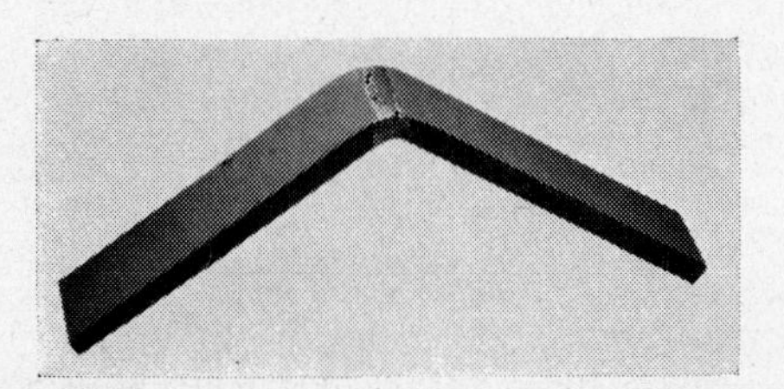

Fig. 15. Guided-bend specimen of apparently perfect weld in 76,000-psi. plate. Weld failed at 40% elongation after bending only a few degrees.

To determine the effect of a notch on the steels being used, a guided-bend specimen was made from the unwelded plate material. In each case a notch $\frac{1}{64}''$ deep having a 60° included angle was machined into the flat plate. The plates were then bent such that the notch was in the center of the bend.

Fig. 11 represents the type of specimen made from plates having 49-, 59-

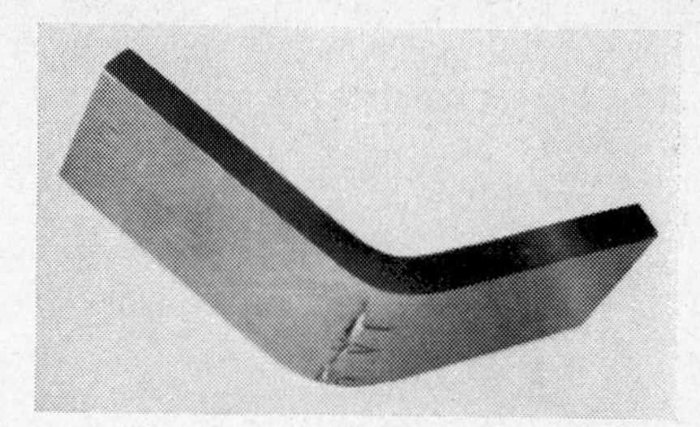

Fig. 16. Guided-bend specimen made of plate having 85,000-psi. tensile strength.

and 67,000-psi. tensile strength. These all bent 180° without failure with the notch in the surface. These steels were not particularly notch sensitive and the notch did not produce failure. An example of a welded specimen tested for notch effect is shown in Fig. 13. Fusion to the root of the weld was not complete and yet the specimen bent 180° without failure.

Fig. 14 represents the type of specimen made from plates having 80-, 85-, 98- and 106,000-psi. tensile strength without welds but with a machined notch. Failure started almost immediately upon the start of the bending. Bending was continued beyond failure to show the type of failure. These steels were sufficiently notch sensitive to have the notch cause failure. A failure such as this in a guided-bend specimen of a test joint cut from a pipeline probably would completely shut down welding on the pipeline.

Fig. 15 represents an apparently perfect weld which failed on bending. The tensile strength of the plates was 76,000 psi. The tensile strength of the weld was 65,000 psi. Elongation of the weld at failure was 40%. The weld had no flaws.

Perhaps some may consider that the $\frac{1}{64}''$ notch across the specimen is too severe. Therefore, the sample shown in Fig. 16 was made. A guided-bend specimen was made from the 85,000-psi. plate. Stencil marks were cut into the plate at the point to be bent. Failure occurred after bending only slightly.

Anyone who has had any experiences with pipelines or pressure-vessel construction knows that stencil marks and scars and digs from handling always appear on the surface. If the pipe or pressure-vessel plate material were bend tested at these marks, it would have a good chance of failing. Yet, no one worries about them. Experience has shown that these flaws or marks do not fail in service.

The question then is should a welding operator be summarily dismissed or "busted out" because he fails the guided-bend test. Failure to pass the guided-bend test may involve the factors previously discussed and considerable care should be taken to make sure that the plate on which the operator is being tested is approximately the same tensile strength as the electrode being used and possesses other features of good weldability.

PART 7.4 — *Welding Symbols*

The American Welding Society Standard Welding Symbols are used on engineering drawings to transmit information from the designer to the fabricator. The system is complete and can be used to convey all needed information. Most of the information can be told with the standard symbols; if more details are required, simple reference notes are used.

The basic component of the standard symbol is the arrow and reference line to which other symbols are added. The significance and use of these components is described in Lesson 1.35. With these parts of the symbol as a foundation, other marks are added to better describe the required welding. Figure 1 is a model symbol showing most of the information which can be included in the complete welding symbol.

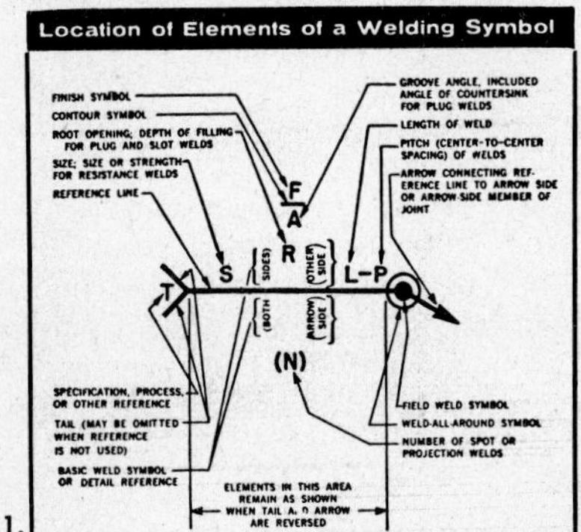

Fig. 1.

The size and length of the desired weld are frequently specified on draw-

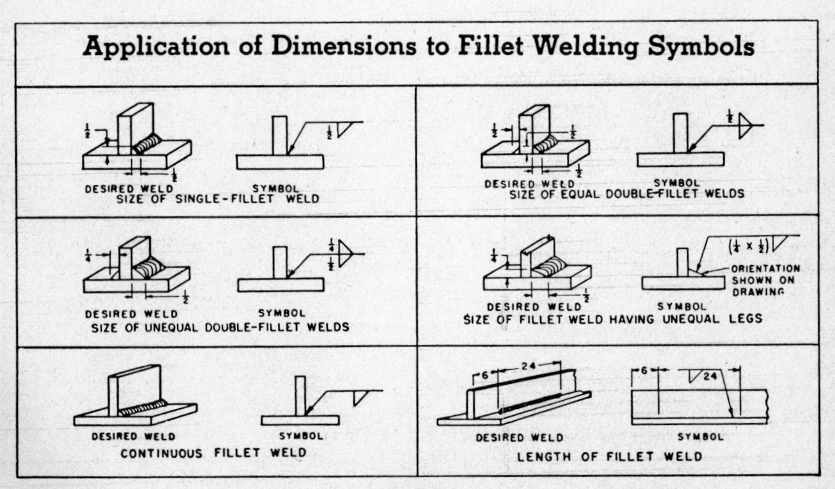

Fig. 2.

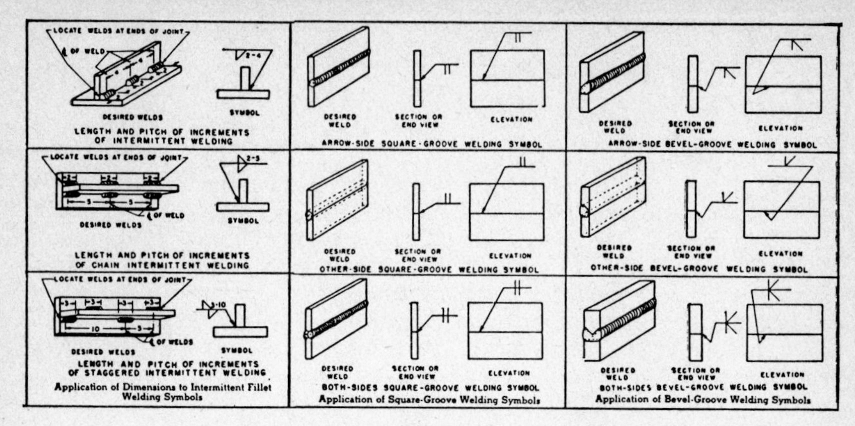

Application of Dimensions to Intermittent Fillet Welding Symbols

Application of Square-Groove Welding Symbols

Application of Bevel-Groove Welding Symbols

Fig. 3.

ings. This information is placed on either side of the basic symbol. On the left side is the size of the weld, and on the right is the length and spacing. Figures 2 and 3A are illustrations of the application of this information to a symbol.

On groove welds it is frequently necessary to specify the edge preparation. The first step is to specify which plate is to be bevelled. If both plates receive similar preparation no notation is necessary, but if only one plate is bevelled it is necessary to indicate which one. This is done by breaking the arrow and pointing it to the plate which is to be prepared. Figures 3C and 4B are illus-

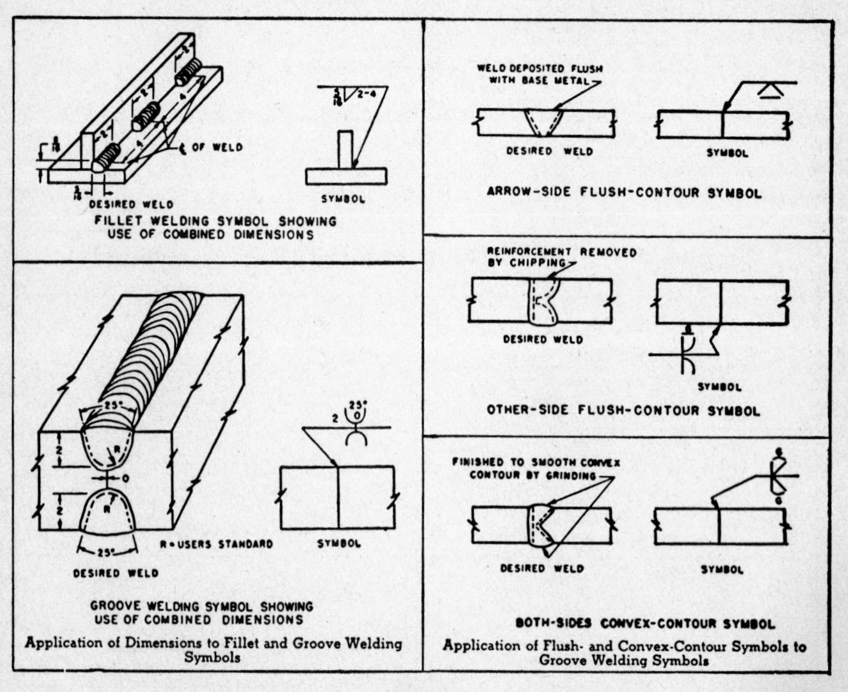

Application of Dimensions to Fillet and Groove Welding Symbols

Application of Flush- and Convex-Contour Symbols to Groove Welding Symbols

Fig. 4.

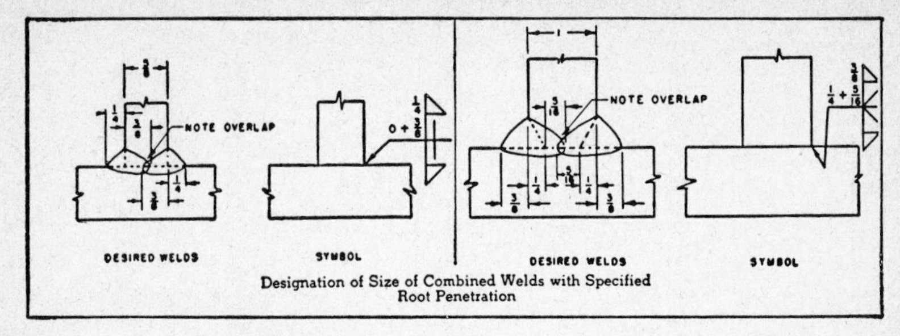

Designation of Size of Combined Welds with Specified Root Penetration

Fig. 5.

trations of this. The type of preparation is indicated by the basic welding symbol, but dimensions may also be added to further define the joint preparation. Figures 4A and 5 show how this is done.

It is possible to show both the preparation and another bead on top by simply placing one symbol on top of the other as shown in Figure 5.

Two supplementary symbols may be placed on the junction of the arrow and the reference line. The first is the "weld all around" symbol which is a circle and means that a weld is to be made completely around the joint. The second is the "field weld" symbol and is a solid dot. It means that the weld is to be made in the field instead of in the shop.

A summary chart of all the welding symbols and examples of their use is included on the following page. Note that this chart also has symbols for resistance welding. It is well to be familiar with these so that there is no confusion with arc welding symbols.

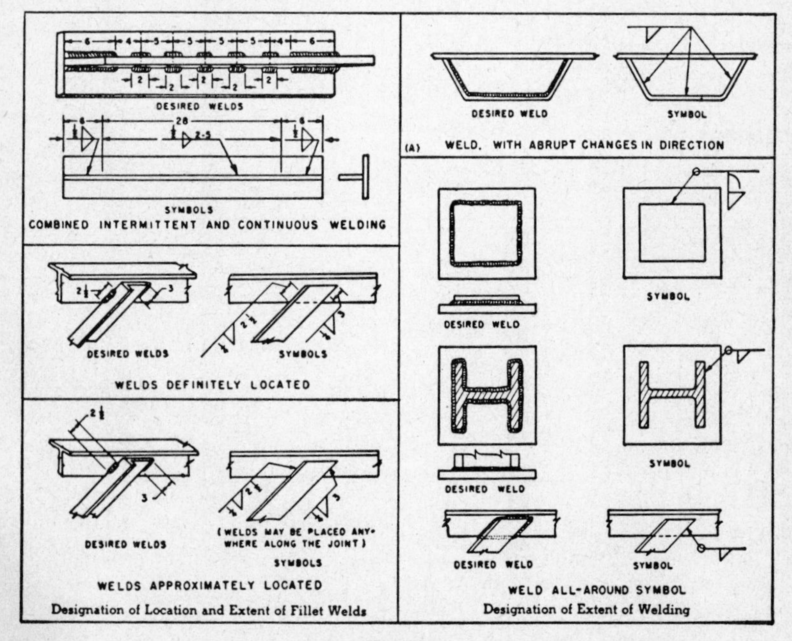

Designation of Location and Extent of Fillet Welds

Designation of Extent of Welding

Fig. 6.

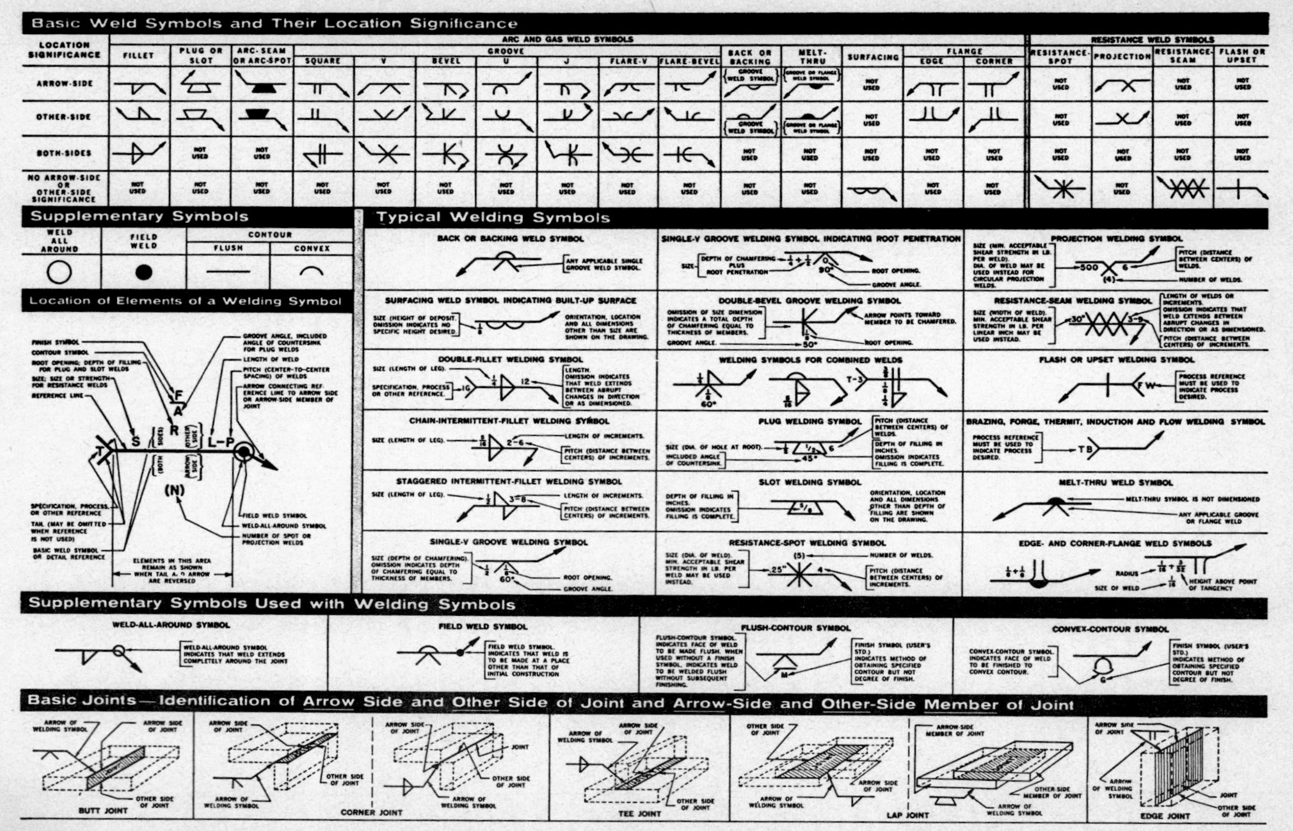
Basic Weld Symbols and Their Location Significance
ARC AND GAS WELD SYMBOLS
RESISTANCE WELD SYMBOLS
LOCATION SIGNIFICANCE
FILLET
PLUG OR SLOT
ARC-SEAM OR ARC-SPOT
GROOVE
SQUARE
V
BEVEL
U
J
FLARE-V
FLARE-BEVEL
BACK OR BACKING
MELT-THRU
SURFACING
FLANGE
EDGE
CORNER
RESISTANCE-SPOT
PROJECTION
RESISTANCE-SEAM
FLASH OR UPSET
ARROW-SIDE
OTHER-SIDE
BOTH-SIDES
NO ARROW-SIDE OR OTHER-SIDE SIGNIFICANCE
NOT USED
Supplementary Symbols
WELD ALL AROUND
FIELD WELD
CONTOUR
FLUSH
CONVEX
Location of Elements of a Welding Symbol
Typical Welding Symbols
BACK OR BACKING WELD SYMBOL
SINGLE-V GROOVE WELDING SYMBOL INDICATING ROOT PENETRATION
PROJECTION WELDING SYMBOL
SURFACING WELD SYMBOL INDICATING BUILT-UP SURFACE
DOUBLE-BEVEL GROOVE WELDING SYMBOL
RESISTANCE-SEAM WELDING SYMBOL
DOUBLE-FILLET WELDING SYMBOL
WELDING SYMBOLS FOR COMBINED WELDS
FLASH OR UPSET WELDING SYMBOL
CHAIN-INTERMITTENT-FILLET WELDING SYMBOL
PLUG WELDING SYMBOL
BRAZING, FORGE, THERMIT, INDUCTION AND FLOW WELDING SYMBOL
STAGGERED INTERMITTENT-FILLET WELDING SYMBOL
SLOT WELDING SYMBOL
MELT-THRU WELD SYMBOL
SINGLE-V GROOVE WELDING SYMBOL
RESISTANCE-SPOT WELDING SYMBOL
EDGE- AND CORNER-FLANGE WELD SYMBOLS
Supplementary Symbols Used with Welding Symbols
WELD-ALL-AROUND SYMBOL
FIELD WELD SYMBOL
FLUSH-CONTOUR SYMBOL
CONVEX-CONTOUR SYMBOL
Basic Joints—Identification of Arrow Side and Other Side of Joint and Arrow-Side and Other-Side Member of Joint
BUTT JOINT
CORNER JOINT
TEE JOINT
LAP JOINT
EDGE JOINT

Fig. 7.

PART 7.5 — *The Lincoln Welding School*

The Lincoln Welding School provides an effective foundation for a career in arc welding. This is not a "school" in the ordinary sense of the word, for the instructions are practical with little theoretical work involved. The welding courses, both Basic and Advanced, enable you to learn not only the fundamental principles of arc welding, but also the latest procedures and techniques.

For the industrious individual with welding ambitions, the Welding School offers a real opportunity. Lincoln's experienced, congenial instructors keep close watch at all times to see that each student progresses. And each student is assigned to his own welding booth and welding machine as his private "shop."

The fees for the following courses are based on the *cost of power and material only.* No charge is made for the instruction given, nor for the use of equipment. A $5.00 deposit is required upon registration. This is applied to the total fee. Protective clothing, shields, gloves, etc., can be purchased at low cost.

BASIC COURSE

Five weeks intensive training . . . 150 hours continuous practice. Subjects include: study of arc welding machine, its performance and control; practice with various electrodes—striking the arc, build-up of plates and shafts; practice with shielded arc electrodes—effect of arc length, current and speed on bead, sizes and uses of various electrodes, butt-lap-tee-fillet welds in flat, vertical and overhead positions; sheet metal welds; "Fleet-Welding" technique; use of iron powder "Jetweld" electrodes; penetration cutting.

ADVANCED SPECIAL COURSES

Pipe Welding Course. Two weeks' training in welding of pipe in all positions.

Alloys Course. Two weeks' training in welding high tensile steels, stainless steel, chrome-moly, cast iron, copper, bronze, aluminum, and hard surfacing.

A copy of the Arc Welding "Bible," the "Procedure Handbook of Arc Welding Design and Practice" (see Part 7.6) will be an invaluable aid for preparation and study.

An outline of the course you prefer will be sent on request. Advance notice of one week is required for registration in the course. Be sure to specify the course in which you wish to enroll. WRITE TO THE LINCOLN ELECTRIC COMPANY, CLEVELAND, OHIO 44117.

PART 7.6 *Aids for Arc Welding Progress*

In the interests of scientific and social advancement through the use of arc welding the Publishers of this book have other books and bulletins on the various phases of arc welding application for sale. The following books are recommended for engineers, designers, production supervisors, shop men, weldors, students and others seeking advancement through knowledge of arc welding.

"Procedure Handbook of Arc Welding Design and Practice"

The 11th Edition Handbook contains up-to-date facts about all aspects of the arc welding process and its many profitable applications, compiled and edited for quick reference and easy understanding.

1200 pages of up-to-date facts.

In eight parts—(1) Welding Processes and Equipment . . . (2) Techniques, Procedures, Speeds and Costs . . . (3) Weldability . . . (4) Basic Design Data for Welded Construction . . . (5) Machine Design with Arc Welded Construction . . . (6) Designing of Arc Welded Structures . . . (7) Applications . . . (8) Reference Data.

Written clearly. Profusely illustrated by over 1300 photos and drawings. Well indexed for quick, detailed reference.

Recognized throughout the world as the authentic reference guide on Arc Welding. Approximately 400,000 copies of previous editions have been sold.

Size 6″ x 9″ x 1⅝″—ideal for use in office, shop or school. Printed on fine paper. Bound in semi-flexible simulated leather, gold embossed.

"The Stabilizer"

"The Stabilizer," published by Lincoln, is virtually the official organ of the vast army of men who weld. Each issue is packed full of practical welding ideas contributed by welders themselves, promoting good welding and the advancement of the industry. Free to employed weldors and supervisors. Give position, name and address of company and home address.

"How To Read Shop Drawings"

A complete revision and organization of the material previously printed in "Simple Blueprint Reading." As produced, this new book for students and shop personnel provides a simplified approach to the easy mastery of reading shop drawings. Designed for use as a working manual, "How To Read Shop Drawings" enables weldors and others engaged in mechanical construction to better understand design, manufacturing, fabrication and construction detailing.

Text contains 187 pages with more than 100 illustrations. Size 8½x11″ durable cloth covered board binding.

PREHEAT CALCULATOR

A slide rule type of calculator for determining the proper preheat and interpass temperature to be used when welding high carbon and alloy steels that need to be preheated for the best quality welds. Temperature can be calculated for any steel whose analysis and thickness is known; eliminates guessing.

FILMS

Movies in color and sound are available to schools, firms and organizations. These include: "Magic Wand of Industry," a general story of what arc welding is and its many uses and "Prevention and Control of Distortion in Arc Welding;" Film size, 16mm. No charge except for transportation. Write Lincoln for further information.

"A New Approach To Industrial Economics"

The ideas expressed in Mr. Lincoln's latest book on industrial economics are based on 53 years of executive leadership. "A New Approach To Industrial Economics" addresses some of the most vital problems facing our country today: labor management cooperation—pricing and price cutting—product development—product marketing—industrial organization—continuous employment—methods of applying incentive—industry and taxation—profits and how to use them—economic waste. Mr. Lincoln's proven answers to these problems are based on the philosophy that industry and business should be conducted for the ultimate benefit of the consumer. Adherence to this principle has made his company the world leader in the welding industry, enabled him to reduce product prices in spite of rising costs, given his employees continuous employment and placed them among the highest paid industrial workers in the world.

"Incentive Management"

The second of three books by Mr. Lincoln, "Incentive Management" discusses a philosophy of human relations in business through which people who work in industry can develop their skills and work together cooperatively to everyone's benefit. Mr. Lincoln explains how incentives can be put into any business, large or small, to increase productivity and make a better product that sells at a lower price. Publisher: The Lincoln Electric Company, 288 pages, hard cover.

THE JAMES F. LINCOLN ARC WELDING FOUNDATION

The James F. Lincoln Arc Welding Foundation was established in 1936 by the Lincoln Electric Company for the scientific and/or practical development of the arc welding process. As a contribution to scientific progress and to promote industrial progress through education the Foundation has produced and published the following books. Copies should be ordered from the James F. Lincoln Arc Welding Foundation, Box 3035, Cleveland, Ohio, 44117.

"Metals and How to Weld Them"—Second Edition

This nationally used reference text by T. B. Jefferson and Gorham Woods has been completely revised and rewritten. Result: the easy reading text clearly explains elementary metallurgy using readily understood terms and illustrations. The book describes the internal structure of metals and its relation to mechanical properties and weldability. Existing arc welding processes are described and their application discussed. Physical characteristics are reviewed and welding procedures outlined for the commonly used metals. An entirely new chapter discusses the "space age developed" exotic metals. 400 pages, 195 illustrations, gold embossed cloth bound cover. Publisher: The James F. Lincoln Arc Welding Foundation.

"Design of Weldments"

This reference handbook describes in detail design techniques used to create machine designs in arc welded steel. Much of this material has had no previous publication. Theoretical analysis and case history studies explain how to design machine components for manufacturing economies and improvement of product performance through efficient use of steel's excellent physical properties. Text covers designing for fatigue, tension, compression, deflection, impact, vibration and torsional load conditions. 464 pages, 923 illustrations, 24 full size nomographs, 8½ x 11 inch page size; bound in gold embossed cloth covered board.

"Arc Welding Lessons for School and Farm Shop"

A book for teaching and learning the skills used with farm arc welding equipment. Contains both informational and operational lessons. Explains how to weld, solder, braze, heat, cut, temper, hard surface with arc welding equipment. Clear step-by-step outline for each lesson given with a typical job explained for practicing the lesson. 342 pages, 550 illustrations, illustrated glossary, 26 pages of welded projects; bound in semi-flexible simulated leather, gold embossed.

"Arc Welding" a Basic Manual of Instruction

This basic manual presents a simple, and easy to read and understand explanation of arc welding and the equipment and skills required to use it. It has large 8½″ x 11″ size pages, punched for notebook filing; illustrated with drawing and photographs. Produced in cooperation with the Vocational Agriculture Service, University of Illinois, and William A. Sellon, Bemidji State College, Bemidji, Minnesota.

"Arc Welded Projects for the School Shop"

This handy student project manual presents full details on how to make 25 shop tested projects. Included are pictures, drawings, bills of material and instructions. Ideally suited to industrial arts training, the projects cover items that are useful in the home or in a workshop. Bibliography lists over 200 other sources of student projects. A few of the projects included in the manual are: Lamps, TV Stand, Portable Grill, Bench Shears, Basketball Goal, Telephone Bench, Wood Lathe and Table Saw.

Written by Wm. A. Sellon (see above). Manual has 46 pages, notebook size, bound and punched.

NOTES

PART 7.7 — *Answers to Review Questions*

Lesson 1.2

1. damages windings, increases shock hazard
2. cut down welder efficiency and arc control
3. No. 10 or 11
4. yes
5. when chipping and grinding
6. "sunburn" and eye "flash"
7. galvanizing, brass, tin, lead
8. welding heat causes explosive fumes to form
9. hold the outstretched palm over it.

Lesson 1.3

1. as strong or stronger than base metal
2. continuous supply of AC or DC sufficient in voltage and amperage to hold an arc
3. motor-generator or selenium rectifier
4. transformer welders
5. National Electrical Manufacturers Association
6. percentage of a 10 minute period that a welder can operate at maximum rated capacity
7. yes, with caution
8. high amperage, low voltage
9. a device to provide a momentary surge of current when striking an arc
10. AC-DC combination welder
11. gasoline or diesel engine driven generator
12. yes.

Lesson 1.4

1. it enables the weldor to handle and control the arc
2. a chemical coating shields the arc and the weld
3. to effect proper fusing action of the weld
4. no, depends upon type of electrode
5. floats impurities out of puddle, shields the arc and puddle, stabilizes the arc
6. oxygen and nitrogen
7. stabilizes and regulates the arc during current alternations
8. to slow cooling, protect hot metal from the air
9. lower tensile strength and impact resistance

Lesson 1.5

1. an arc is formed
2. scratching and tapping
3. scratching
4. scratching
5. snap it backward to break it loose, or release from the holder
6. no, the electrode will "flash"
7. yes
8. yes
9. withdraw electrode tip to form a long arc
10. $\frac{1}{16}$″ to $\frac{1}{8}$″
11. 20 to 25 degrees

Lesson 1.6

1. sound beads are the basis of sound welds
2. perpendicular, inclined 20 to 25 degrees
3. yes
4. yes
5. chip slag with chipping hammer and brush clean
6. 11 or 12 inches
7. $1\frac{1}{2}$″
8. the depth of fusion below the surface of the base metal
9. amperage setting, electrode angle, arc length, travel speed

Lesson 1.7

1. welding speed
2. lower rate of metal deposition,

more electrode changes, more working time
3. electrode diameter should not exceed base metal thickness
4. 3/16″
5. conditions under which electrode is used are too varied
6. about the middle of the range
7. maximum electrode size and maximum amperage

Lesson 1.8

1. the spot where the force of the arc has left an unfilled depression
2. a depression or lump to spoil uniformity
3. the bead covers up the striking marks
4. stress points in the weld which may result in cracks
5. extinguish the arc, let it cool, clean, and refill
6. running back over the bead, or jumping to the end and welding back to the bead

Lesson 1.9

1. an oscillating motion, crosswise to the direction of the bead
2. give a wider bead, float out slag, secure better penetration
3. yes
4. six times the electrode diameter
5. No, in all positions.

Lesson 1.10

1. whipping is lengthwise movement, weaving is crosswise to the bead
2. keep metal "hot" for penetration or "cool" for build-up
3. forward motion
4. to avoid depositing metal until puddle partially solidifies
5. no, backward on "hot" whip, forward on "cool" whip

Lesson 1.11

1. building up by overlaying weld passes
2. rebuilding worn parts or correcting machining errors
3. yes
4. running alternate layers of passes 90 degrees to each other
5. no, straight passes may be used
6. cutting through and etching with acid
7. put acid into water, one to three
8. yes
9. to minimize distortion
10. no, let them cool slowly

Lesson 1.13

1. negative
2. negative
3. no
4. DC negative
5. negative
6. high cellulose
7. break up oxide films

Lesson 1.14

1. at the start and finish of joints and in corners and deep grooves
2. yes
3. no
4. reduce current; this lowers welding speed. Use AC welding current
5. yes. The strength of the flux is reduced because it collapses and builds up 120 times a second
6. stainless, non-ferrous, low hydrogen in small sizes, and hardsurfacing electrodes
7. AC
8. 250-300 is sufficient on most jobs
9. weldors reduce current when arc blow is present. Reduced current means slower welding
10. yes

Lesson 1.15

1. gets larger, expands
2. gets smaller, contracts
3. decreased
4. yes
5. depth or size of throat
6. yes
7. 30 degrees
8. 1/32″ to 1/16″
9. yes
10. use more passes

11. lateral distortion
12. yes
13. yes
14. stretches and work hardens metal
15. use jigs and fixtures

Lesson 1.16

1. to identify and specify electrode for jobs
2. by color code or the manufacturer's container
3. the American Welding Society
4. National Electrical Manufacturers Association
5. "fast freeze," "fast fill," "fill-freeze" and "hard to weld steels"
6. E-6010 and E-6011
7. E-6016 and E-6018
8. tensile strength

Lesson 1.17

1. from the particles of iron in the coating
2. Lincoln's "Jetweld"
3. downhand and horizontal fillets
4. yes
5. yes
6. higher rate of deposition, well-shaped welds, minimum spatter, easy slag removal, stable arc
7. high rate of metal deposition
8. yes, with correct application
9. higher
10. yes
11. wider and shallower
12. fillet
13. yes
14. AC
15. drag technique

Lesson 1.18

1. fillet weld
2. welding speed and quality of the weld are increased
3. "Fleetweld 7"
4. straight bead
5. greater welding speed
6. either a straight bead or a weave
7. smooth, uniform, even penetration into each plate and into the corner

Lesson 1.19

1. lap weld
2. very short to contact
3. heat is directed on the thicker plate
4. greater welding speed
5. depends upon the resultant bead shape
6. smooth bead, equal penetration in each member and into the corner

Lesson 1.20

1. square, vee, groove
2. welding from both sides
3. for maximum strength
4. complete penetration must be secured on the first pass
5. enough to build the weld bead slightly above base metal surface
6. yes
7. higher amperages for greater speed; larger electrodes may be used without burn-through
8. steel or copper

Lesson 1.21

1. by the force of the arc
2. electrode movement and arc length
3. whipping motion, amperage adjustment
4. yes

Lesson 1.22

1. horizontal corner weld
2. square, bevel one plate, bevel both plates
3. about the same or slightly lower
4. shorten arc, reduce amperage
5. short

Lesson 1.23

1. welding up and down
2. on plate 3/16" and under
3. for plate 1/4" and over
4. down welding
5. fast enough to keep ahead of the molten slag
6. yes

Lesson 1.24

1. yes

2. by weaving with one of several patterns
3. to make different sizes or types of deposits
4. shorten arc, reduce amperage, change electrode angle, hesitate on weave pass

Lesson 1.25

1. vertical down welding is generally used on lighter metals
2. to hold them in place before and during welding
3. to provide root spacing for penetration
4. the base metal is too thin

Lesson 1.26

1. on tanks and containers
2. downhand whenever possible
3. no, vertical down welding
4. any thickness may be corner welded
5. open, half open, closed
6. back-step welding, copper back-up strips, adequate tack welds
7. not usually

Lesson 1.27

1. 12 to 16 gauge
2. flange welds
3. no
4. carbon arc or arc torch
5. downhand
6. downward

Lesson 1.28

1. less than 1/8"
2. mild steel
3. mild steel with a zinc coating
4. mild steel with an oxide coating
5. distortion and correct amperage adjustment
6. lap weld
7. 16 gauge
8. soft arc, spray type deposit
9. use back-up strip, preferably copper
10. tilting metal 10 to 15 degrees and welding downward
11. heat may be more carefully controlled
12. clean metal, retard oxidation, aid fusion
13. fumes are toxic and should not be breathed in

Lesson 1.29

1. 1/4" or thicker
2. more passes are needed, less penetration obtained
3. up welding
4. allow the puddle to remain small and solidify

Lesson 1.30

1. for maximum penetration and minimum number of passes
2. to more easily control the molten puddle and produce a uniform bead
3. to carry more metal needed in filler and cover passes

Lesson 1.31

1. square (welded from both sides), vee, or groove
2. yes
3. easier to get into the bottom of the vee

Lesson 1.32

1. holding the electrode steady and keeping the molten metal in the puddle
2. using both hands and resting one arm against the body or an object
3. to allow the sparks and spatter to roll off easier
4. perpendicular, inclined 5 to 10 degrees in the direction of travel
5. keep a very short arc, use a whipping motion

Lesson 1.33

1. yes
2. multiple stringer or weave passes
3. stringer bead
4. fused evenly into each plate and into the corner, no undercut or overlap on members
5. two

Lesson 1.34

1. very short